D0768629
WETLAND
ECOSYSTEM
TEAM
UNIVERSITY OF WASHINGTON

Landscape Ecological Analysis

Springer
New York
Berlin
Heidelberg
Barcelona
Hong Kong
London
Milan
Paris
Singapore
Tokyo

Jeffrey M. Klopatek
Robert H. Gardner
Editors

Landscape Ecological Analysis

Issues and Applications

With 69 Illustrations

Jeffrey M. Klopatek
Department of Plant Biology
Arizona State University
Tempe, AZ 85287-1601, USA

Robert H. Gardner
Appalachian Laboratory
University of Maryland
Frostburg, MD 21532, USA

Library of Congress Cataloging-in-Publication Data

Landscape ecological analysis: issues and applications / edited by
Jeffrey M. Klopatek, Robert H. Gardner.
p. cm.
Includes bibliographical references and index.
ISBN 0-387-98325-2 (hc.: alk. paper)
1. Landscape ecolgy. I. Klopatek, Jeffrey M. II. Gardner, R.H.
QH541.15.L35L33 1999
577—dc21 98-33561

Printed on acid-free paper.

Production coordinated by Chernow Editorial Services, Inc., and managed by Steven Pisano; manufacturing supervised by Jacqui Ashri.
Typeset by Best-set Typesetter Ltd., Hong Kong.
Printed and bound by Braun-Brumfield, Inc., Ann Arbor, MI.
Printed in the United States of America.

9 8 7 6 5 4 3 2 1

ISBN 0-387-98325-2 Springer-Verlag New York Berlin Heidelberg SPIN 10556867

Preface

The emergence of landscape ecology during the past decade as an important scientific discipline has been both remarkable and gratifying. Remarkable because the rate of new publications, the growth of university courses and seminars, and the expansion of new positions in landscape ecology has been greater than anyone anticipated 10 or more years ago. Gratifying because the development of the science and associated methodologies has provided a new perspective on an array of critical problems, such as land use history and change, habitat fragmentation, natural and anthropogenic disturbances, biodiversity and conservation design, and ecosystem dynamics. In addition, studies in landscape ecology are forcing a reevaluation of the difficult issues of scale and the problem of extrapolation of point data to assess changes over broad areas. Although these issues are not the sole domain of landscape studies, new problems associated with global change (IPCC 1995) and subsequent effects on biodiversity (Chapin et al. 1998) continue to fuel the development of broad-scale studies typical of landscape ecology.

The eclectic nature of landscape ecology has resulted in multiple viewpoints concerning the scope and domain of the science (Wiens 1992). In a recent *Bulletin of the International Association for Landscape Ecology* (1998), this definition was offered: "Landscape ecology is the study of spatial variation in landscapes at a variety of scales. It includes the biophysical and societal causes and consequences of landscape heterogeneity. Above all, it is broadly interdisciplinary." The need to define (or refine) what constitutes landscape ecology more than 12 years after Forman and Godron's (1986) seminal text on the subject is symptomatic of a young and evolving discipline—but it is also indicative of the broad, interdisciplinary net cast by landscape studies. It is not surprising to find a diversity of opinions among individuals regarding the significance of developments within landscape ecology over the last 10 years. Our own informal survey at recent society meetings indicates that many believe that technical advances, such as the development and application of GIS, remote sensing, and methods for spatial analysis, are most important; others believe that landscape studies are most relevant when addressing issues of ecosystem management, conservation design, and landscape planning; still others feel that the characterization of the interactive effects of spatial heterogeneity of resources and subsequent effects on ecological processes are the critical issues now being addressed by this new discipline. Clearly, all these attributes are embedded within landscape studies, so it is not surprising that semantic confusion continues for this expansive, yet exacting, science.

The purpose of this book is to provide the reader with a current perspective on the rapidly evolving science of landscape ecology. To do so requires a broad coverage of both theoretical and practical issues and viewpoints. Although discussions continue regarding the nature and content of landscape ecology, it is clear that a single perspective integrating this evolving science has yet to emerge. The first section of this book, Issues and Challenges in Landscape Ecology, discusses these philosophical questions and their practical consequence. The first chapter by Paul Risser stresses the need for landscape ecology to be viewed and practiced within the context set by rapidly changing environmental and economic conditions produced by human population growth, resource depletion, and resulting changes in societal conditions. Chapter 2 by Richard Hobbs points out the critical need for increased participation of landscape ecology (and landscape ecologists!) in landscape planning decisions. Examples from Australia are used by Hobbs to demonstrate that to be successful, landscape ecology requires mutual interaction between research and management to insure adequate information transfer, decision support, and conflict resolution. Ann Bartuska presents a specific example of forest ecosystem management within a landscape framework that includes political, physiographic, and societal components. Richard Foreman examines landscape ecology and its need for practical applications as society produces new two-dimensional structures and patterns (e.g., roads, patches of natural vegetation within suburban developments, etc.) that affect ecological flows within these altered landscapes. In all the above articles the concept illustrated is the utility of landscape ecology as an interdisciplinary science for assessment and solutions.

Computer applications, remote sensing, and GIS developments continue to make significant contributions in landscape ecological analysis. Indeed, we can do much more today with a common desktop PC today than could be done with the most powerful mainframes a decade ago. The next section of the book, Modeling Applications, highlights three new computer applications designed to assist management and planning activities. Ecosystem management requires both new tools and technologies to analyze the problems and derive workable solutions. Robert Coulson and co-authors employ knowledge engineering and object-oriented progamming to develop an expert system that addresses ecosystem management from planning, problem-solving, and decision-making perspectives. A specific example of landscape ecological modeling as applied to ecosystem management is presented by Michael Childress and co-authors. Their hierarchical, mechanism-oriented model simulates ecological dynamics from fine to broad scales. Models of this type can be used as management tools is a variety of ecosystems under different management scenarios and at various spatial scales. Landscape models that relate ecological processes to spatial pattern are finding more and more use as a strategic planning tool. Eric Gustafson has developed a simulation model that is combined with GIS software to produce a stochastic implementation of management decisions. Output from this application can be used to apply principles of landscape ecology to arrive at decisions for managing landscapes.

The difficulties of satisfying multiple objectives faced in policy and planning activities is addressed by three chapters in the next section of the book, Planning Strategies. Landscape ecology is continuing to take advantage of the immense data available from remote sensing. Denis White and co-authors combine remotely sensed data with a number of national and regional databases to develop a multiscale, hierarchical method with the use of regression tree analysis to assess biodiversity. Their method provides a framework for decentralizing resource management decisions to more local levels, while maintaining the larger spatial perspectives required for sustainable resource use. Conservation biology and landscape ecology are inextricably related. This need to address questions of conservation biology at landscape scales is discussed by Stephen Polasky and Andrew Solow. Because economics often drive the conservation strategies, Polasky and Solow pursue the development of a technique that optimizes biological diversity under a constrained economic status. The final chapter of this section by Jack Ahern attacks the problem of sustainable landscape planning. Ahern uses experience and methods derived by European landscape architects to integrate the disciplines of ecology and planning into landscape ecology. He argues that the landscape scale is the only viable unit for sustainable planning.

The rapid development of landscape ecology forces concepts and methodologies to continually adjust. This is most evident in improvements in methods and applications involving spatial analysis of both pattern and process at landscape scales. Changes in accessibility and sophistication of today's software compared to that used in the earlier volume on quantitative methods by Turner and Gardner (1990) clearly marks the progress being made. Concepts and caveats concerning the use of software for spatial analysis are addressed in the next section of the book, Concepts, Methodological Implications, and Numerical Techniques. The relationship between pattern and processes operating at different scales is fundamental not only in landscape ecology, but at all levels of biological organization. The term scale and spatial and temporal scaling issues have virtually exploded in the ecological literature in the past decade (e.g., Ehleringer and Field 1993). Mark Withers and Vernon Meentemeyer offer a thorough treatise on the use (and misuse) of the term scale and incorportate it as an essential element in landscape ecological studies. Progress in remote sensing technology is yielding extensive matrices of land cover data and associated attributes. As computer hardware and software (particularly GIS) have made tremendous advances, so has the attempts to perform spatial analysis of the data. Marie-Joseé Fortin provides a thorough primer on the use of spatial statistics. The chapter by Robert Gardner presents a computer program that not only analyzes spatial pattern but also provides a suite of methods for generating neutral landscape models. Bruce Milne, Alan Johnson, and Steven Matyk take some of the mystery, but none of the magic, out of the use of fractal geometry for quantitative analysis of spatial patterns. This program also provides interactive learning exercises allowing a novice to quickly reach expert status in the principles and concepts of fractal patterns and analysis.

It is not surprising that there has been a significant lag between the development of the science of landscape ecology and classroom instructional materials. Each of the above chapters provides excellent sources of new materials for classroom exercises. The next section, The Teaching of Landscape Ecology, contains a single chapter by Scott Pearson, Monica Turner, and Dean Urban that integrates materials each author has successfully used in the classroom. These exercises will be a valuable asset to those developing new courses and wishing to demonstrate principles and applications of landscape ecology to both undergraduate and graduate students.

The urgency and difficulty of the issues (both empirical and theoretical) facing the young science of landscape ecology has, as John Wiens has written in the final chapter, created an identity crisis. Because this crisis is due to the interdisciplinary approach to problems addressed by landscape ecologists, it defies simple resolution. What is important, as Wiens eloquently argues, is that the science should be based on "clear logic, sound design, careful measurement, quantitatively rigorous and objective analysis, and thoughtful interpretation." When these conditions are met, a pluralistic approach that embraces complexity is a necessary and useful approach for advancing knowledge of pattern and process at landscape scales.

The chapters of this book offer a broad perspective of the science of landscape ecology. As this science continues to evolve we hope that this volume will serve as a catalyst to further stimulate the development of the science and the utility of its applications.

Acknowledgments

The chapters presented here stem principally from a series of workshops held at the 11th and 12th annual meetings at the U.S. International Association of Landscape Ecology in Galveston, Texas, and Durham, North Carolina, respectively, in 1996 and 1997. Jeff Klopatek, as Program Chair for the Galveston meetings, organized several workshops designed as interactive sessions on applications of landscape ecology ranging from ecosystem management to instructional techniques. The plenary session speakers for the Galveston meetings were Paul Risser, Ann Bartuska, and Richard Hobbs, who were invited to present current issues and challenges for landscape ecologists. The results of these workshops and plenary sessions, along with the contributions of a number of other authors, were the stimulus for this volume to present both the present state and future direction of landscape ecology.

The contributions of each author in this book are greatly appreciated. The efforts to condense, explain, and illustrate their presentations have required a significant amount of time. All of the chapters were peer-reviewed, and we are grateful to the following individuals for their thoughtful comments and suggestions: John Bissonette, David Breshears, Curt Flather, Joyce Francis, Tim Keitt, Joyce King, Robert Lackey, Sam Schreiner, Fitz Steiner, Sandra Turner, Dean Urban, Jianguo Wu.

JEFFREY M. KLOPATEK
ROBERT H. GARDNER

References

Chapin, F.S. III, O.E. Sala, I.C. Burke, J.P. Grime, D.U. Hooper, W.K. Lauenroth, A. Lombard, H.A. Mooney, A.R. Mosier, S. Naeem, S.W. Pacala, J. Roy, W.L. Steffen, and D. Tilman. 1998. Ecosystem consequence of changing biodiversity. Bioscience 48:45–52.

Ehleringer, J.R., and C.E. Field (eds.). 1993. Scaling physiologic processes: leaf to globe. Academic Press, New York.

Forman, R.T.T., and M. Godron. 1986. Landscape ecology. John Wiley & Sons, New York.

Intergovernmental Panel on Climate Change (IPCC). 1995. Climate change 1995: Impacts, adaptations, and mitigation. Summary for Policy Makers. World Meteor. Org. and U.N. Environ. Prog., Montreal.

Turner, M.G., and R.H. Gardner (eds.). 1990. Quantitative Methods in Landscape Ecology. The Analysis and Interpretation of Landscape Heterogeneity. Springer-Verlag, New York.

Wiens, J.A. 1992. What is landscape ecology, really? Landscape Ecology 7:149–150.

Contents

Preface v
Contributors xv

Part I Issues and Challenges in Landscape Ecology

1. Landscape Ecology: Does the Science Only Need to Change at the Margin? 3
   Paul G. Risser

2. Clark Kent or Superman: Where Is the Phone Booth for Landscape Ecology? 11
   Richard J. Hobbs

3. Cross-Boundary Issues to Manage for Healthy Forest Ecosystems 24
   Ann M. Bartuska

4. Horizontal Processes, Roads, Suburbs, Societal Objectives, and Landscape Ecology 35
   Richard T.T. Forman

Part II Modeling Applications

5. A Knowledge System Environment for Ecosystem Management 57
   Robert N. Coulson, Hannu Saarenmaa, Walter C. Daugherity, E.J. Rykiel, Jr., Michael C. Saunders, and Jeffrey W. Fitzgerald

6. A Multiscale Ecological Model for Allocation of Training Activities on U.S. Army Installations 80
   W. Michael Childress, Terry McLendon, and David L. Price

7. HARVEST: A Timber Harvest Allocation Model for Simulating Management Alternatives 109
Eric J. Gustafson

Part III Planning Strategies

8. A Hierarchical Framework for Conserving Biodiversity 127
Denis White, Eric M. Preston, Kathryn E. Freemark, and A. Ross Kiester

9. Conserving Biological Diversity with Scarce Resources 154
Stephen Polasky and Andrew R. Solow

10. Spatial Concepts, Planning Strategies, and Future Scenarios: A Framework Method for Integrating Landscape Ecology and Landscape Planning 175
Jack Ahern

Part IV Concepts, Methodological Implications, and Numerical Techniques

11. Concepts of Scale in Landscape Ecology 205
Mark Andrew Withers and Vernon Meentemeyer

12. Spatial Statistics in Landscape Ecology 253
Marie-Josée Fortin

13. RULE: Map Generation and a Spatial Analysis Program 280
Robert H. Gardner

14. ClaraT: Instructional Software for Fractal Pattern Generation and Analysis 304
Bruce T. Milne, Alan R. Johnson, and Steven Matyk

Part V The Teaching of Landscape Ecology

15. Effective Exercises in Teaching Landscape Ecology 335
Scott M. Pearson, Monica G. Turner, and Dean L. Urban

Part VI Synthesis

16. The Science and Practice of Landscape Ecology 371
John A. Wiens

Index 385

Contributors

JACK AHERN
Department of Landscape Architecture and Regional Planning, University of Massachusetts, Amherst, Amherst, MA 01003, USA.

ANN M. BARTUSKA
Director, Forest Health Protection, USDA Forest Service, Washington, DC 20250, USA.

W. MICHAEL CHILDRESS
Shepard Miller, Inc., Fort Collins, CO 80525, USA.

ROBERT N. COULSON
Department of Entomology, Texas A&M University, College Station, TX 77842, USA.

WALTER C. DAUGHERITY
Department of Computer Science, Texas A&M University, College Station, TX 77842, USA.

JEFFREY W. FITZGERALD
Department of Entomology, Texas A&M University, College Station, TX 77842, USA.

RICHARD T.T. FORMAN
Graduate School of Design, Harvard University, Cambridge, MA 02138, USA.

MARIE-JOSÉE FORTIN
Départment de Géographie, Université de Montréal, Montréal, Québec H3C 3J7, Canada.

KATHRYN E. FREEMARK
National Wildlife Research Center, Canadian Wildlife Service, Environment Canada, Ottawa K1A 0H3, Canada.

ROBERT H. GARDNER
Appalachian Laboratory, University of Maryland, Frostburg, MD 21532, USA.

ERIC J. GUSTAFSON
U.S. Forest Service, North Central Forest Experiment Station, Rhinelander, WI 54501, USA.

RICHARD J. HOBBS
Division of Wildlife and Ecology, CSIRO, Wembly WA 6014, Australia.

ALAN R. JOHNSON
Department of Geography, New Mexico State University, Las Cruces, NM 88003, USA.

A. ROSS KIESTER
USDA Forest Service, Corvallis, OR 97331, USA.

STEVEN MATYK
Department of Biology, University of New Mexico, Albuquerque, NM 87131, USA.

TERRY MCLENDON
Department of Biological Sciences, University of Texas at El Paso, El Paso, TX 79968, USA.

VERNON MEENTEMEYER
Department of Geography, University of Georgia, Athens, GA 30602, USA.

BRUCE T. MILNE
Department of Biology, University of New Mexico, Albuquerque, NM 87131, USA.

SCOTT M. PEARSON
Department of Biology, Mars Hill College, Mars Hill, NC 28754, USA.

STEPHEN POLASKY
Department of Agriculture and Resource Economics, Oregon State University, Corvallis, OR 97331-3601, USA.

Eric M. Preston
U.S. Environmental Protection Agency, Corvallis, OR 97333, USA.

David L. Price
U.S. Army Corps of Engineers, Construction Engineering Research Laboratory, Champaign, IL 61826, USA.

Paul G. Risser
Office of the President, Oregon State University, Corvallis, OR 97331-2128, USA.

E.J. Rykiel, Jr.
Department of Environmental Science, Washington State University, Pullman, WA 99164, USA.

Hannu Saarenmaa
European Environmental Agency, DK-1050 Copenhagen, Denmark.

Michael C. Saunders
Department of Entomology, Pennsylvania State University, University Park, PA 16802, USA.

Andrew R. Solow
Marine Policy Center, Woods Hole Oceanographic Institution, Woods Hole, MA 02543, USA.

Monica G. Turner
Department of Zoology, University of Wisconsin, Madison, WI 53706, USA.

Dean L. Urban
School of the Environment, Duke University, Durham, NC 27708, USA.

Denis White
U.S. Environmental Protection Agency, Corvallis, OR 97331, USA.

John A. Wiens
Department of Biology and Graduate Degree Program in Ecology, Colorado State University, Fort Collins, CO 80523, USA.

Mark Andrew Withers
Department of Geography, University of Georgia, Athens, GA 30602, USA.

Part I
Issues and Challenges in Landscape Ecology

1
Landscape Ecology: Does the Science Only Need to Change at the Margin?

Paul G. Risser

In 1983, 25 scientists from the United States and elsewhere gathered for a workshop designed to define the concept of landscape ecology as it applied to the United States, to draw conclusions about the future of landscape ecology as a developing discipline, and to describe the key scientific questions (Risser et al. 1984). The central questions at that time included the following:

1. How are fluxes of organisms, material, and energy related to landscape heterogeneity?
2. What formative processes, both historical and present, are responsible for the existing pattern in a landscape?
3. How does landscape heterogeneity affect the spread of disturbance?
4. How can natural resource management be enhanced by a landscape ecology approach?

The essence of these four questions was included in a rather long and complicated operational definition of landscape ecology. In addition, the workshop concluded that further advancement of landscape ecology depended on (1) generating fundamental concepts for understanding the relationships between spatial and temporal scales, (2) making comparisons of managed and natural systems at multiple space and time scales, (3) building connections between population-level and ecosystem-level processes, and (4) involving a broad spectrum of natural, physical, and social scientists. It was also concluded that methods should be developed for managing and analyzing large amounts of data about landscape characteristics and processes, both as tools for scientists and as ways to share data with resource managers.

A more recent review of landscape ecology (Pickett and Cadenasso 1995) offered a simpler definition, namely, as *the study of the reciprocal effects of spatial pattern on ecological processes*. In this discussion, as well as an earlier one (Risser et al. 1984), there was a major focus on the structure of the landscape, especially as this structure relates to patches of different types of ecosystems and the flows between and among these patches. Pickett and

Cadenasso (1995) emphasized that the landscape is not static, but rather a "shifting mosaic" of patches and processes. Despite the refinement toward a more dynamic view of landscapes, much of the discussion in the 1995 paper is very similar to the ideas included in the 1984 paper (Risser 1995).

Progress in Landscape Ecology over the Past 15 Years

Landscape ecology today still tends to focus on the same issues described 15 years ago, although progress has been made. The pervasive consideration of landscape-level processes in many ecological studies is a powerful indication that the field of ecology itself now routinely and formally includes the concepts of landscapes into its experimental design. New approaches to examining landscape processes occur at many levels of field study, and landscape concepts are now invoked in most discussions of policies relating to resource management and preservation.

The concept of biodiversity exemplifies the challenge of bringing scientific information and values to a public policy issue, especially one that has large economic consequences. Landscape ecologists have contributed significantly to the challenge of providing guidance to those who wish to preserve biological diversity by protecting species habitats or those ecosystems relatively rich in biological diversity. There are now several models with "population viability analyses" that attempt to estimate the long-term viability and reproductive success of individual populations (Grumbine 1990). In addition, landscape ecologists have used population viability analyses and various mapping techniques to identify high-priority locations for preservation, including those that include taxonomic uniqueness, rarity, number of listed threatened endangered species, uniqueness of ecosystem type, biodiversity "hot spots" where there are concentrations of endemic species, magnitude and severity of threat, and some estimate of the overall chance of preservation success (Mengel and Tier 1993; Prendergast 1993). However, rarely do these analyses consider the actual costs of preservation as a process itself, and almost never is there a comparison of lost opportunity costs. This lack of complete economic comparison is simple to understand: ecological goods, services, and values cannot be currently calculated in ways that will withstand rigorous scrutiny and that will be evaluated with favor in comparisons to economic benefits and costs (Daily et al. 1997). So, because of the absence of a comparable economic system and because many natural features, including species and ecosystems, are considered to have personal and social values outside conventional economic systems, the cost analyses of preserving sections of the landscape are rarely calculated. So, even though landscape ecology concepts are now routinely invoked in discussions of biodiversity, the characterizing landscape variables are not different from those routinely used in conventional ecological and systematic studies. As will be noted later, more realistic landscape studies will invoke more relevant economic considerations.

The processes that characterize and impact landscapes occur at many spatial and temporal scales. Some processes occur at both broader and finer scales than the landscape, although the landscape itself is usually defined in the context of the question or issue being studied. Landscape ecologists have made some progress in scaling up ecological processes from fine scale measurements, and conversely, translating broader spatial processes into smaller scales more compatible with customary ecological field measurements (Root and Schnider 1995). For example, it is quickly obvious that there would be significant inaccuracies in simply scaling up the gas exchange rates from individual leaves to forests across the landscape, because leaves have many different physiological states and operate in a wide variety of microhabitats. Similarly, if one were to attempt to scale down, for example, the broad scale Holdridge life zone concept to fine-scale local habitats, transitional states would be missed and successional and migration processes would be ignored. Thus, there is a significant challenge in translating ecological processes and phenomena from one spatial and temporal scale to another (Rosswall et al. 1988). The most significant advances have involved linking geographic information system spatial data bases with ecological process models (Starfield and Chapin 1996). Further advances may involve cyclical or iterative modeling where processes are modeled at different scales (Root and Schnider 1995).

Ecosystem Management

Much of the current focus on "ecosystem management" can be derived from earlier landscape ecology studies and from incorporating the broader landscape perspective in formulating natural resource management philosophies and approaches. Ecosystem management embodies many ideas (Cortner et al. 1997), but among the most central are the following:

1. Both understanding and management principles are focused at broader landscape levels rather than on single ecosystem types, and there is a recognition that ecological processes operate at multiple scales.
2. Humans and their needs and impacts are integral parts in the consideration of ecosystems to be managed.
3. It is not just the commercial products that are considered when planning the management of landscape. Decisions must incorporate a range of ecosystem goods, services, and values.
4. The values of ecosystems include not only the natural values, but also economic, social, and cultural values.
5. In making decisions about public natural resources, the public and stakeholders (e.g., landowners) are a part of the process.

6. Ecosystem management includes the notion of adaptive management, that is, management approaches will adapt as conditions change and as different requirements are made of the natural resources.
7. Because of the integral connections among different parts of the landscape, it is frequently not possible to manage public lands without taking into consideration the conditions and effects of private lands in the same or related landscapes.

To its credit, the field of landscape ecology anticipated many of these concepts, which form the backbone of ecosystem management. And because national environmental issues have become so important politically, economically, and socially, landscape ecologists now have the opportunity to participate more broadly in guiding management policies and practices.

This involvement of landscape ecology in ecosystem management raises several new considerations about the relationship between science and policy development. These questions have frequently been raised in the arena of risk assessment, but not in the realm of landscape ecology. For example, there is a legitimate question about the degree to which scientists should be involved in making policy and participating in the decision-making processes related to landscape management. A useful way of examining the issue is to break the process in to three parts: (1) formulation of the question or problem; (2) conduct of the science or scientific interpretation of the data; and (3) promulgation of a policy, regulation or management process. Traditionally scientists have been involved in the second component—that of conducting the science. However, the landscape perspective, with its focus on interactions of humans with ecological processes across heterogeneous regions, suggests that scientists should also be involved in formulating the question and framing the issues. This is because the potential interactions are numerous and complex, thus requiring an understanding of these processes before deciding upon the most important considerations to be addressed in the analysis. In the final step, namely of establishing policy, the scientists may have no particular contribution, because policies represent a statement about society values. However, even in this last step, scientists have the responsibility to predict the consequences of adopting and implementing any potential policy. So, because landscape ecology deals with issues at scales of human concern and because so many crucial natural resources issues must be addressed at landscape spatial scales, landscape ecology has been thrust into a prominent role in decision making and policy formulation.

Ecological Economics

In the last few years, there have been numerous discussions about "ecological economics" and the economics of sustainable ecological systems. To place the discussion in stark contrasts, some mainstream economists have argued that with respect to the earth's carrying capacity, there are no set limits to economic growth and that the earth's carrying capacity is a function of human ingenuity, preferences, knowledge, and technological advances (Sagoff 1995). On the other hand, ecological economists reject the idea that technology and resource substitution (ingenuity) can continuously outrun depletion of natural resources and pollution. They believe that sources of raw materials and sinks for wastes (natural capital) are relatively fixed in the biosphere and therefore limit the growth of global economy. To this contention, mainstream economists argue that with "the application of knowledge to knowledge, we can always obtain the other resources" and that the "reserves" of natural resources themselves are actually functions of technology. For example, new technologies allow humans to extract more resources or allow humans to find substitutes for resources that become scarce. The power of knowledge, mainstream economists would argue, continually reduces the amounts of resources needed to produce a constant or increasing flow of consumer goods and services (Sagoff 1995). Indeed, there is little empirical evidence that natural resource scarcity will restrict human production in the broadest sense, at least in the next few decades, and in fact, the absence of economic growth and strength is the principal cause of some ecologically destructive practices leading to deforestation, soil erosion, and loss of biological diversity.

These two extreme views do not recognize many uncertainties, such as whether or not incomes, preferences and technology will remain the same over future generations. Castle et al. (1996) place this recognition as a third view, namely that "uncertainty is the dominant economic condition describing the relationship of natural to human capital and resources in the distant future. The nature of the substitution relation between human and natural capital in the distant future is unknown." This view argues that cultural capital (including human legal structures and institutions) is an important part of the total capital stock, that not all natural capital is likely to be or can be preserved, and recognizes the importance of evolutionary processes in both the economic and ecological spheres. Obviously, the adoption of any one of these three options has great implications for the appropriate policies to be used in managing landscapes. Further, it is important to recognize that although ecological processes are surely important in landscape behavior, human activities on landscapes (e.g. logging, grazing) are human processes driven by economics and politics (Castle et al. 1996).

Currently there is no way to resolve the three points of view. Yet this discussion, and the implications of choice among the options, is of great importance to landscape ecology. If one believes that natural capital can be largely replaced by human capital (e.g., human ingenuity),landscape ecologists should focus on the incorporation of technology in landscape studies. If there is less substitution, then landscape ecologists need to focus on decisions about which resources must be preserved and how landscape processes can be used to predict the most optimum strategy. If the third, uncertainty option is selected, then landscape ecologists should focus on adaptive management techniques geared to benefit from experience and on human institutions that can synthesize information from political and economic considerations as well as ecological considerations.

Social Forces

There is a great danger that landscape ecology will focus on traditional science, retaining an avid interest in quite legitimate research questions. In doing so, however, it may become of only marginal importance, as landscape ecologists will not be connected with the social forces that shape society. Thus, because decisions are made within a social and political setting, landscape ecology must become an integral part of that setting.

In "The Coming Anarchy," a controversial article, R.D. Kaplan (1994) wrote that the environment will be the national security issue of the twenty-first century. He argues that to understand the next 50 years, one must understand environmental scarcity, cultural and racial clashes, geographic destiny, and the transformations caused by war. In essence, the argument is that as natural resources become more scarce, especially in developing countries where the human population is increasing most rapidly, there will be increased conflicts over these resources. These clashes, driven by geographic differences in natural resource productivity exacerbated by cultural differences, will be most severe in countries where central governments have decreasing abilities to govern. Therefore, there will be a rise in tribal and regional domains, a growing persuasiveness of war, and less attention will be devoted to implementing and managing sustainable productive ecosystems, and to the protection and preservation of natural resources (Kaplan 1996).

This pessimistic scenario may not evolve and these conditions may be limited to isolated localities rather than being widespread. However, this possible set of circumstances points to the need for a much broader concept of landscape ecology. For example, the current rather benign landscape ecology maps of species distributions, habitat and land use types, and preservation corridors will be insufficient to address the real issues of landscape ecology of the future. If the science is really to be useful, landscape ecology maps must include such variables as human cultural group identities, political system characteristics, flows of refuges from cultural conflict and resource scarcity, influence zones of private security forces, and flows of national currencies and their international market forces. The incorporation of social sciences into landscape ecology will move well past simply conducting surveys of personal preferences and values as is done today, into a hardened analysis of political systems operating against a complex ecological, geographical, and cultural background.

Conclusions

At first glance, it could be assumed that the field of landscape ecology has been a great success. A number of the technical issues have been addressed (e.g., managing large data sets, scaling processes among different spatial and temporal scales), and the whole concept of "ecosystem management" includes many of the tenants of landscape ecology. Indeed, a quick review of the contents of most professional scientific journals in the ecological field reveals many titles incorporating the term "landscape" or at least an indication that the study is at the landscape level or scale. However, one is also struck with the notion that most of the fundamental questions in the field of landscape ecology have not changed in the last 15 years. And therein lies the problem. Landscape ecology began as a perspective in the ecological sciences, largely because it was obvious that there were significant interactions among ecosystems and because preserves for biological diversity required a larger spatial framework for protection. The relevant scientific questions for landscape ecology derive from these two sources, and the questions remain the same today.

To be successful in enriching our understanding of the landscape and to assist in making meaningful contributions to natural resource management, landscape ecology as a discipline must change. And this change cannot be marginal—it must undergo a far more sweeping change. To be sure, landscape ecology must continue to study *the reciprocal effects of spatial pattern on ecological processes*, and models must be developed to organize and extend ecological understanding of landscapes. However, these basic scientific studies of landscapes must now be radically expanded to include rigorous economic analysis, not just of the local landscape, but of intersecting economic systems including the global scale. Cultural differences and social systems of the inhabitants and users of the landscape will become important variables in the landscape geographical information systems. The characteristics of political systems and of the politicians themselves are variables just as or more important than soil type or the average size of various habitat types. Thus, landscape ecology projects in the future will include expertise in economics, politics, cultural anthropology, and several other disciplines. The involvement of these other disciplines will not be just cursory rhetoric or superficial applications of current techniques. Rather, these other disciplines will need to develop new approaches to landscape ecology, and they must be involved in all three parts, namely, formulation of the question or problem; conduct of the science or scientific interpretation of the data; and promulgation of a policy, regulation, or management process. Landscape ecology is at the crossroads. As a field it can continue along the conventional pathway or it can reconstruct itself by focusing on questions in a social, political, and cultural context as fully as it focuses on traditional ecological questions. If the field takes the latter (and correct) approach, then it will need to involve expertise in these areas just as deeply as the field now involves ecologists, soil scientists and climatologists. It is a very exciting prospect.

References

Castle, E., R.P. Berrens, and S. Polasky. 1996. The economics of sustainability. National Resources Journal 36(4):475–490.

Cortner, J.C., J.C. Gordon, P.G. Risser, D.E. Teeguarden, and J.W. Thomas. 1998. Ecosystem management: evolving model for stewardship of the nation's natural resources. Proceedings of the Ecological Stewardship Conference (Tucson, Arizona, December 1995) In press.

Daily, G.C., S. Alexander, P.R. Ehrlich, L. Goulder, J. Lubchenco, P. Matson, H.A. Mooney, S. Postel, S.H. Schnider, P. Tilman, and G.M. Woodwell. 1997. Ecosystem Services: benefits supplied to human societies by natural ecosystems. Ecological Society of America, Washington, DC.

Grumbine, R.E. 1990. Viable populations, reserve size, and Federal lands management. Conservation Biology 4:127–134.

Kaplan, R.D. 1994. The coming anarchy. Atlantic Monthly, Feb.:44–76.

Kaplan, R.D. 1996. The Ends of the Earth. Random House, New York.

Mengel, M., and C. Tier. 1993. A simple direct method for finding persistence times of populations and applications to conservation problems. Proceedings of the National Academy of Sciences (USA) 90:1083–1086.

Pickett, S.T.A., and M.L. Cadenasso. 1995. Landscape ecology: spatial heterogeneity in ecological systems. Science 269:331–334.

Pendergast, J.R., R.M. Quinn, J.H. Lawton, B.C. Eversham, and D.W. Gibbons. 1993. Rare species, the coincidence of diversity hotspots and conservation strategies. Nature 365:335–337.

Risser, P.G. 1995. The Allerton Park workshop revisited–A Commentary. Landscape Ecology 10:129–132.

Risser, P.G., J.R. Karr, and R.T.T. Forman. 1984. Landscape ecology: directions and approaches. Illinois Natural History Survey Special Publication, Number 2.

Rosswall, T., R.G. Woodmansee, and P.G. Risser. 1988. Scales and Global Change. Spatial and Temporal Variability in Biospheric and Geospheric Processes: A Summary. In: SCOPE 35, pp. 1–10. T. Rosswall, R.G. Woodmansee, and P.G. Risser (eds.). John Wiley and Sons, Chichester, UK.

Root, T.L., and S.H. Schnider. 1995. Ecology and climate: research strategies and implications. Science 269:334–341.

Sagoff, M. 1995. Carrying capacity and ecological economics. BioScience 45:610–662.

Starfield, A.M., and F.S. Chapin III. 1996. Model of transient changes in arctic and boreal vegetation in response to climate and land use change. Ecological Applications 6:842–864.

2
Clark Kent or Superman: Where Is the Phone Booth for Landscape Ecology?

RICHARD J. HOBBS

We live in a rapidly changing world in which the pressures of increasing human populations and their impacts on the world's ecosystems force us to consider complex problems concerning land use policy and management. We are increasingly realizing the need to deal with these issues at the relevant scale, which in many cases is at the landscape or regional scale. The underlying reason for this is the prevalence of patterns and processes which operate at these larger scales. The science of landscape ecology has been developing in response to this recognition that examination of the traditional scales studied in ecology (i.e., the quadrat, plot, and stand) can only provide part of the overall picture and may often omit some of the most important broader-scale and cross-boundary patterns and dynamics. Landscape ecology is thus poised to play a central role in finding solutions to many of today's land management and conservation problems. It is appropriate, therefore, to assess the current status of landscape ecology, not only in terms of its theoretical constructs and scientific rigor, but also in terms of its ability to provide principles and methodologies applicable to management and planning.

This chapter examines various aspects of landscape ecology in this context, and suggests that the ingredients of a thriving and successful landscape ecology include the following:

1. Effective tools,
2. Integration and cooperation, and
3. Relevance and participation.

Effective tools take the form of theoretical constructs, methodologies, and technologies that advance the science of landscape ecology by facilitating the study of landscape-scale patterns and processes and by allowing the development and testing of sound hypotheses and models. Integration and cooperation refer to the need for integration of information and approaches from many different disciplines to deal with complex environmental problems and issues and the cooperation among disciplines needed to accomplish this. Relevance refers to the relationship of these hypotheses and models to the types of problem frequently encountered by land managers and planners. Participation relates to the willingness of landscape ecologists to be involved in the planning and management process and actively participate in goal setting and decision making.

Effective Tools

For landscape ecology to contribute usefully to management, planning, and decision making, it must have a set of tools that contains the basic scientific components of a sound conceptual framework and effective methodologies and analytical techniques, together with mechanisms for applying these.

My recent assessment (Hobbs 1996) of the development of landscape ecology in terms of its methodologies and analytical techniques repeated a survey carried out by Wiens (1992). Wiens' initial survey examined the content of the first five volumes of the journal, *Landscape Ecology*, and gave an indication that landscape ecology concentrated mainly on the spatial scale of hectares to square kilometers and consisted of a relatively high proportion of purely descriptive and nonquantitative studies. Of particular note was the fact that few studies contained any experimentation. The results for volumes five through ten indicated a shift away from descriptive, nonquantitative studies toward more quantification and use of statistics (Fig. 2.1). In addition, a greater emphasis on modeling and development of methodology was apparent. Notably, however, there was no increase in the level of experimentation. There was little change in the relative importance of particular subject areas, although Weins' categories concentrated exclusively on biophysical areas.

It appears, therefore, that landscape ecology is developing away from being mostly descriptive to being more quantitative, with an increasing battery of statistical and modeling methodologies. The development of the theoretical basis for the science can be assessed by comparing the books produced by Richard Forman in 1986 and 1995 (Forman and Godron 1986; Forman 1995). While the former, together with Naveh and Lieberman (1984), provided the basis for a theoretical framework for landscape ecology, it lacked substance in many areas. The latter book, on the other hand, incorporates a rich variety of literature published in the intervening ten years and builds from this a considerably stronger framework. Although there is still a long way to go in many areas and not everyone will agree with Forman's formulations, it is clear that the conceptual basis for landscape ecology is solidifying rapidly.

In terms of methodologies, the lack of experimentation still prevalent in the discipline is viewed by some as a problem. Considerable debate in ecology in general has centered on the contribution that different methodologies may make. Methods of obtaining information in ecology include observation, experimentation, and modeling. Much early ecology was

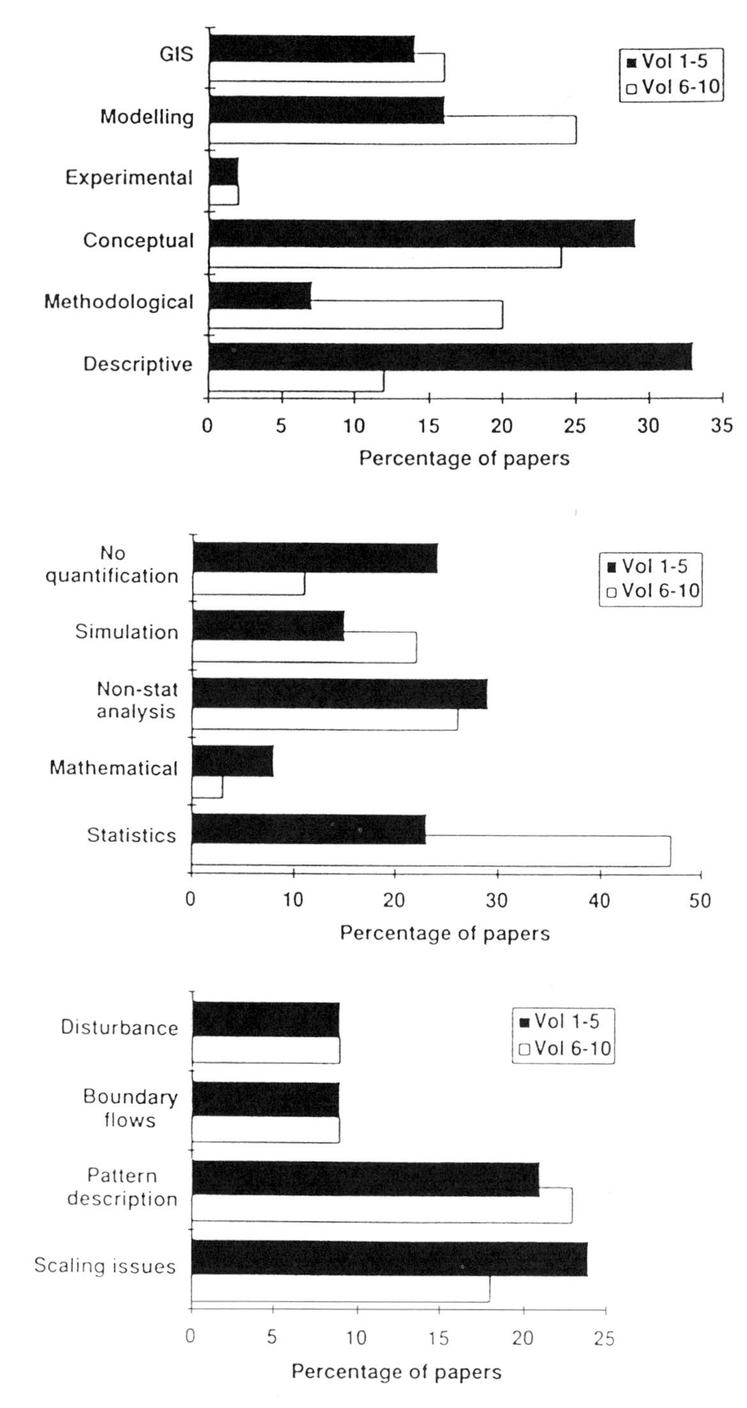

FIGURE 2.1. An analysis of all papers contained in Volumes 1–10 of the journal *Landscape Ecology* examining the approaches taken (*top*), quantification methods used (*middle*) and topics considered (*bottom*). Data for Volumes 1–5 are from Wiens (1992), who set the range of categories examined; data from Volumes 6–10 follow the same criteria as used by Wiens. Reproduced from Hobbs (1996), with kind permission from Elsevier-NL Sara Burgerhartstraat, 1055 KV Amsterdam, The Netherlands.

purely observational and descriptive, but increasing emphasis has been placed on the role of experimentation and the importance of falsificationism. Indeed, some ecologists argue that only by using replicated, statistically robust experimentation and falsifiable hypotheses can we be truly scientific (Underwood 1990; Underwood and Chapman 1995). Thus, landscape ecology, with geology and astronomy, must be unscientific!

However, other commentators have discussed the scientific method in relation to ecology and have pointed out that experimentation and strict falsificationism are only part of the scientific process (Pickett et al. 1994). The development of ecological understanding depends on a rich tapestry of methodologies, all of which can play important roles depending on the questions being asked and the systems being studied. Indeed, an undue emphasis on experimentation severely limits the possibility of ever progressing far in the study of complex or broad-scale systems and/or long-term processes.

Dogmatic adherence to classical experimentation as the primary scientific method leads almost inevitably to fine-scale reductionist approaches to problems. This is undoubtedly appropriate in some instances. The Rothampstead and Cedar Creek experiments have, for instance, provided large amounts of valuable information on plant community responses to varying environmental and management regimes (Johnston 1991; Tilman and Downing 1994). It can, however, be argued that few ecological experiments are carried out at broad enough spatial scales or over long enough time periods to provide a clear answer to many important ecological questions (Tilman 1989; Magnuson 1990, 1995). Extrapolation of experimental results from fine to broad spatial scales can be fraught with problems and open to misinterpretation. For instance, Murphy (1989) strongly questioned the validity of results from fine-scale experimental work on fragmentation carried out by Robinson and Quinn (1988). Thus, while fine-scale experimentation may yield some useful information of relevance to broad-scale processes, its validity in broad-scale contexts must always be carefully assessed. This is also true of the proposal to study landscape processes using fine-scale model systems (Wiens et al. 1993).

Having pointed out the problems of using fine-scale experimentation to study broad-scale processes, it is also appropriate to point out the problems associated with broad-scale experimentation. As we move to broader scales, control over individual variables becomes increasingly difficult, true replication becomes almost impossible, and the results of an experiment may become almost impossible to interpret. Hargrove and Pickering (1992) have discussed in detail the problems associated with experimentation at broad spatial scales and have pointed out that pseudoreplication, regarded as a major offense in classical ecology (Hurlbert 1984), is almost inevitable in landscape ecology. Elsewhere, I have also pointed out the problems associated with achieving true replication, even within a classical experimental design, where spatial processes such as disturbances are patchily distributed in space and time (Hobbs and Mooney 1995). Nicholls and Margules (1991) have also provided a clear illustration of the problems related to setting up rigorous experiments on landscape-scale processes, in particular the effect of confounding factors in swamping out any treatment response.

Where, then, does this leave experimentation in landscape ecology? Is landscape ecology a science like geology and astronomy, in which experimentation is essentially impossible? Clearly, some broad-scale experimentation has been very successful, for instance in watershed studies such as at Hubbard Brook (Bormann and Likens 1977). Note, however, that even these well-known experiments would not satisfy the purists because they are unreplicated. However, the fact that much useful information has been derived from them indicates that a relaxation of rigor in experimental design does not immediately consign a study to the scrap heap. It does, however, recognize the practical limitations placed on broad-scale studies and indicates that some sort of compromise between rigor and reality can be reached. So-called "natural experiments", in which natural events such as broad-scale disturbances occur, provide invaluable opportunities and should not be ignored. Similarly, management actions often provide opportunities for a quasi-experimental approach. Indeed, there is a need for greater interaction between research and management, in which management activities are viewed as experiments and managers become part of the research process (Walters and Holling 1990; Underwood 1995). Adaptive management and Adaptive Environmental Assessment and Management (AEAM) models provide fruitful areas for developing this approach.

Experiments are also possible on model systems, and these may provide insights into landscape processes, *if* they are conducted and interpreted in the context of broad-scale studies and a battery of other approaches. Herein lies the crux of the solution to the problem. Experimentation must take its place along with the suite of other methodologies available in an integrated research approach. These methodologies include spatial statistics, correlative studies, and modeling. Pattern recognition and description is essential before any sensible hypotheses can be developed. Tools, such as GIS and remote sensing, increasingly allow the study and interpretation of broad-scale pattern and process. Techniques of pattern description rely on the identification of the relevant aspects of landscape structure, which need to be quantified. Considerable effort has gone into the development of an array of descriptive statistics (Turner and Gardner 1991), but it is debatable whether many of these have been adequately tested as yet. Further, we need to guard against the tendency to describe pattern for description's sake, Pattern description is relevant only in the context of the particular question being asked, and generalized descriptors may have little meaning in relation to specific processes or functional aspects (Cale and Hobbs 1994). It is important not to lose sight of the goal of matching pattern with process (Turner 1989) in our preoccupation with ever more powerful GIS and statistical tools and increasing capacity to produce multicolored images. It is clear that our ability to interpret such images in terms of landscape function lags far behind our ability to create them (Turner et al. 1995; Litvaitis et al. 1996).

Correlative studies can go a long way to producing working hypotheses relating pattern to process and are often the main avenue for obtaining information at broad spatial scales. Here, again, the important elements for success are the identification of the relevant variables to be measured and the recognition of potentially important covarying factors. Modeling is currently the most effective (and often the only) way of doing the types of experiments we would like to do in real landscapes but cannot. Here, too, modeling must be carried out in the context of the real world and in relation to other approaches. Ecology is full of mathematically robust models that are entirely spurious in terms of the real world. Although simplification and abstraction are essential for modeling to be successful, continual cross-checking with reality also assures that the models may actually prove to be useful as well as elegant.

From the above, it is apparent that no one methodology will be sufficient or preeminent in the quest to answer questions about complex large-scale systems. Ensuring an adequate mix of appropriate methods and continual cross-checking of results from each is an important task for landscape ecology. Rather than bemoan the fact that traditional avenues of study are inappropriate, we must be innovative in our approach and develop new methodologies appropriate to the task at hand. Further, we must be prepared to defend this approach against those who would deny its validity.

Integration and Cooperation

One view of landscape ecology is that it is simply ordinary ecology writ large. In other words, it is the study of the kinds of things that ecology usually studies (populations, communities, etc.), but with a better-defined spatial component and often at a scale much broader than that often considered in classical ecology. To some extent this is true, because landscape ecology is concerned with examining structure and function (or pattern and process) at a landscape scale and how these change in relation to changing conditions. The search for pattern and how this relates to process runs through much of ecology. Increasingly, attempts to understand patterns and processes at other levels of organization are looking to landscape-level phenomena. For instance, in population ecology, metapopulation dynamics (Hanski and Gilpin 1991; Harrison 1994) and source-sink relationships (Pulliam 1988) both implicitly assume the importance of landscape patterns or processes. Landscape phenomena are increasingly being put forward as important components of a new paradigm in ecology (Wu and Loucks 1995).

However, landscape ecology is potentially more than simply broad-scale ecology. The adequate study of landscape scale processes involves input from a number of different disciplines. Depending on the question being asked, the relevant factors involved might include ecological aspects of population, community and ecosystem patterns and processes, behavior, climate, hydrology, soil, geomorphology, limnology, ecotoxicology, and many others. Further factors of importance may include socioeconomic and political aspects of the history and patterns of land use and tenure, aesthetics, and human values and perceptions. Landscape ecology thus draws on a wide array of expertise to answer complex, broad (and often multi-) scale questions.

The process of integration of different scientific fields and of building linkages between the sciences and the social sciences is not easy. The prevailing trend in science has been towards increasing fragmentation and specialization, and this is nowhere more true than in ecology. Disciplinary rivalries, the development of "invisible colleges" promoting the development of particular subdisciplines, current reward systems in science, and narrowly defined funding opportunities have positively reinforced this trend (Hobbs and Saunders 1995). It continues to be reinforced by the way in which students are trained at an undergraduate and graduate level, particularly by the strictures of Ph.D. programs that encourage or even demand narrow specialization (Lubchenco 1995; Slobodkin 1995). Scientific training also poorly equips students to deal with the increasing need to communicate with others outside their own discipline (Hobbs 1998). Lack of understanding of other points of view and approaches often precludes effective interaction, and conversely, failure to make our science available in an understandable format frequently leads to it being relegated to the sidelines.

Landscape ecology provides a ready-made arena for overcoming these problems, but still has to address the problems effectively. Landscape ecology encompasses a wide array of disciplines and interests and provides a concrete meeting point for the biophysical and social sciences. For it to achieve its potential as a truly integrative science, however, there needs to be a real recognition of the value and contribution of the different subdisciplines involved.

In particular, many biophysical scientists still pay scant attention to the socioeconomic aspects of their studies. This socioeconomic context is essential for ensuring that scientific questions and outcomes are relevant to the real world and hence liable to be useful in a management and planning context. However, as McHarg (1995) notes, "I have yet to find a physical scientist who is very well informed about biological science. I have yet to find a biological scientist who's even familiar with physical science. I have yet to find either of them who knew very much about social science. I've yet to meet any scientists who know very much about planning. And if you talk about design, they run away screaming." On the other hand, landscape planning and architecture could benefit greatly from increased scientific input to ensure functionality of planned landscapes. Too often form and aesthetics take prevalence over functional aspects. The point is that the approaches are complementary, and their combination could be highly synergistic. At the moment, the majority of interactions, where they occur, seem to be either neutral or antagonistic. The cementing of the link between the biophysical and social components of landscape ecology is essential if landscape ecology is to be applied.

Relevance and Participation

As indicated elsewhere, landscape ecology lies at the interface of many applied disciplines (Hobbs 1996). There is increasing recognition that issues related to global change, conservation biology, and restoration ecology need to be tackled at a landscape level and that land-use planning decisions frequently involve landscape-scale considerations. By the very nature of its subject matter, landscape ecology is an applied science. To what extent, then, is it actually applied? How much landscape ecology currently finds its way into land-use planning decisions? Or into landscape design? The answer is depressingly little.

There is a perception among many in other disciplines that, although the landscape may be the relevant scale at which to study and manage things, landscape ecology has not come up with much that will help. Despite commentaries on the need to link landscape ecology with planning (e.g., Golley and Bellot 1995; Rookwood 1995), much more needs to be done by landscape ecologists of foster this linkage. In the end, the products of landscape ecology (i.e., theory, methodology, etc.) will be assessed not on their intrinsic interest or popularity in the scientific literature, but on the impact they have on the planning and management of real landscapes. There is still a great need for good empirical and theoretical research, especially in the area of linking pattern with process, but there is an equally great need to link this firmly with application.

I have reached this conclusion from the perspective of the work with which I have been associated in the agricultural area of Western Australia. Here, conversion of natural vegetation to agricultural land over the past century has led to dramatic changes in landscape structure and function through alteration of hydrological, ecosystem and biotic processes (Hobbs 1993a; Hobbs et al. 1993b; George et al. 1995; Saunders and Ingram 1995). The rapidity of change and the prospect of dramatic and continuing declines in agricultural production and conservation values over the next few decades has led us to attempt to develop a landscape-ecological "rescue mission," which involves reversing current trends of landscape change.

This approach aims to provide spatially-explicit design principles which will slow down current hydrological changes (rising saline water tables) and halt the current trends of biotic impoverishment, while at the same time allowing a sustainable and economically viable agriculture to continue in the area (Hobbs and Saunders 1991, 1993; Hobbs et al. 1993a). There are many components to this approach, which requires broad-scale reconstruction (Saunders et al. 1993), cooperation among numerous agencies, and considerable community participation (Saunders et al. 1995). To be successful, it also requires that research and management interact strongly and that good mechanisms are available for information transfer, decision making, and conflict resolution. Land allocation models, decision support systems, and AEAM models are all currently under development in the region, expressly to facilitate the interaction between research and management.

The problems in the Western Australian wheatbelt are acute and require immediate action. This has, perhaps, focused our activities more than would have been the case otherwise. However, this situation differs little from that found in many other parts of the world. Acute, broad-scale environmental problems can be found just about everywhere. Chapter 1 discusses developing nations and the pressures placed on the environment by rapidly rising populations, resource depletion, territorial conflicts, and mass migrations. Developed nations, including the United States, are also not immune, as urbanization, water supply problems, ecosystem fragmentation, and land-use conflicts increase. Is it therefore not appropriate to adopt a sense of urgency in landscape ecology as a whole?

Landscape ecologists are faced with a philosophical choice. Are we to be mere recorders of landscape change or do we want to be active in determining the direction and rate of that change? Many scientists view science as an objective pursuit isolated from social and political pressures. With that view, science will be simply a recording of the inevitable. However, many commentators have argued that science is intimately linked with social and political processes (e.g., Levins and Lewontin 1985), and recent failures of scientists to recognize and accept that link have jeopardized the status and potential contribution of science to world problems (Allaby 1995). Nowhere is this more true than in landscape ecology.

The title of this chapter refers to the mythical superhero Superman and his mildmannered alter ego, Clark Kent. Clark was apparently neither equipped for nor interested in solving societal problems. However, whenever such a problem arose, he would find a phone booth and change rapidly into his Superman outfit and deal effectively with the crisis. Landscape ecology currently faces the choice of stepping into the phone booth or not. It can either remain a mild-mannered academic pursuit which has little or no impact on the real world or it can equip itself for action and take a proactive role in shaping the landscapes of the future. I strongly advocate the latter approach.

Summary

Landscape ecology is poised to play a central role in finding solutions to many of today's land management and conservation problems. It is becoming increasingly recognized that the landscape is the relevant scale at which to address these problems. It is appropriate, therefore, to assess the current status of landscape ecology in terms of its ability to provide principles and methodologies applicable to management and planning.

Considerable progress has been made in developing the quantitative and theoretical base of landscape ecology, but it is often debatable whether its practical relevance has been considered or assessed. Methodologies of landscape description have proliferated, but many remain untested and few are related to landscape functional characteristics. Studies of landscape change have also increased in number and sophistication, but often little attempt has been made to determine how changes impact landscape function. A clear understanding of how landscape structure and change affect landscape function is nevertheless essential if we are to apply landscape ecology to real world problems.

Although landscape ecology brings together researchers and practitioners from a wide array of backgrounds and claims to be a integrative science, there is still much to be done to forge closer linkages between different areas—in particular, between the biophysical and social components and among research, design, and planning.

To capitalize on the potential offered by landscape ecology, landscape ecologists must continue to develop methodologies and theoretical constructs relevant to the broad scales at which they commonly work. This may mean radical departures from traditional scientific approaches. At the same time, however, they must ensure that the methods and constructs they develop are applicable in the real world, and can be linked with management, design and planning. Landscapes are real things where people live and work and where broad-scale ecological processes operate. We must ensure that our science treats them as such and enhances our ability to plan and manage our landscapes for the future.

Acknowledgments. I wish to thank Jeff Klopatek and IALE(US) for the opportunity to present a plenary address at the US(IALE) meeting in Galveston in March 1996. The ideas presented here developed from discussions with many colleagues, and I wish to thank them for their inputs. Particular thanks go to Peter Cale, Richard Forman, Robert Lambeck, Ted Lefroy, Monica Turner, Denis Saunders, and John Wiens.

References

Allaby, M. 1995. Facing the Future: The Case for Science. Bloomsbury, London.

Bormann, F.H., and G.E. Likens. 1977. Pattern and Process in a Forested Ecosystem. Springer-Verlag, New York.

Cale, P., and R.J. Hobbs. 1994. Landscape heterogeneity indices: problems of scale and applicability, with particular reference to animal habitat description Pacific Conservation Biology 1:183–193.

Forman, R.T.T. 1995. Land Mosiacs: The Ecology of Landscapes and Regions. Cambridge University Press, Cambridge, UK.

Forman, R.T.T., and M. Godron. 1986. Landscape Ecology. Wiley & Sons, New York.

George, R.J., D.J. McFarlane, and R.J. Speed. 1995. The consequences of a changing hydrologic environment for native vegetation in south Western Australia. In: Nature Conservation 4: The Role of Networks, pp. 9–22. D.A. Saunders, J. Craig, and L. Mattiske (eds.). Surrey Beatty and Sons, Chipping Norton, Australia.

Golley, F.B., and J. Bellot. 1995. Interactions of landscape ecology, planning and design. Landscape and Urban Planning 21:3–11.

Hanski, I., and M. Gilpin. 1991. Metapopulation dynamics: brief history and conceptual domain. Biological Journal of the Linnean Society 42:3–16.

Hargrove, W.W., and J. Pickering. 1992. Pseudoreplication: a sine qua non for regional ecology. Landscape Ecology 6:251–258.

Harrison, S. 1994. Metapopulations and conservation. In: P.J. Edwards, R.M. May, and N.R. Webb (eds.). Large-Scale Ecology and Conservation Biology, pp. 111–128. Blackwell, Oxford.

Hobbs, R.J. 1993. Effects of landscape fragmentation on ecosystem processes in the Western Australian wheatbelt. Biological Conservation 64:193–201.

Hobbs, R.J. 1996. Future landscapes and the future of landscape ecology. Landscape and Urban Planning 37:1–9.

Hobbs, R.J. 1998. Ecologists in public. In: Ecology for Everyone, pp. 20–25. R.T. Wills and R.J. Hobbs (eds.). Surrey Beatty and Sons, Chipping Norton, Australia.

Hobbs, R.J., and H.A. Mooney. 1995. Spatial and temporal variability in California annual grassland: results from a long-term study. Journal of Vegetation Science 6:43–57.

Hobbs, R.J., and D.A. Saunders. 1991. Reintegrating fragmented landscapes: a preliminary framework for the Western Australian wheatbelt. Journal of Environmental Management 33:161–167.

Hobbs, R.J., and D.A. Saunders (ed.). 1993. Reintegrating Fragmented Landscapes. Towards Sustainable Production and Conservation. Springer-Verlag, New York.

Hobbs, R.J., and D.A. Saunders 1995. Conversing with aliens: do scientists communicate with each other well enough to solve complex environmental problems? In: Nature Conservation 4: The Role of Networks, pp. 195–198. D.A. Saunders, J. Craig, and L. Mattiske (eds.). Surrey Beatty and Sons, Chipping Norton, Australia.

Hobbs, R.J., D.A. Saunders, and G.W. Arnold. 1993a. Integrated landscape ecology: a Western Australian perspective. Biological Conservation 64:231–238.

Hobbs, R.J., D.A. Saunders, L.A. Lobry de Bruyn, and A.R. Main. 1993b. Changes in biota. In: Reintegrating Fragmented Landscapes. Towards Sustainable Production and Nature Conservation, pp. 65–106. R.J. Hobbs and D.A. Saunders (eds.). Springer-Verlag, New York.

Hurlbert, S.H. 1984. Pseudoreplication and the design of ecological field experiments. Ecological Monographs 54:187–211.

Johnston, A.E. 1991. Benefits of long-term ecosystem research: some examples from Rothamsted. In: Long-term Ecological Research: An International Perspective, pp. 89–114. P.G. Risser (ed.). John Wiley and Sons, Chichester, UK.

Levins, R., and R. Lewontin. 1985. The Dialectical Biologist. Harvard University Press, Cambridge, MA.

Litviatis, J.A., D.F. Smith, R. Villafuerte, J. Oehler, and J. Carlson. 1996. Landscape ecology today. Conservation Biology 10:306–308.

Lubchenco, J. 1995. The relevance of ecology: the societal context and disciplinary implications of linkages across levels of ecological organization. In: Linking Species and Ecosystems, pp. 297–305. C.G. Jones and J.H. Lawton (eds.). Chapman and Hall, New York.

Magnuson, J.J. 1990. Long-term ecological research and the invisible present. Bio Science 40:495–501.

Magnuson, J.J. 1995. The invisible present. In: Ecological Time Series, pp. 448–464. T.M. Powell and J.H. Steele (eds.). Chapman and Hall, New York.

McHarg, I. 1995. Ian McHarg reflects on the past, present and future of GIS. GIS World 8:46–49.

Murphy, D.D. 1989. Conservation and confusion: wrong species, wrong scale, wrong conclusions. Conservation Biology 3:82–84.

Naveh, Z., and A.S. Lieberman. 1984. Landscape Ecology: Theory and Application. Springer-Verlag, New York.

Nicholls, A.O., and C.R. Margules. 1991. The design of studies to demonstrate the biological importance of corridors. In: Nature Conservation 2: The Role of Corridors, pp. 49–61. D.A. Saunders and R.J. Hobbs (eds.). Surrey Beatty, Chipping Noton, Australia.

Pickett, S.T.A., J. Kolasa, and C.G. Jones. 1994. Ecological Understanding: The Nature of Theory and the Theory of Nature. Academic Press, New York.

Pulliam, H.R. 1988. Sources, sinks and population regulation. American Naturalist 132:652.

Robinson, E.R., and J.F. Quinn. 1988. Extinction, turnover, and species diversity in an experimentally fragmented California annual grassland. Oecologia 76:71–82.

Rookwood, P. 1995. Landscape planning for biodiversity. Landscape and Urban Planning 31:379–385.

Saunders, D.A., J. Craig, and L. Mattiske (eds.). 1995. Nature Conservation 4: The Role of Networks. Surrey Beatty and Sons, Chipping Norton, Australia.

Saunders, D.A., R.J. Hobbs, and P.R. Ehrlich (eds.). 1993. Nature Conservation 3: Reconstruction of Fragmented Ecosystems, Global and Regional Perspectives Surrey Beatty and Sons, Chipping Norton, NSW.

Saunders, D.A., and J. Ingram. 1995. Birds of Southwestern Australia: An Atlas of Changes in the Distribution and Abundance of the Wheatbelt Fauna. Surrey Beatty and Sons, Chipping Norton, Australia.

Slobodkin, L.B. 1995. Linking species and ecosystems through training of students. In: Linking Species and Ecosystems, pp. 306–312. C.G. Jones and J.H. Lawton (eds.). Chapman and Hall, New York.

Tilman, D. 1989. Ecological experimentation: strengths and conceptual problems. In: Ling-Term Studies in Ecology: Approaches and Alternatives, pp. 136–157. G.E. Likens (ed.). Springer Verlag, New York.
Tilman, D., and J.A. Downing. 1994. Biodiversity and stability in grasslands. Nature 367:363–365.
Turner, M.G. 1989. Landscape ecology: The effect of pattern on process. Annual Review of Ecology and Systematics 20:171–197.
Turner, M.G., and R.H. Gardner (ed.). 1991. Quantative Methods in Landscape Ecology. The Analysis and Interpretation of Landscape Heterogeneity. Springer-Verlag, New York.
Turner, M.G., R.H. Gardner, and R.V. O'Neill. 1995. Ecological dynamics at broad scales. Ecosystems and landscapes. BioScience Suppliment 1995:S29–S35.
Underwood, A.J. 1990. Experiments in ecology and management: their logics, functions and interpretations. Australian Journal of Ecology 15:365–389.
Underwood, A.J. 1995. Ecological research and (and research into) environmental management. Ecological Applications 5:232–247.
Underwood, A.J., and M.G. Chapman. 1995. Introduction to coastal habitats. In: Coastal Marine Ecology of Temperate Australia, pp. 1–15. A.J. Underwood and M.G. Chapman (eds.). University of New South Wales Press, Sydney.
Walters, C.J., and C.S. Holling. 1990. Large-scale management experiments and learning by doing. Ecology 71:2060–2068.
Wiens, J.A. 1992. What is landscape ecology, really? Landscape Ecology 7:149–150.
Wiens, J.A., N.C. Stenseth, B. Van Horne, and R.A. Ims. 1993. Ecological mechanisms and landscape ecology. Oikos 66:369–380.
Wu, J., and O.L. Loucks. 1995. From balance of nature to heirarchical patch dynamics: a paradigm shift in ecology. Quarterly Review of Biology 70:439–466.

3 Cross-Boundary Issues to Manage for Healthy Forest Ecosystems

Ann M. Bartuska

Since the mid-1980s, the frequency of severe and extensive wildfires, involving millions of hectares, has increased. Beginning with the 1988 fires and the concern about Yellowstone National Park through the 1994 season, in which 34 lives were lost and 1.9 billion ha burned, the extremes in fire activity have become the norm. Then, in 1996, the fires began in the southern United States much earlier then normal and raged unceasingly in the west well into September, covering over 2.5 million ha and costing well over $1 billion. All this activity has made one issue very clear—fires do not respect property lines or political boundaries. These events have brought the issue of forest ecosystem health into a central position in the debate on the use and management of natural resources and the many values society places on these resources. And by the very nature of the issue, the concept of the landscape as organizational unit has gained recognition from the public, the media, and the policy makers.

This chapter specifically illustrates the issues surrounding the "forest health" debate as of March 1996, when the Forest Service was in the middle of a protracted process to define and describe a policy for forest health. The absence of such a policy was noted in the Western Forest Health Initiative, a 1994 report that made 38 recommendations for the improvement of internal policies and procedures regarding forest health. This discussion is not a treatise on the science of forest ecosystem health or a comprehensive review of the literature, but rather an attempt to bring forward some of the parameters that have and will continue to define the scope of the forest health issue and our ability to respond to the issue in a meaningful way through management. A great deal of rhetoric certainly surrounds the term forest health—not least of which is the debate on whether the country is in the midst of an ecosystem "crisis" (Peters et al. 1996). The debate cannot be resolved here. Rather, this chapter focuses on how forests reached their current condition and several management approaches that should be considered in a restoration of these ecosystems to an ecologically sustainable condition.

The Situation

The root of much of the unraveling ecological condition of many of the western forests of the United States is an aggressive policy of fire suppression, combined with the practice of selective harvesting and overgrazing. But how did we get there? Evidence from tree ring analysis shows that many forests of the interior west have been characterized by short-interval, low-intensity fires every 12 to 20 years, with periodic large-scale, stand-replacing fires (Sampson and Adams 1994). The low-intensity fires encouraged open-stand conditions, acting as a natural thinning agent by killing many seedlings and saplings, a condition typical of the ponderosa pine and white pine types, and a concomitant suppression of insect and disease activity. The stand-replacing fires have helped shape the landscape, creating a mosaic of seral stages. Ecologically, the system was dynamic and sustainable across the landscape.

Human settlers from the eastern United States moved into this landscape establishing homesteads, ranches, and the beginnings of an active timber industry. At the turn of the century the United States was growing by leaps and bounds. Housing and factories were springing up at record numbers, the government was providing incentives for people to expand into the midwest and west, and railroads began crisscrossing the country. Wood was the resource most in demand during this expansion. As the eastern forests had been all but eliminated by timber harvesting by the early 1900s, we turned to the west. The Forest Service was authorized as an agency at this time—an agency whose Organic Administration Act of 1897 stated "*No National Forest shall be established except to improve and protect the forest within the boundaries or for the purpose of securing favorable condition of water flows, and to furnish a continuous supply of timber for the use and necessities of citizens of the United States.*" Congress saw the clear need to protect the forests for a growing nation. One can see how uncontrolled wildfires would not be consistent with this need. Then in 1910, wildfires swept the West with an intensity greater than those of 1994 and 1996. In Idaho, 400 million ha burned in four days and numerous human lives and much property were lost. These events led to the implementation of an aggressive fire suppression policy throughout the United States, and land managers did an excellent job meeting their obligations. From 1930 to 1970, the acreage burned by wildfires decreased from approximately 20 million to less than 2 million ha. Our approach made sense from a societal perspective, and the best available science was used to address the problem. What was not taken into account were the long-term ecological effects, but this is not surprising recognizing that ecology was a very new and emerging science in the early 1900s.

The landscape effects are clear. The absence of short-interval fires has led to forests characterized by increased density of shade-tolerant trees (e.g., Douglas fir and white fir) rather then the more open-stand structure of ponderosa and white pine. These species have many more lower branches, which combined with the stand density, has led to increased woody biomass—and a lot more fuel for fires. The density also predisposes many stands to insects and diseases, as the trees become stressed as a result of competition effects for nutrients and water. The absence of stand-replacing fires further compounds the problem. Much larger areas of the landscape have become homogeneous with respect to stand ages, because of the loss of the high-intensity fires. Insect and disease infestations typically increase as stands mature. It is not surprising, therefore, that the aging forest structure has led to broad-scale insect epidemics. Recent epidemic levels of bark beetle in Alaska and of southern pine beetle are natural, cyclical events in these ecosystems; however, the geographic extent of the infestations has caused concern. Heterogeneity of stands across the landscape would not have prevented the high beetle activity, but it may have limited the extent.

One other factor considered concerning forest health—the introduction of nonnative species into the United States. Our current forest ecosystems have been shaped as much by these invaders as by fires and harvesting policies. In the eastern United States, chestnut blight eliminated one of the most ecologically and socially important tree species—the American chestnut. With the loss of the chestnut, various oak species became the dominant tree species through much of the range. The subsequent introduction of the European gypsy moth has further shaped the forest; the mortality of several oak species (*Quercus rubra*, *Q. coccinea*, *Q. alba*, *Q. prinus*) in many parts of the East can still be observed in the numerous "gray ghosts" that remain standing throughout the range. Other recent introductions have resulted in localized losses of key tree species—flowering dogwood at higher elevations throughout the Appalachians because of dogwood anthracnose, butternut throughout its range because of the butternut canker, and most recently, the increasingly rapid loss of eastern hemlock because of the hemlock wooly adelgid.

The isolation of the western forests has limited, but not entirely eliminated, the introduction of nonnative insects and diseases. Western white pine was a dominant tree species in the fire-maintained intermountain west. White pine blister rust was introduced in 1909 in British Columbia and rapidly spread throughout the Intermountain West. Its impacts have been most severe on younger trees, limiting, for example, the western white pine population (Byler et al. 1994). Unfortunately, many of the nonnative pests originally introduced into the eastern United States (e.g., dogwood anthracnose) have now made their way to the West and are becoming established (Campbell and Liegel 1996).

Arguably, the most significant threat to the western forests is not from introduced insects and diseases, but from plants. Nonnative species, such as scotch broom, leafy spurge, and spotted knapweed have taken hold in much of the West. While initially a concern on grazing lands because of competition with forage, these species have been recognized as severely impacting native biodiversity wherever they are introduced. Even the heart of many wilderness areas are at risk. Many interior forest meadows in Montana and Idaho wilderness areas have noxious weed infestations, introduced by way of the forage used on pack mule and horse excursions.

Management and a Consideration of Boundaries

In 1992, the Forest Service adopted a policy of ecosystem management; by 1994, eighteen Federal agencies, as well as state organizations and non-governmental organizations had identified ecosystem management principles to be used as guidelines for management actions. While each agency or organization defined "ecosystem management" in slightly (or largely) different terms (Christensen et al. 1996), at least two characteristics were common: (1) that management must be built on ecological science and an understanding of how ecosystems function and (2) that humans are integral components of ecosystems. It is in this context that the protection and restoration of forest ecosystem health must be addressed. Concepts of ecological integrity and resiliency are fundamental to defining and managing for ecosystem health (Costanza 1993).

Much of the debate around forest health stems from the inability to define it in strictly ecological or biocentric terms. The very concept of ecosystem health is derived from human values—healthy enough for what? For whose purpose? Throughout the previous discussion on the current condition of forest ecosystems, human intervention and influence shaped the issue. The bark beetle epidemic of Alaska is of great concern because many small communities believe they are at risk because of fires; the southern pine beetle outbreak was less an ecological consideration then an economic one, driven by the potential commercial loss of timber in the southern markets. Any resolution to the restoration of healthy, forested ecosystems will require considerations of ecological, societal, and political parameters. I would like to suggest some steps to this resolution through a consideration of the boundary issues that delimit the controversy.

Disciplinary Boundaries

Healthy forest ecosystems occur through consideration of multiple processes and components of an ecosystem. In fire-dependent ecological communities, there is broad agreement on steps that can be taken to restore ecosystem health. The desired condition would provide for the return of short-interval fires as a primary controlling mechanism in the ecosystem. To get there, existing fuel conditions must be reduced. A prudent management program should include vegetation management to thin overcrowded stands and to remove shade-tolerant species, enhanced fuels management to reduce the existing fuel load, and then, a reintroduction of prescribed fire. These actions must be done with full recognition of watershed stability, air quality, and the presence of threatened and endangered species. Biological responses to periodic fires is not the same as to vegetation management through timber harvesting. In the longleaf communities of the southern United States, red-cockaded woodpecker populations can be improved by removing the midstory canopy; restoration of the ground cover, insect populations, and the associated wildlife can only be accomplished by also reintroducing fire (Farrar 1990).

In the above example, knowledge in silviculture, fire dynamics, fire operations, botany, ornithology, systems ecology, and even history—to understand the presettlement conditions—are essential pieces to solve this particular problem. As simple and obvious as that sounds, the integration of the above topics rarely occurs.

Another boundary is the aquatic versus terrestrial demarcation of scientific thought. It is surprising how frequently a fisheries biologist reacts negatively to hearing the phrase "forest health" because of an assumption that streams, wetlands, and riparian communities—and their fish—are not included. If forest health is a goal of ecosystem management, then forest health implicitly incorporates these factors. If it does not, then that is more likely the result of the disciplinary boxes in which many resource specialists operate.

The final disciplinary hurdle may be the highest, effectively integrating ecological, cultural, and economic theory and principles. The report by Christensen et al. (1996) on the scientific basis of ecosystem management explicitly recognizes that humans and ecosystems are intricately linked, yet ecologists and other natural resource professionals continuously try to establish some "natural" baseline of condition and then incorporate human activity. The entire issue of forest ecosystem health is a societal construct and is an ideal proving ground for more fully integrating the natural systems and human systems disciplines to describe the current and future condition of the landscape and to establish realistic resource objectives.

Ecosystem Boundaries

The Southern Appalachian ecoregion—the longleaf pine ecosystem—the Lake Tahoe Basin—the Applegate watershed—each of these systems has been the subject of forest ecosystem health scientific study and management action. Each is on a different scale. Each has been delineated—defined as an ecosystem—although including different ecosystem types, to accomplish particular objectives, mostly driven by the need to establish ecological and social sustainability. What is common to each is that the delineation of the area became necessary to provide a framework for management actions. What also is common is that the generally accepted methods of establishing boundaries, primarily jurisdictional line of counties, states, National Forests, and other public land lines, were determined to be inadequate. Many in the general public and among policy makers find confusion in the term "ecosystem" because it can not readily be drawn on a map, yet when attempting to resolve a natural resource problem across a landscape, these same individuals implicitly adopt an ecosystem approach. This confusion is not surprising, as the definition of an ecosystem has several meanings in the ecological literature, making the transition to the public policy arena that much more difficult (Blew 1996). Even with problems caused by this imprecision, "ecosystem" will continue to be used because it is the only integrative concept that readily works.

Problem solving at a landscape level brings with it social science as well as ecological challenges. The breakdown of jurisdictional boundaries means that private lands become included. Problem solving then becomes a group process, with all relevant parties at the table to identify a common goal for management. Decision making at the ecosystem level becomes a study in social dynamics.

Societal Boundaries

Although we may be able to understand how ecosystems function at a landscape scale, reaching decisions on management goals and objectives across that landscape can be problematic. The mixture of public (Federal and state), tribals, and privately owned land occurs in many different combinations. There are areas throughout the West that are best described as a "checkerboard," reflecting a mixing of public and private lands. In the eastern United States, 95% of the forestland is in private ownership (there are approximately 9 million landowners), so landscape-level problem solving can become extremely complex. Lack of action on public lands can have marked impacts on the adjacent private lands, whether it be an escaped prescribed fire or an insect infestation that jumps boundaries. If the management objectives for the adjacent lands in these areas are not the same, conflicts may and do arise. Clearly, action must be taken collaboratively to meet common goals.

Nowhere does this issue reach greater prominence then in the wildland–urban/rural interface. Demographic shifts are bringing people into and adjacent to public lands in greater numbers, not only with second homes but also with primary residence. In places like the Tahoe basin of California and Nevada, subdivisions of expensive homes have been developed adjacent to lands that are in an unhealthy condition (i.e., dense stands with a high degree of mortality). This is not just a western condition: similar developments have been constructed and similar interface problems exist from Florida to North Carolina to Long Island, New York.

The tension that frequently exists arises from principles set down at the founding of our country—the right to and the sanctity of private property. This concept is problematic when resource issues do not recognize those same boundaries, and the solutions to these problems often require active management on more then just public lands. Once again, the proven method of breaking this barrier is through joint problem solving. Recent efforts in the restoration of the Everglades early recognized that the entire south Florida ecoregion must be included for any solution to be effective (Harwell et al. 1996), and Federal, State, tribal, and community members have come together to craft that solution.

What Is Next?

Since 1994, the media and the popular literature have focused the debate on whether there is a forest health "crisis." Undoubtedly, the United States has areas where ecological conditions have been altered and the current landscape is not socially desirable. Although many of the parameters which should be considered to create and manage a healthy forest condition are known, our ability to accomplish these objectives is limited. The United States has over 280 million ha of forest land, approximately 50% of which is in the west, including Alaska. There are approximately 16 million ha in the short-interval, low-intensity fire type that could be treated to restore a normal fire regime. This is a huge area, with a concomitant cost, estimated to be $5 billion (Ross Gorte, Congressional Research Service, personal communication), that is prohibitive. Actions can and should be taken, with full recognition of the complexity of ecosystems and the multiple values humans place on those ecosystems.

As a result of the 1994 wildfires, the USDA Forest Service paid greater attention to its policy on forest health. A definition of forest health has been developed through a collaboration with other Federal and state agencies. FOREST ECOSYSTEM HEALTH is a condition of forested ecosystems characterized at the landscape scale by ecological integrity and, within the capability of the ecosystems, sustainability of multiple benefits, products, and values.

Building on the concepts and ecosystem management goals embedded in the definition, we have begun to identify guidelines for action. In so doing, we have brought in the various disciplinary, organizational, and societal perspectives that make this issue so complex.

Collaboration

From the first steps of problem identification through the establishment of management objectives to the implementation of project plans, partnerships with other Federal agencies, states, tribal governments, and communities must be established. The implementation of ecosystem management calls for the development of a common vision for the landscape in question as an absolutely critical first step (Christensen 1996; Harwell et al. 1996). The up-front time this takes, the lengthy meetings, the endless dialogue does have long-term benefits. Those involved in the process become partners in achieving a solution and work gets done!

Knowledge

Where are the problem areas and how serious is the problem? This understanding becomes the basis for prioritization. Currently, no one data base identifies the locations of the healthy and unhealthy forest ecosystems. The Forest Health Monitoring program, a collaboration between the USDA Forest Service and the states, does provide a systematic, statistically based system for identifying trends for specific indicators of forest condition. However, it presently covers only approximately 60% of all forest land. Through this program and other surveys available at regional and local levels, high-priority areas for active management should be targeted.

Just as the economic health of the country is analyzed and reported on annually, ecosystem health should be reported in a similar way. The USDA Forest Service annually publishes an insect and disease condition report for the United States, but this report only looks at one of the health parameters. A more comprehensive look should consider ecological characteristics, establish current condition and trends over time, and highlight "hotspots" where management or policy attention can be focused. The information must stand up to scientific scrutiny, and to be effective, it must be reported simply and clearly, in a form accessible to a diverse readership (USDA 1996).

Planning and Implementation

It is critical that a rigorous priority setting process be implemented to identify what systems are most at risk and which of these can benefit most from proactive management. One possible triage scheme was originally developed for the Western Forest Health Initiative in direct response to the western fires of that year. The priority rankings are:

- Priority 1: Reduce the hazard of catastrophic loss of key ecosystem structure, composition, and processes.
- Priority 2: Restore critical ecosystem processes.
- Priority 3: Restore stressed sites.

To implement these priorities, data on the distribution of forested ecosystems and their condition is necessary. Access to geographic information systems, combined with a set of key ecosystem characteristics that are routinely described—preferably quantitatively—will enable resource managers to effectively focus actions.

The intent was to use these priorities in selecting projects for action on the National Forests and, although not mandated, they have been used to varying degrees. But there was a second step to the priority setting process, which really addressed only ecological conditions. An additional set of selection criteria, given below, were identified to address social issues, in part in recognition that health is a human, not an ecological, value.

1. Protection of human life and property.
2. Restoration of stressed sites and landscapes.
3. Projects that directly maintain, protect, or restore habitat or sites required by threatened or endangered species.
4. Projects that protect investments or intensively developed sites.

Because the objective of priority setting is to motivate action, these criteria recognize attributes that people really care about. One has only to look at the newspaper reports of the 1994 and 1996 fires to see the pattern: firefighters were more often engaged when communities were at risk. Similarly, focusing resources on special places, whether a protected glade with a high proportion of rare and endangered plants or a frequently used recreation site, is of importance to the broader public—whose representatives, remember, are sitting at the table helping define management actions.

Monitoring and Evaluation

One reason why managers can be proactive for short-interval fire ecosystems of the United States is that they understand the fire return interval, learned from fire scar history. Through photographic records, we have a visual image of what these landscapes looked like in the mid-1800s, which can be used to identify the plant community structure of the time. This information helps define the goal for resource managers; an effective monitoring program will help determine if that goal is being reached. It is critical that monitoring becomes an established part of the way we do business, fully recognizing the spatial scale for the indicators which are measured. Measuring all possible ecological characteristics is impossible, so more use of key criteria and indicators, as has been promoted through the Santiago agreement for sustainable forest management (Anonymous 1995), must become the rule. An effective monitoring program is one that includes a periodic evaluation or assessment of changing conditions.

The use of new and emerging technologies and analytical tools should facilitate an effective monitoring and evaluation effort. Remote sensing, simulation models GIS, combined with global positioning systems and statistically rigorous monitoring protocols must become routine. Once the information is gathered, summarized and synthesized, the next step is to communicate and use the findings.

Communication, Communication, Communication!

The tradition of scientists and technical specialists working in isolation, of making decisions on the best science without public involvement, is over. Legally, the National Environmental Policy Act of 1969 requires public involvement in the management of Federal lands, but recent legal challenges to resource management decisions have emphasized that it is never too soon to bring interested and affected parties to the table. Effective two-way communication is more likely to lead to effective collaboration and problem solving. If there is a new public policy mantra, it might be "communicate early and often!" As scientists, a greater responsiveness and recognition of broadening communication suggests a different approach to publishing new information. Along with peer-reviewed publications, scientists should be encouraged to translate their findings into a form that can be readily understood by an intelligent but untrained public that includes policy makers. Recent efforts by the Ecological Society of America to increase outreach and translation of ecological theory for use by Congress and other policy makers is an explicit recognition of the greater society in which we operate.

Summary

Following the principles of ecosystem management, a shared understanding of the issue is the starting point for dialogue and action. I have tried to define the issue and the problem, if it is agreed there is a problem, to be solved. Some probably do not agree with all the words and concepts embodied in that definition and may have felt something was missing. And yet, for communication purposes, scientists and managers are expected to translate complex issues into simple sentences, which, from my experience, cannot be done to the satisfaction of all parties. If we begin with what we can agree on—concepts of ecological resiliency, landscape restoration (not stands), and recovery from disturbance—then we are 80% of the way toward a working solution. The process is all about breaking through boundaries—barriers to communication, to integrated management, to disciplinary boxes—and ultimately achieving ecologically sustainable forest ecosystems.

References

Anonymous. 1995. Sustaining the world's forests: the Santiago agreement. Journal of Forestry 93:18–21.

Blew, R.D. 1996. On the definition of ecosystem. Ecological Society of America Bulletin 77:171–173.

Byler, J.W., R.G. Krebill, S.K. Hagle, and S.J. Kegley. 1994. Health of the Cedar-Hemlock-Western white pine forests of Idaho. In: Proceedings of the Interior Cedar-Hemlock-White Pine Forests: Ecology and Management, Washington State University, Pullman.

Campbell, S., and L. Liegel (tech. coords). 1996. Disturbance and Forest Health in Oregon and Washington. Gen. Tech. Rep. PNW-GTR-381. USDA Forest Service, Portand, OR.

Christensen, N.L., A.M. Bartuska, J.H. Brown, S. Carpenter, C. D'Antonio, R. Francis, J.F. Franklin, J.A. MacMahon, R.F. Noss, D.J. Parsons, C.H. Peterson, M.G. Turner, and R.G. Woodmansee. 1996. The report of the Ecological Society of America Committee on the Scientific Basis for Ecosystem Management. Ecological Applications 6(3):665–691.

Costanza, R. 1993. Toward an operational definition of ecosystem health. In: Ecosystem Health, pp. 239–256. R. Costanza, B.G. Norton, and B.D. Haskell (eds.). Island Press, Washington, DC.

Farrar, R.M. (ed.). 1990. Management of Longleaf Pine. Proceedings of a symposium. Gen. Tech. Rep. SO-75. USDA Forest Service, New Orleans, LA.

Harwell, M.A., J.F. Long, A.M. Bartuska, J.H. Gentile, C.C. Harwell, V. Myers, and J.C. Ogden. 1996. Ecosystem management to achieve ecological sustainability: the case of South Florida. Environmental Management 20(4):497–521.

Peters, R.L., E. Frost, and F. Pace. 1996. Managing for Forest Ecosystem Health: A Reassessment of the "Forest Health Crisis." Defenders of Wildlife, Washington, DC.

Sampson, R.N., and D.L. Adams (eds.). 1994. Assessing Forest Ecosystem Health in the Inland West. Proceedings of the American Forests workshop. Haworth Press, Binghamton, NY.

USDA Forest Service. 1996. America's Forests: 1996 Health Update. AIB-727, Wsahington, DC.

4
Horizontal Processes, Roads, Suburbs, Societal Objectives, and Landscape Ecology

RICHARD T.T. FORMAN

The growth rate and acceleration of landscape ecology in a scant fifteen years have been extraordinary. I believe two reasons explain this phenomenon. First, the field has opened a frontier of research scholarship and theory development focused on spatial pattern intertwined with processes and changes. Second, by focusing on the broad human scale and including all landscapes, it offers direct applications for solving vexing environmental and societal issues.

Such growth invites controversy. It is like an expanding archery target. The faded message on an old rusty arrow said landscape ecology was nothing but island biogeography. Three other old arrows claimed it was ecosystem science, total human systems, or physical geography. Newer arrows suggest that it is only conservation biology, nature conservation, land-use mapping, or just nothing. Another says it is all abstract theory. Yet another finds it totally applied. More arrows will hit the growing target. Fine. If I could hit the bulls-eye, my arrow would say that during these early years the primary thrust was development of basic theory, models, concepts, and principles, together with enough empirical evidence to remain on reasonably solid ground. In addition, landscape ecology meetings from the beginning were ecumenical, welcoming people from numerous basic and applied disciplines. So as the applications in diverse fields now begin to mushroom, the lag between development of theory and its application is impressively short.

Therefore, in this chapter I will highlight some principles and opportunities for application that apply essentially to any landscape. To emphasize their wide utility, however, the focus is on places with a heavy human imprint (i.e., places where roads, homes, and natural processes share space). This begins with a distinctive set of natural processes and their role in conservation, planning, and management. Next is road ecology and its importance to transportation planning. Then the role of patches of nature in suburbia is pinpointed for planning and conservation. Finally, the unique role of landscape ecology for addressing society's manifold concurrent and conflicting issues is highlighted.

Horizontal Natural Processes

Processes on the land involve flows, transport, and movement through space. Some are mainly vertical, such as succession, evapotranspiration, water percolation, nutrient uptake by plants, photosynthesis, and treefalls. In contrast, this chapter focuses on horizontal processes—those that move across the land (in mountainous terrain, a vertical component is of course present). Such flows and movements generally cross two or more local ecosystems or land uses, and thus are central in landscape ecology. Some are basically human driven, including vehicle usage, cycling, walking, trains, canal boats, truck (lorry) transport of goods, electricity transmission, and piping of sewage, water, oil, and natural gas. Human-driven processes tend to be constrained to straight lines. Hikers, horseback riders, and off-road vehicles may take convoluted routes, but, compared with the other processes, the total amount of movement involved is minor. Basically the straight routes of human-driven processes are efficient for getting from here to there; they also serve to protect the surrounding matrix from much human impact.

The focus here, however, is on horizontal natural processes or ecological flows (Turner 1987; Harris et al. 1996; Forman and Hersperger 1997). Important examples with significant horizontal components are surface water (in streams and rivers), groundwater, fire, animal foraging, animal dispersal, migration, pollination, plant dispersal, wind erosion and deposition, water erosion and sedimentation, fish movements, sheet flow of water, and aerosol, nutrient, and gaseous transport. The routes of natural processes are overwhelmingly curvilinear—even convoluted (Fig. 4.1a). This is because the transport vectors or mechanisms (i.e., wind, water and animals) (Forman 1995) typically take curvy or convoluted routes. Note however that the routes are not random; patterns predominate. Surface water flow in streams is commonly dendritic and coalescing. Groundwater flow generally appears to be only slightly curvilinear, whereas foraging animals often trace highly convoluted routes. Pollinators often alternate relatively straight trajectories with very convoluted stretches.

These horizontal natural processes are the common day-to-day flows and movements resulting from or involving species adaptation. Periodically, major forces build to produce disturbances such as hurricanes, floods, large wildfires, and pest explosions (Turner 1987; Forman 1987). The routes of these large flows tend to be straighter and much wider.

Why are these horizontal ecological flows so important to planning, conservation, design, and management? Humans see objects and arrange them in spatial patterns to achieve certain goals. For example, the designer or manager sees trees, grassy areas, a pond, and buildings, and arranges or manages them for aesthetic or wildlife population objectives. Similarly, the planner or conservationist sees woods, fields, roads, and housing developments, and works out a logical spatial pattern to accomplish various goals.

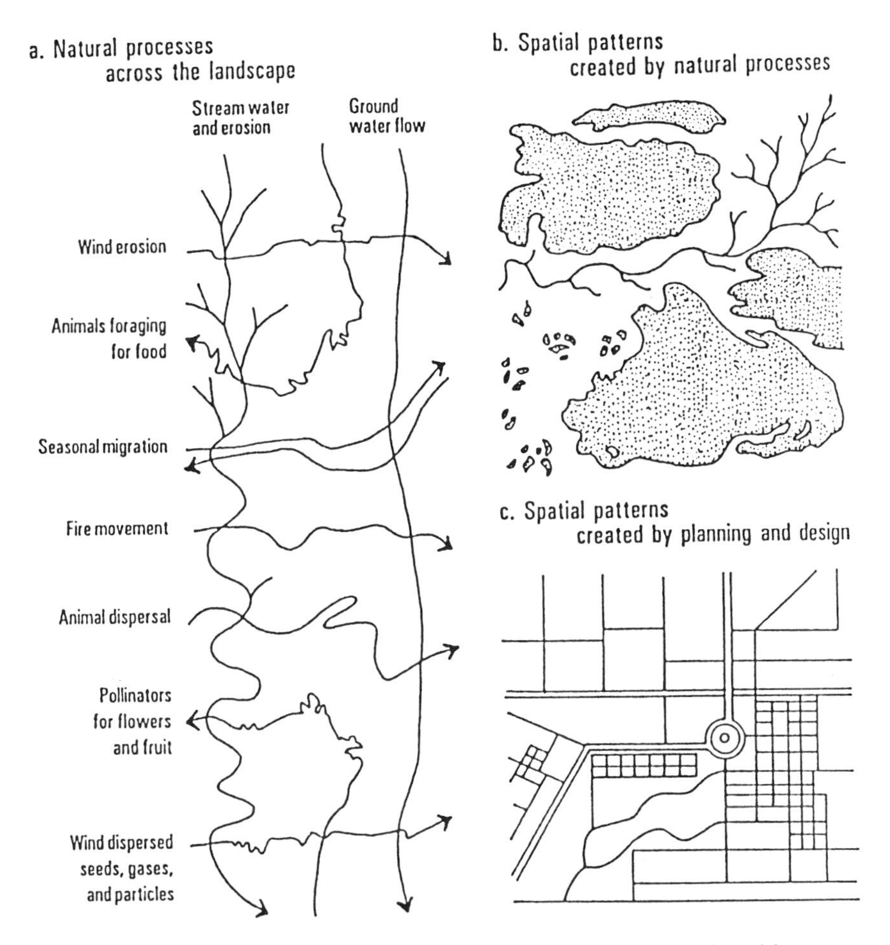

FIGURE 4.1. Horizontal natural processes and spatial patterns produced by nature versus planning/design. See also Forman and Hersperger (1997).

Rather than simply identifying and arranging objects, an alternative approach emphasizes the importance of providing for horizontal natural processes. Focusing first on horizontal processes and then on arranging objects produces a different—perhaps radically different—spatial solution or pattern.

This approach has two major advantages. First, the sizes and shapes of the objects themselves will often mimic those of nature. Humans tend to make geometric patterns (Fig. 4.1c) (Hough 1984; Spirn 1984; Diamond and Noonan 1996), whereas horizontal natural processes tend to make irregular, curvilinear, aggregated, size-variable, and richly textured spatial patterns (Fig. 4.1b) (Forman 1995; Forman and Hersperger 1997). The flows and movements form patches, corridors, and boundaries. Thus, a plan or design or management action that provides for, rather than interrupts, horizontal ecological flows contains distinctive naturally shaped objects. It also arranges them into an ecologically logical spatial pattern that mimics nature.

Second, the resulting solution or spatial arrangement should be more sustainable. It is more apt to last over human generations. Objects that are sized and shaped like natural forms, plus an arrangement of objects similar to that in landscapes with little human impact, are more likely to be compatible with and maintained by natural processes. Furthermore, artificial barriers are present. In short, maintenance budgets necessary to hold back natural processes and to repair their effects are low.

A caveat is in order. Making natural sizes and shapes of objects is but a first step. Natural processes are directional, so the "angle of orientation" is critical (Forman 1995; Forman and Godron 1986). For example, an elongated patch in a landscape with strong wind is most effective if the long axis of the patch is reasonably parallel with the preponderant wind direction. Tailoring the sizes and shapes of objects to directional flows and movements is a key second step. Arranging the objects into an ecologically appropriate spatial pattern or solution is the third step.

Consider briefly the ecological design of suburbs. Think of your favorite city. Where would you go to find the "hot spots" of biodiversity, the sites with highest species richness? Often these are the unmanaged spots by water bodies, the forgotton corners by railroads and highways, or the overlooked edges of landfills and industries. Alternatively, where would you find green spots with the lowest biodiversity? Good candidates are the intensively managed grass–tree–bench ecosystem, the intensively designed downtown plaza, and the manicured residential lawn. Look at the implications. When not based on ecologically sound principles, intensive design, planning, and management commonly produce biological impoverishment. In contrast, biological richness typically emerges in places essentially overlooked by society.

Now consider at a landscape scale the difference between a planned versus an unplanned suburban area (Forman 1995). Here, planning or design refers to the area as a whole; individual objects within the unplanned area such as schools, cul-de-sacs, and malls may be highly designed. Although an occasional town, such as Woodlands, Texas (Spirn 1984), has been planned with a strong emphasis on both ecological and human dimensions, typical town and suburb planning has given overwhelming emphasis to transportation, shopping, residential development, and so forth. The planned suburban area is commonly characterized by squares, rectangles, grids, occasional circles with radiating lines, double lines, and smooth curves (Fig. 4.1c). The unplanned suburban area on the other hand is a complex mixture of these geometric and natural forms. The geometric shapes, however, are often poorly arranged on the land relative to horizontal natural processes. Which is ecologically better, the typical planned or unplanned suburb? Both fall far below their potential (e.g., in terms of water quality, fish populations, species richness, interior species, and the like). However, overall the typical unplanned area seems better than the planned. Again the basic reason is that the unplanned suburb or town contains many small places with somewhat natural conditions . . . the sites largely overlooked by society. They escaped intensive design or management.

Clearly the unplanned area is not a general model to follow. Equally evident is the huge gap between the poor ecological conditions of the typical planned area and the robust environment that could be produced by serious ecological planning. Until ecological planning moves beyond its formative stage, the following three planning and design actions would at least improve planned areas so that they are ecologically better than unplanned areas: (1) provide for horizontal natural processes; (2) incorporate more of the spatial patterns produced by natural processes; and (3) include one or more "bio-rich places," where native species richness is exuberant, in every plan and design. These are implementable now. In effect, these actions would ecologically improve our landscapes and make them more sustainable over human generations.

Road Ecology

In considering horizontal natural processes and landscapes with heavy human imprints, it is impossible to overlook the abundance and importance of roads. The use of landscape ecology in transportation planning is especially appropriate. Roads (and highways) cut across the landscape and intersect many local ecosystems. Many roads (including their roadsides/verges) are significant barriers or filters to horizontal natural processes such as animal movement. The conduit, habitat, source, and sink functions of roads (Bennett 1991; Forman 1995) are sometimes also of major ecological significance. The next section introduces the zone of ecological effects along a road, and then illustrates the use of landscape ecology and road ecology in transportation planning.

Road-Effect Zone

Numerous ecological effects are produced by the road infrastructure and associated vehicle usage. Different ecological effects extend outward from the road over a range of distances, which vary from meters to thousands of meters (Reck and Kaule 1993; Forman 1995). These different ecological effects extending varying distances from a road define a road-effect zone (Forman et al. 1997; Forman and Alexander 1998) (see Fig. 4.2).

The road-effect zone not only varies in width along a road, but also exhibits highly convoluted margins. Furthermore, at any given point the road-effect zone is usually asymmetric with significant ecological impacts

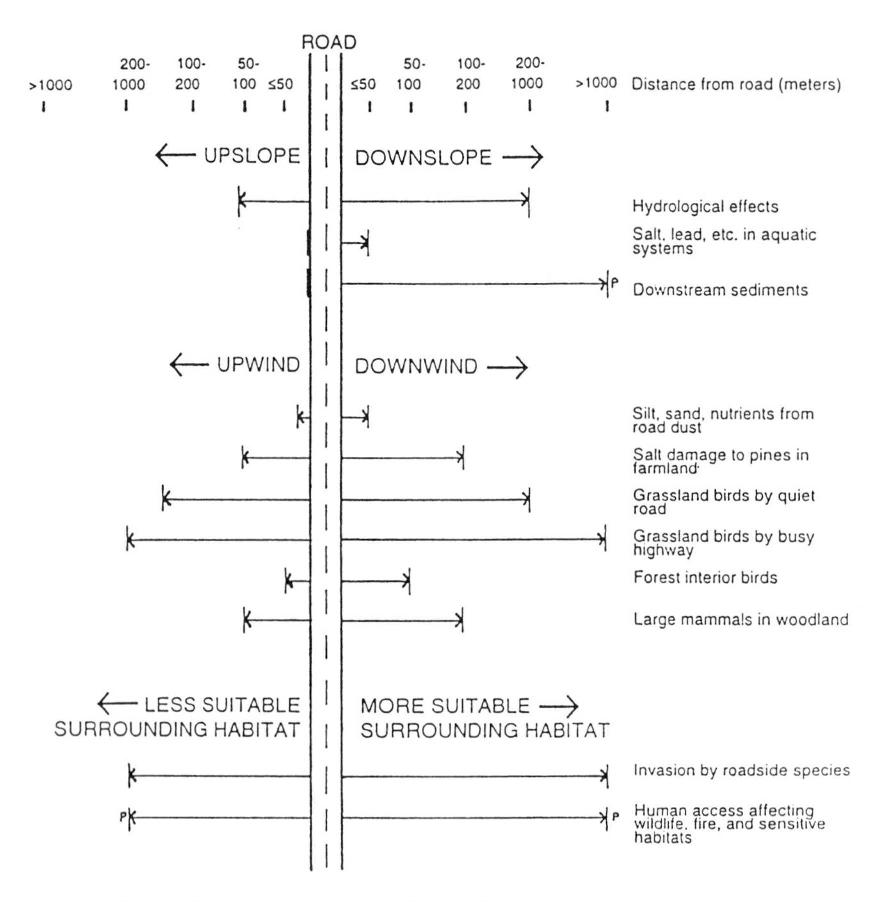

FIGURE 4.2. Road-effect zone defined by ecological effects extending different distances from a road. Most distances are based on specific studies (Forman 1995). However, distance extending to left is arbitrarily half of that to the right. "P" indicates an effect primarily at specific points where other corridors cross the road. See also Forman et al. (1997) and Forman and Alexander (1998).

extending further on one side than on the other. Three extrinsic factors determine the degree of asymmetry. Topography, with upslope–downslope differences, primarily affects hydrology and water-transported materials. Wind direction, with upwind–downwind differences, mainly affects wind-transported items, including particulate matter and traffic noise. Finally, the surrounding habitat commonly differs in quality or suitability on opposite sides of a road. This difference principally affects the spread of species, such as invasive roadside plants, as well as humans that enter the matrix and cause disturbance.

Roads, however, not only cross patches and the matrix, but also bisect corridors. These corridors, such as hedgerows, streams, wildlife routes, and powerlines, are important ecological conduits. At the points where conduits intersect the road, the road effects may extend much further outward. For example, sand from a bridge can be carried and deposited a long distance downstream. A path provides access for picknickers and hunters, which may impact wildlife over a considerable distance. In short, the road-effect zone varies in width, is asymmetric, and has convoluted margins with protruding fingers. The zone is commonly more than 200 m wide (Forman et al. 1997).

The cause of the ecological effects may be the road infrastructure itself or vehicle usage of the road (Forman and Hersperger 1996). Atmospheric effects are largely related to vehicle usage, whereas hydrologic and soil effects are primarily associated with the size and structure of the road itself. Some species and habitat effects are mainly the result of vehicles (e.g., traffic noise affecting vertebrates [Reijnen et al. 1995]), and some to the combination of roads and vehicles (disruption of certain wildlife corridors). However, most species and habitat effects are caused by the road infrastructure. This underlines the ecological importance of, for example, the four million miles of public roads in the United States. Road networks are prominent features of almost all landscapes, and their ecological effects permeate and spread widely.

Yet road ecology literature is skimpy, essentially embryonic. It is a wide open frontier promising a short lag between development of theory and its application to ecologically sound land planning.

Transportation planning for roads and highways has generally considered as narrow a zone along the highway as possible. Thus the engineering and design dimensions of the road and roadside have been essentially the only interest. In contrast, the road-effect zone points to the importance of considering the road in a wide spatial context. The ecological effects of roads with vehicles permeate numerous ecosystems and land uses of the broad landscape.

Thus a key application principle is to perforate road barriers. Roads need to be porous enough to provide for the horizontal natural processes (i.e., the flows and movements across the landscape) (Harris et al. 1996). Tunnels, underpasses, and overpasses are easily incorporated into new road construction (Evink et al. 1996; Forman and Hersperger 1996; Forman et al. 1997). Upgrading or widening projects are also opportune times to perforate road barriers. Indeed, the Dutch have shown that existing roads can be successfully perforated for ecological benefits (Natuur over Wegen 1995; Canters 1997). The costs are minuscule compared with the costs of road construction.

Many other mitigation techniques are available, including earth and vegetation berms to protect animal communities from traffic noise, perforated "jersey barriers" in the center of highways, fencing and shrubs to channel animals to particular spots, and so forth. With rare exceptions such as the Florida panther and key deer (*Felis concolor coryi*, *Odocoileus virginianus clavium*) roadkills (faunal casualties) do not significantly reduce animal population sizes (Evink et al. 1996; Forman et al. 1997). However, habitat fragmentation caused by road barriers separates populations into smaller subpopulations where the frequency of local extinction is higher (Charlesworth and Charlesworth 1987; Soule 1987). The barrier effect, therefore, is apparently of much greater ecological significance than the roadkill effect (Forman and Alexander 1998). Even if roadkill rates increase, providing for horizontal natural processes by perforating road barriers is a good societal investment in an ecological future.

Planning Framework for Addressing Road and Nature Conflicts

Most roads were constructed before the explosion in ecological knowledge, and hence before society's recognition of its dependence on nature. Therefore a procedure developed in The Netherlands is outlined (with slight modification) to identify the primary conflicts between existing roads and nature, and thus target locations for possible mitigation (Morel and Specken 1992; van Bohemen et al. 1994; Natuur over Wegen 1995). The approach also applies to avoiding environmental problems in new road construction.

In essence, the network of nature's patterns and processes are first mapped (Forman and Hersperger 1996). Then the road network is superimposed. Intersections of the two networks are examined to identify "bottlenecks," where natural patterns or processes are significantly interrupted. An array of existing solutions is then used to alleviate a bottleneck.

Ideally species populations, dispersal routes, groundwater flows, and the like are mapped for a state or region. But generally such detailed information is unavailable for large areas. So a spatial, more-integrative landscape-ecology approach is taken. Here the large patches of natural vegetation are mapped. These are surrogates for aquifer protection, large-home-range species, sustainable populations of interior species, and so forth (Forman 1995). Next the major corridors or routes of animal movements and water flows across the landscape are added (Harris and Scheck 1991; Saunders and Hobbs 1991; Binford and Buchenau 1993). These large patches and major corridors form the primary ecological network or infrastructure of the landscape.

The remaining land is then differentiated into two types based on how rapidly or easily the land use could be converted into natural vegetation (Hein van Bohemen, pers. comm.; Forman and Hersperger 1996). "More suitable areas," such as many cultivated fields, pastures, golf courses, and mowed parkland, theoretically could be readily transformed into natural vegetation. "Less suitable areas," such as commercial, dense residential, industrial, and urban areas, normally could only be converted to natural vegetation over long periods. In addition, more suitable areas are often compatible for movement by some species characteristic of native vegetation. However, less suitable areas typically provide extreme resistance to movement of most such species (Knaapen et al. 1992; Forman 1995). In short, nature's network is mapped (the ecological network of nodes and corridors) and then juxtaposed with more suitable and less suitable areas.

Next the road network is superimposed on the ecological network (Forman and Hersperger 1996). Locations where the two networks cross are identified as potential "bottlenecks,"—places where major ecological flows or patterns are interrupted by roads. For each bottleneck identified, the surrounding landscape area (a few tens of kilometers long and several kilometers wide) is examined in detail (Natuur over Wegen 1995; Pfister and Keller 1995). Patches, corridors, more suitable areas, and less suitable areas are mapped at this relatively fine scale using aerial photographs, topographic maps, site visits, and other information. At this scale detailed information on species populations, dispersal routes, groundwater flows, and so forth are more readily available or can be determined.

Next the array of techniques for avoidance, mitigation, and compensation of ecological impacts is used (van Bohemen et al. 1994; Natuur over Wegen 1995). The first consideration is "avoidance" to prevent negative ecological impacts altogether. For example, a road might be removed, constructed differently, built in another place, or not built at all. If this is impossible, the second step is "mitigation." How can ecological impacts be reduced or minimized (e.g., through restricted access, reduced vehicle speed, wildlife tunnels, fencing and the like)? If significant ecological impacts still remain, the third step is "compensation." How can an ecological impact be compensated to provide an equivalent amount of ecological enhancement to a nearby area? For example, a wildlife corridor may be widened, a naturally functioning wetland constructed, or a patch of natural vegetation enlarged. No net loss of ecological value is the guiding principle in compensation (H. van Bohemen, pers. comm.). Often several techniques are used for a bottleneck on a new road, in widening an existing road, or in alleviating impacts of previous construction.

An example of this transportation planning process has been outlined for a 77,000 km^2 area in the central and eastern portion of The Netherlands (Forman and Hersperger 1996). The "National Ecological Network" of large patches and major corridors was mapped, and the highway network superimposed on it to identify bottlenecks (Morel and Specken 1992; van Bohemen et al. 1994). In the central portion with a large national park and national forest, the following mitigation structures have been built or are planned: six wildlife overpasses or large tunnels, and 18 locations for pipes, tunnels, or wildlife culverts. Eight bottlenecks required compensation. In the surrounding fragmented landscape, one wildlife overpass or large tunnel, 35 bottleneck areas for smaller structures, and four compensations have been completed or are planned. The government has established and is meeting targets set for the percent of bottlenecks to be solved within five- and ten-year periods.

In summary, the planning framework to address conflicts between nature and roads makes the ecological network or green infrastructure of the landscape explicit, in order to identify road-caused bottlenecks. The array of avoidance, mitigation and compensation techniques available to scientists and engineers is used to eliminate or minimize the ecological impacts.

Large and Small Patches of Nature in Suburbia

Suburban Nature

Nature in suburban landscapes differs markedly from that in forested, agricultural, and desert landscapes. In suburbia, species richness at medium to broad spatial scales is very high because of the abundance of nonnative exotic species (Sukopp et al. 1990; Gilbert 1991). Generalist species and edge species predominate almost everywhere. A paucity of specialists, of interior species, and of large-home-range vertebrates means that rather few species are dependent on large patches of natural vegetation (Tilghman 1987). Few rare native species are present (although rare exotic species are numerous), because although some previous species remain as rarities they are generally common in nearby rural landscapes.

Additional distinctive ecological attributes of suburbia include elevated inputs of mineral nutrients and toxins that accumulate in soil (Gilbert 1991; Craul 1992). Plant productivity is high in most green areas. Soil erosion and water and substance runoff to water bodies is striking. Many streams are channelized or piped so water runoff is rapid and downstream floods characteristic. Most water bodies are eutrophicated and otherwise polluted, so fish populations are reduced.

Natural (or somewhat natural) vegetation in the suburban landscape is generally recognizable in three categories. The first two, large patches and major corridors, are typically prominent and easily recognized. The third is an "interstitial vegetation" of small patches and narrow corridors appearing in complex patterns. These back lot lines, street tree plantings, vest pocket parks, edges of school yards, channelized streams, and much more offer rich opportunities for serious ecological design and planning. However, here the discussion focuses on large patches of natural vegetation, supplemented with brief consideration of small patches, for planning, conservation, and management purposes.

Large Patches of Natural Vegetation

Eight ecological values of large natural vegetation patches in rural forested, agricultural, and desert landscapes are documented (Forman 1995): (1) water quality protection for aquifer and lake; (2) connectivity of a low-order stream network; (3) habitat to sustain populations of patch interior species; (4) core habitat and escape cover for large-home-range vertebrates; (5) source of species dispersing through the matrix; (6) microhabitat proximities for multihabitat species; (7) near-natural disturbance regimes enhancing species dependent on disturbance; and (8) buffer against extinction during environmental change.

This range of ecological benefits which can be provided in no other way makes large patches indispensable (Forman and Collinge 1996, 1997). Therefore large patches are generally the most important ecological feature in landscapes and the top priority for planning and conservation.

While these patterns and principles are general for all landscapes, some significant differences exist for suburban and urban landscapes. Six key ecological values of large natural-vegetation patches in suburbia are now presented (Table 4.1).

First, water quality is protected by large vegetated patches for groundwater and other water bodies. Cities often get clean water from aquifers or other sources from afar, but many suburbs depend on groundwater for drinking and other uses. Lakes in suburbs are often used for recreation, sometimes with the goal of being swimmable and fishable.

Second, large natural-vegetation patches in suburbia are key hydrologic sponges that reduce flooding. Compared with no patch or a small patch, the large patch normally results in much less downslope water runoff and lower later peak flows. It is hypothesized that the same result occurs, though on a quantitatively more modest scale, when a large patch is compared to the same vegetated area subdivided into smaller patches. This is based on the assumption that in the edge area of a patch more water moves outward in surface and/or subsurface flow to the surrounding nonvegetated matrix, compared with that in the interior patch area where more water percolates deeper into the soil. Thus a large natural-vegetation patch has less perimeter length than several small patches and is considered to be a more effective sponge.

TABLE 4.1. Primary ecological values of natural vegetation patches in suburban landscapes.

Large patch
1. Water quality protection for groundwater and surface water.
2. Hydrologic sponge that reduces flooding.
3. Interior portion maintains the best facsimile of native plant and animal communities.
4. Major source of species that disperse across the landscape.
5. Magnet rest stop for birds migrating across suburbia.
6. Microclimatic amelioration of adjacent area, especially downwind.

Small patch
1. Stepping stone for species dispersing across the landscape.

Source: Adapted from Forman and Hersperger (1997).

Third, large patches in suburbia maintain the best facsimile of native plant and animal communities in their interior (Table 4.1). Small patches and corridors typically contain numerous exotic species and are strongly affected by human disturbance (Matlack 1993). Even large patches, though, are generally somewhat impoverished in native species because of human disturbance and relative isolation from rural sources of species. Consequently, even though few rare and endangered species will be protected in suburbia, the inherent value of retaining some representatives of natural ecosystems in a landscape is considerable.

Fourth, large patches of natural vegetation are major sources of species that disperse across the suburban landscape. Rural landscapes are sources for the suburban fringe. However, for suburbia, and even the adjacent city, the limited number of specialists, interior species and large-home-range species present is concentrated in large patches (Burgess and Sharpe 1981; Ranney et al. 1981). Such species doubtless disperse at low rates to the surrounding interstitial nature in small patches and narrow corridors, where the species must generally persist only short periods. The presence of these species in the numerous interstitial bits of nature means that such populations are larger in size and more widely distributed in suburbia. Overall then, patches are important as sources to maintain this richness of scarce native species.

Fifth, these large patches in a sea of suburbia are magnets for birds migrating across the landscape (Table 4.1). The natural vegetation is a familiar habitat type where birds can rest and fill up on food. The surrounding matrix has an unfamiliar vegetation structure mixed with buildings, an abundance of exotic vegetation with unfamiliar fruits and relatively few insects, unfamiliar predators such as house cats, and all kinds of human disturbance. Large patches are oases in an inhospitable matrix.

Finally, large natural-vegetation patches ameliorate microclimate in their vicinity, especially downwind. For example, summer temperatures near large city parks are several degrees celsius cooler for a distance of hundreds of meters (Hough 1984; Spirn 1984; Gilbert 1991). Similarly, based on oasis studies, evapotranspiration in the park provides moisture downwind for similar distances (Laikhtman 1964; Bill et al. 1978; Balser et al. 1981). The impaction of particles and aerosols by the vegetation, plus the absence of human sources of pollution, makes the air cleaner downwind of large patches (Geiger 1965; White and Turner 1970; Art et al. 1975).

Note the differences between the ecological values in suburbia and rural areas. Similarities include protecting water resources and habitats and serving as species sources. The special suburban roles are as a sponge against flooding, an oasis for migrants, and a microclimatic ameliorator. Protecting large patches of natural vegetation in suburbia is still a top planning and conservation priority, but for somewhat different reasons than in rural landscapes.

Small Patches

Five ecological values are recognized for small natural-vegetation patches in rural areas (Forman 1995): (1) habitat and stepping stones for species dispersal and for recolonization after local extinction of interior species; (2) high species densities and high population sizes of edge species; (3) matrix heterogeneity that decreases fetch (run) and erosion, and provides escape cover from predators; (4) habitat for occasional small-patch-restricted species; and (5) protection of scattered small habitats and rare species.

Only one of these commonly appears important in suburban landscapes. Small patches are stepping stones for species dispersal across the landscape (Table 4.1). Because of the abundance of narrow corridors, such as along suburban property boundaries, as well as the scarcity of species present of major conservation interest, this ecological benefit of small patches is probably less than is the norm in other landscapes.

The educational and political value of small natural-vegetation patches probably exceeds the ecological benefit. That is, in numerous nations where >40% of the population lives in metropolitan areas, the interstitial vegetation of small patches and corridors is the major provider of nature in these people's daily life. Without this nature, suburbia is a sea of hard surface and sterile greenery. These tiny ecosystems teach and inspire people about species and natural processes, and human dependence on them. Such a linkage with nature is equally important to provide political support for the protection of nature in rural and remote areas, where a shrinking proportion of the population lives. In a real sense the future of natural landscapes and the global environment depend on the suburban interstitial nature being a vibrant linkage between land and people.

Landscape Ecology and Societal Objectives

Most of society's major objectives meet, too often in conflict, at the landscape and regional scales (Diamond and Noonan 1996). Prevailing thought is that effective solutions are impossible because the area is too large (e.g., regional planning is unpopular), local governments and the public focus on home rule (e.g., in towns or counties), and the issues are too numerous, difficult, and persistent. I no longer believe this. Landscape ecology has now emerged with many attributes that could break the logjam.

Those in this field are familiar with the focus on large heterogeneous areas (regions, landscapes, and portions thereof), landscape structure, function, and change, natural- and human-dominated areas, a spatial language that enhances cross-discipline communication, a tight linkage between theory and application, and an integrated focus on water, soil, and biodiversity issues. However, societal objectives for the land are much more diverse. Meshing them goes way beyond the landscape ecology thought of today, yet this body of theory appears to be an entree into meshing such diverse objectives. To illustrate, I will first encapsulate the primary objectives of ten "fields" or "communities" in society. All of these objectives are usually perceived anthropogenically, that is, as providing resources for people.

1. Transportation community: provide mobility in a safe and efficient manner.
2. Agriculture: sustain crop production.
3. Conservation biology: save rare species and biodiversity.
4. Game management: maintain wildlife populations for hunting.
5. Range management: maintain forage and soil conditions for livestock production.
6. Forestry: manage and harvest forests for wood products.
7. Water resources: maintain water quantity and high water quality.
8. Design: enhance aesthetics in the built environment.
9. Town (and suburban) planning: mesh people's homes with other places for daily living.
10. Recreation community: provide healthy places and activities for leisure time.

Secondary objectives exist in every field. One secondary objective of all fields is to "minimize environmental degradation" or "accomplish the primary objective consistent with maintaining ecological integrity or a healthy environment." This distinctive commonality to all fields is where landscape ecology becomes so important and an opportunity for society. The fields all focus on the landscape scale. Landscape ecology provides an analytical framework and a communication mechanism to understand spatial patterns, horizontal processes, and change in the land mosaic over time.

Examine the preceding list of primary objectives, and see if any two are incompatible at the landscape scale. The answer is no. Are any three incompatible? No. In fact, all ten theoretically seem compatible, despite the many existing conflicts. Spatial arrangement of land uses is central. Yet, as seen earlier in this article, horizontal natural processes must help determine spatial pattern.

Take this analysis another step. Are any of the primary objectives listed incompatible with a "healthy" environment? No. A landscape ecologist could readily design a plan compatible for both transportation and ecological objectives, or for both game management and ecology, or agriculture and ecology (Dramstad et al. 1996). It would not maximize either component. It would be an optimization, where bits are lost in tradeoffs, but neither component as a whole is significantly reduced or degraded. Could such a plan be made compatible for two primary objectives plus ecology? Three plus ecology? All ten plus ecology? I think so. This appears to be a tractable challenge for landscape ecology, for each field or community itself, and for society as a whole. A resulting vision that makes all the objectives spatially compatible would truncate the current high level of conflicts and pessimism for solutions and would be ecologically logical at the core. Changing the trajectory of landscape change in this direction would be stunning.

Two additional key elements are qualitatively different from, yet permeate, all objectives in the preceding list: economics and culture. Economics should be there, yet too often it is placed first in priority by society. Probably it is better for societal objectives to be wisely meshed first based on their individual inherent values and secondly on their emergent combined value. Then determine how, if, or when the economics work to attain the integrated solution. Economics often change rapidly. What will not work today will tomorrow, and vice versa. At present there appears to be no vision of how society's major objectives can compatibly and ecologically mesh in a landscape or region. Outline a vision and then add economics.

Culture, on the other hand, cuts through the list of society's priorities in nearly the same way as ecology. (I use human culture in the traditional sense of language, arts, aesthetics, morals, literature, and education). Culture should be a "secondary" objective common to all fields. "Minimize cultural degradation" or "accomplish the primary objective consistent with maintaining cultural integrity or maintaining a healthy culture." All the questions posed and answers given for ecology in this section refer almost equally to culture. When pushed, I would place ecology somewhat ahead of culture in priority for sustainability. But the important point is that culture seems more compatible with ecology than almost all the listed primary objectives, and thus the landscape ecological approach for meshing the objectives seems especially promising. Furthermore, because culture and landscapes commonly change at about the same rate (i.e., over human generations) (Forman 1995), the resulting solution is clearly on the trajectory toward a sustainable environment.

Summary

Horizontal natural processes produce curvilinear, often convoluted, routes across the land. Planning, conservation, design, and management normally arrange pieces of the landscape into patterns. A better approach first provides for horizontal natural processes and then arranges the objects. This results in more natural forms and greater sustainability to persist over generations. In the built environment, intensive planning or design using symmetric geometric patterns commonly produces biological impoverishment, whereas biological richness emerges in places overlooked by society. Three actions would make planned/designed areas better: (1) provide for horizontal natural processes; (2) use more spatial patterns such as produced by nature; and (3) include one or more bio-rich places where native species richness is exhuberant in every design and plan.

Ecological impacts from a road traversing numerous ecosystems and land uses spread outward to form a road-effect zone. The zone is usually wide and asymmetric, with convoluted margins and occasional fingers where roads cross other corridors. The barrier effect of roads leading to small fragmented populations is probably quite significant. Thus perforating road barriers is a key applied principle. A planning framework for transportation using landscape ecology begins by mapping the ecological network or infrastructure (i.e., the large natural-vegetation patches and the major wildlife and water flow corridors). Next the road network is superimposed on the ecological network to identify bottlenecks. Then an array of avoidance, mitigation, and compensation techniques is used to eliminate bottlenecks. In this transportation planning, road barriers are perforated to maintain horizontal natural processes or ecological flows.

Six major ecological values of large natural-vegetation patches in the suburban landscape include water, species, and microclimatic benefits. Although these values differ significantly from those in rural landscapes, large patches of natural vegetation are a top priority for planning and conservation in suburbia as elsewhere. Of the values of small natural-vegetation patches, only the stepping-stone role for species movement across the matrix is of major importance in suburbia. Overall, the educational and political value of linking nature and people in metropolitan regions exceeds the ecological value of suburban nature.

For ten disciplines illustrated that focus on land use, all have a secondary objective of minimizing environmental degradation. Landscape ecology provides a unique analytical framework and spatial broad-scale language for society to spatially mesh and accomplish its objectives. Society's collective land-use objectives appear to be spatially compatible with ecological objectives. Therefore, in the face of pervasive land-use conflicts and pessimism, such a landscape ecological optimization provides a vision for the future.

Acknowledgments. I warmly thank: Larry D. Harris for highlighting the importance of horizontal natural processes; Hein van Bohemen for insight into ecological transportation planning in The Netherlands; the National Research Council "Committee to Study the Future of Transportation in the United States in the Context of a Sustainable Environment" for numerous discussions of road ecology; my Suburban Ecology classes at Harvard for discussions on suburban patches of nature; Lauren E. Alexander, Robert H. Gardner, and Jeffrey M. Klopatek for kindly reviewing the manuscript; and Anna M. Hersperger for insight in several areas.

References

Art, H.W., F.H. Bormann, G.K. Voight, and G.M. Woodwell. 1974. Barrier island forest ecosystem: role of meteorologic nutrient inputs. Science 184:60–62.

Balser, D., A. Bielak, G. De Boer, T. Tobias, G. Adindu, and R.S. Dorney. 1981. Nature reserve designation in a cultural landscape, incorporating island biogeographic theory. Landscape Planning 8:329–347.

Bennett, A.F. 1991. Roads, roadsides and wildlife conservation: a review. In: Nature Conservation 2: The Role of Corridors, pp. 99–117. D.A. Saunders and R.J. Hobbs (eds.). Surrey Beatty, Chipping Norton, Australia.

Bill, R.G. Jr., R.A. Sutherland, J.F. Bartholic, and E. Chen. 1978. Observations of the convective plume of a lake under cold-air advective conditions. Boundary-Layer Meteorology 14:543–556.

Binford, M.W., and M. Buchenau. 1993. Riparian greenways and water resources. In: Ecology of Greenways: Design and Function of Linear Conservation Areas, pp. 69–104. D.S. Smith and P.C. Hellmund (eds.). University of Minnesota Press, Minneapolis.

Burgess, R.L., and D.M. Sharpe (eds.). 1981. Forest Island Dynamics in Man-Dominated Landscapes. Springer-Verlag, New York.

Canters, K. (ed.). 1997. Habitat Fragmentation and Infrastructure. Ministry of Transport, Public Works and Water Management, Delft, Netherlands.

Charlesworth, D., and B. Charlesworth. 1987. Inbreeding depression and its evolutionary consequences. Annual Review of Ecology and Systematics 18:237–268.

Craul, P.J. 1992. Urban Soil in Landscape Design. John Wiley & Sons, New York.

Diamond, H.L., and P.F. Noonan. 1996. Land Use in America. Island Press, Washington, DC.

Dramstad, W., J.D. Olson, and R.T.T. Forman. 1996. Landscape Ecology Principles in Landscape Architecture and Land-use Planning. Island Press, Washington, DC.

Evink, G.L., P. Garrett, D. Zeigler, and J. Berry (eds.). 1996. Trends in Addressing Transportation Related Wildlife Mortality. Report FL-ER-58-96, Florida Department of Transportation, Tallahassee.

Forman, R.T.T. 1987. The ethics of isolation, the spread of disturbance, and landscape ecology. In: Landscape Heterogeneity and Disturbance, pp. 213–229. M.G. Turner (ed.). Springer-Verlag, New York.

Forman, R.T.T. 1995. Land Mosaics: The Ecology of Landscapes and Regions. Cambridge University Press, Cambridge, UK.

Forman, R.T.T., and L.E. Alexander. 1998. Roads and their major ecological effects. Annual Review of Ecology and Systematics 29:207–231.

Forman, R.T.T., and S.K. Collinge. 1996. The "spatial solution" to conserving biodiversity in landscapes and regions. In: Conservation of Faunal Diversity in Forested Landscapes, pp. 537–568. R.M. DeGraaf and R.I. Miller (eds.). Chapman and Hall, London.

Forman, R.T.T., and S.K. Collinge. 1997. Nature conserved in changing landscapes with and without spatial planning. Landscape and Urban Planning 37:129–135.

Forman, R.T.T., and M. Godron. 1986. Landscape Ecology. John Wiley & Sons, New York.

Forman, R.T.T., and A.M. Hersperger. 1996. Road ecology and road density in different landscapes, with international planning and mitigation solutions, pp. 1–22. Report FL-ER-58-96, Florida Department of Transportation, Tallahassee.

Forman, R.T.T., and A.M. Hersperger. 1997. Landscape ecology and planning: a powerful combination. Urbanistica 108:61–66.

Forman, R.T.T., D.S. Friedman, D. Fitzhenry, J.D. Martin, A.S. Chen, and L.E. Alexander. 1997. Ecological effects of roads: toward three summary indices and an overview for North America. In: Habitat Fragmentation and Infrastructure, pp. 40–54. K. Canters (ed.). Ministry of Transport, Public Works and Water Management, Delft, Netherlands.

Geiger, R. 1965. The Climate Near the Ground. Harvard University Press, Cambridge, MA.

Gilbert, O.L. 1991. The Ecology of Urban Habitats. Chapman and Hall, London.

Harris, L.D., T.S. Hoctor, and S.E. Gergel. 1996. Landscape processes and their significance to biodiversity conservation. In: Population Dynamics in Ecological Space and Time, pp. 319–347. O. Rhodes, Jr., R. Chesser, and M. Smith (eds.). University of Chicago Press, Chicago.

Harris, L.D., and J. Scheck. 1991. From implications to applications: the dispersal corridor principle applied to the conservation of biological diversity. In: Nature Conservation 2: The Role of Corridors, pp. 189–220. D.A. Saunders and R.J. Hobbs (eds.). Surrey Beatty, Chipping Norton, Australia.

Hough, M. 1984. City Form and Natural Processes: Towards an Urban Vernacular. Van Nostrand Reinhold, New York.

Knaapen, J.P., M. Scheffer, and B. Harms. 1992. Estimating habitat isolation in landscape planning. Landscape and Urban Planning 23:1–16.

Laikhtman, D.L. 1964. Physics of the Boundary Layer of the Atmosphere. Israel Program for Scientific Translations, Jerusalem.

Matlack, G.R. 1993. Sociological edge effects: spatial distribution of human impact in suburban forest fragments. Environmental Management 17:829–835.

Morel, G.A., and B.P.M. Specken. 1992. Versnippering van de ecologische hoofdstructuur door de natte infrastructuur. Project Versnippering Deel 4 (H. Duel, schrijver). Directoraat-Generaal Rijkswaterstaat, Dienst Weg- en Waterbouwkunde, Delft, The Netherlands.

Natuur over Wegen (Nature Across Motorways). 1995. Rijkswaterstaat (Directorate-General for Public Works and Water Management), Dienst Weg- en Waterbouwkunde (Road and Hydraulic Engineering Department), Delft, The Netherlands.

Pfister, von H.P., and V. Keller. 1995. Strassen und Wildtiere: Sind Grunbrucken eine Losung? Bauen fur die Landsirtschaft 32:26–30.

Ranney, J.W., M.C. Bruner, and J.B. Levenson. 1981. The importance of edge in the structure and dynamics of forest islands. In: Forest Island Dynamics in Man-dominated Landscapes, pp. 67–96. R.L. Burgess and D.M. Sharpe (eds.). Springer-Verlag, New York.

Reck, H., and G. Kaule. 1993. Strassen und Lebensräume: Ermittlung und Beurteilung strassenbedingter Auswirkungen auf Pflanzen, Tiere, und ihre Lebensräume. Forschung Strassenbau und Strassenverkehrstechnik, Heft 654. Herausgegeben vom Bundesminister für Verkehr, Bonn-Bad Gödesberg, Germany.

Reijnen, R., R. Foppen, C. ter Braak, and J. Thissen. 1995. The effects of car traffic on breeding bird populations in woodland. III. Reduction of density in relation to the proximity of main roads. Journal of Applied Ecology 32:187–202.

Saunders, D.A., and R.J. Hobbs (eds.). Nature Conservation 2: The Role of Corridors. Surrey Beatty, Chipping Norton, Australia.

Soule, M. 1987. Viable Populations for Conservation. Cambridge University Press, Cambridge, UK.

Spirn, A.W. 1984. The Granite Garden: Urban Nature and Human Design. Basic Books, New York.

Sukopp, H., S. Hejny, and I. Kowarik (eds.). 1990. Urban Ecology: Plants and Plant Communities in Urban Environments. SPB Academic Publishing, The Hague, The Netherlands.

Tilghman, N. 1987. Characteristics of urban woodlands affecting breeding bird diversity and abundance. Landscape and Urban Planning 14:481–495.

Turner, M.G. (ed.). 1987. Landscape Heterogeneity and Disturbance. Springer-Verlag, New York.

van Bohemen, H., C. Padmos, and H. de Vries. 1994. Versnippering-ontsnippering: Beleid en onderzoek bij verkeer en waterstaat. Landschap 1994(3):15–25.

White, E.J., and F. Turner. 1970. A method of estimating income of nutrients in catch of air borne particles by a woodland canopy. Journal of Applied Ecology 7:441–461.

Part II
Modeling Applications

5 A Knowledge System Environment for Ecosystem Management[1]

ROBERT N. COULSON, HANNU SAARENMAA,
WALTER C. DAUGHERITY, E.J. RYKIEL JR.,
MICHAEL C. SAUNDERS, AND JEFFREY W. FITZGERALD

Ecosystem management is a collective term used to embrace a philosophy and set of methodologies associated with land-use manipulation or modification (Grumbine 1994; Kaufmann et al. 1995; Sedjo 1995; Lackey 1996). Broad-based issues dealt with under the umbrella of ecosystem management include biodiversity (Wilson 1992; Groombridge 1995; Heywood and Watson 1995), sustainability of ecological systems (Lubchenco et al. 1991; Risser et al. 1991; Covington and DeBano 1993; Levin 1993; Bormann et al. 1994a), maintenance of ecosystem health (Costanza et al. 1993; USDA Forest Service 1996), preservation of ecosystem integrity (Monnig and Byler 1992; Woodley et al. 1993), conservation and stewardship (Sample 1991; Callicott 1994; Alpert 1995), public participation (Wondolleck 1988; Knopp and Caldbeck 1990; Loikkanen 1995), and landscape management (Lucas 1991; Diaz and Apostol 1993; Urban 1993; Boyce 1995; Forman 1995a, 1995b).

The subjects considered within the perview of ecosystem management are certainly diverse, but they all have in common a substantial existing knowledge base associated with them. This knowledge base evolves continuously as a function of new research and expanded human experience. Given this circumstance, it follows that a fundamental question in ecosystem management, regardless of the specific domain focus, centers on how to use existing knowledge for land-use planning, problem solving, and decision making. Answering this question requires consideration of several related issues: (1) ecosystem management includes an agenda of scientific, social, legal, political, and technical topics, (2) knowledge about ecosystems is scattered among a variety of domain specialities, (3) efficient use of the existing knowledge base available for a specific problem in ecosystem management generally necessitates application of integrative computer-based

[1]This work was supported in part by a contract with Battelle, Pacific Northwest Laboratory, and the USDA Forest Service, Forest Health Protection. The opinions expressed herein are those of the authors.

technologies that can accommodate both quantitative as well as qualitative information, and (4) integrative computer-based decision support systems represent a new and untested approach for agencies and organizations involved in ecosystem management.

Accordingly, our goals in this chapter are to examine how computer-based technologies can be used to address complex problems in ecosystem management and to consider how integrative systems can be employed in practical settings. There are four specific objectives. First, we will examine the concept of ecosystem management. Second, we will describe an approach, referred to as knowledge engineering, that can be used to organize solutions for ecosystem management problems. Third, we will examine a type of integrative computer-based application, referred to as a knowledge system environment, that can be tailored to address ecosystem management from planning, problem solving, and decision making perspectives. Fourth we will consider several important issues associated with efficacy, deployment, and implementation of knowledge-based computer systems for ecosystem management.

Ecosystem Management: An Evolving Concept

The general notion of ecosystem management was recognized and well formulated through technical writings published in the 1960s and 1970s, for example, Watt (1968), Van Dyne (1969), Holling (1978), and Bormann and Likens (1979). Recent reviews of the concept and practice include Sample (1991), Covington and DeBano (1993), Callicut (1994), Grumbine (1994), Kaufmann et al. (1995), Lackey (1995), Thompson (1995), Christensen et al. (1996), and Vogt et al. (1997). Although many scientists and practitioners advocated a holistic approach, there was no broad-based support for the use of ecosystem concepts in land-use management until 1993. At that time President William J. Clinton convened the Pacific Northwest Forest Conference. This summit, which took place in Portland, Oregon, was a response to controversy over forest management on public lands in the Pacific Northwest. The event was significant in that it provided authorization to enact the use of a new paradigm for land-use management in the United States, that is, ecosystem management.

As there was no consensus as to the scope and bounds of ecosystem management, a flurry of activity followed in which the goal was to develop a workable concept that could be implemented by the agencies responsible for federal land-use management, that is, the National Park Service, the Bureau of Land Management, the Fish and Wildlife Service (within the Department of the Interior), and the Forest Service (within the Department of Agriculture). These agencies are charged with management of approximately 30% of the landmass of the United States. One immediate product of this activity was the FEMAT (Forest Ecosystem Management

Team) report, which addressed requirements for ecosystem management. Commentary on this report and other broader issues of ecosystem management followed, for example Slocombe (1993), Bormann et al. (1994a, 1994b), Caldwell et al. (1994), Franklin (1994), Noss et al. (1995), Salwasser (1994), Thomas (1994), and Kaufman et al. (1995). In 1994, the U.S. General Accounting Office (GAO) published a report that summarized the status of federal initiatives for implementation ecosystem management. Also identified were additional needed actions and barriers to government-wide implementation. This same process took place simultaneously in Europe. Emphasis was placed on defining the concept, criteria, and indicators of sustainable development for forestry (Ministerial Conference on the Protection of Forests in Europe, 1993).[2]

Although the topic of ecosystem management has been the focus of considerable critique, the concept is still evolving. From an examination of the literature, we identified four subjects on which further commentary could contribute to better understanding.

First, there is little consensus about the meaning of the term ecosystem management. Examples representing a spectrum of definitions are presented in the Appendix. A fundamental message contained in all the definitions of the concept is that ecosystem management includes consideration of scientific, social, legal, political, and technical issues. The range in meaning of the definitions illustrates how difficult a task it is to merge a scientific concept (ecosystem ecology) with a human activity (land-use management) (Vogt et al. 1997).

Second, a somewhat surprising feature of the literature on ecosystem management is that only modest emphasis is placed on fundamental principles of ecosystem ecology (Golley 1993) or ecological science. The notable exception is Christensen et al. (1996). A logical beginning for a definition of ecosystem management, founded on a basic textbook definition of the term ecosystem, might be: the orchestrated modification or manipulation of primary production, consumption, decomposition, or abiotic storage. Putting such a definition into practice would require integration of a substantial knowledge base on (1) the structure and function of the natural ecological system, (2) the current land use system, and (3) the socioeconomic system governing the land unit in question.

Third, at least part of the explanation for the lack of consensus on the boundary for the practice of ecosystem management centers on the casual use of the term ecosystem. (i.e., in the literature the term is used imprecisely). The common definition (e.g., the biotic community plus the abiotic environment) is inherently vague about scale and boundary. In fact, ecosystems are often described simply as environments around special organ-

---

[2] Documents of the Helsinki process are available on the Internet WWW at http://www.efi.joensuu.fi/publications/helsinki/.

isms of interests (e.g., the forest ecosystem, the cotton ecosystem, the apple orchard ecosystem) or land areas of interests (e.g., the Yellowstone Ecosystem, the Everglades Ecosystem). Because ecosystem management is "place-based" and involves discrete human activities, it is imperative that boundary and scale be defined in precise ways. In practice this definition is often quite difficult. In some instances natural boundaries (such as those delineating watersheds) are present and management units are obvious. In other instances, property rights of individuals create highly fragmented "ecosystems" that are not amenable to corporate management. The issue of practicing ecosystem managment at various spatial and temporal scales has been addressed in the literature (e.g., McNab and Avers 1994; Forman 1995b). However, there is little consensus as to the resolution (grain size) and range (extent) (Schneider 1994) needed to achieve specific ecosystem management goals and objectives. Examples of multiscale analyses of proposed land-use management practices are rare.

Fourth, characteristic patterns of landscapes presumably are a result of the operation of ecological processes, that is, "ecological processes generate patterns, and by studying these patterns we can make useful inferences about the underlying processes" (the pattern/process paradigm, quoted from Urban 1993). How to identify and manipulate the processes to create a desired pattern is very much a part of landscape ecology research (Holling 1992). One aspect of this research considers ways and means for integrating information on land units and land attributes (Zonneveld 1990). At this stage in the evolution of the concept and practice of ecosystem management, it is not clear what the goals are. Is the desired outcome of management to create landscape patterns that are (relative to a natural condition or nominal state) more than illusions, delusions, or deceptions?

Knowledge Engineering

Ecosystem management deals with subjects that typically have large and disparate knowledge bases. The data and information that form the knowledge base for a specific problem often come from several different domain specialties, such as, ecology, geography, sociology, and economics. The knowledge base can exist in several forms: (1) tabular information (usually stored in a database management system), (2) spatially referenced data themes (usually associated with a geographic information system—GIS), (3) numerical output from simulation models and mathematical evaluation functions, (4) unstructured paper and hypertext documents, and (5) heuristics of experts (based on corporate experiences of humans). Although the knowledge base for most problems in ecosystem management is substantial, it is also incomplete and in a state of evolution.

Knowledge engineering is an activity that embraces a set of concepts and methodologies dealing with (1) acquisition of knowledge, (2) analysis and

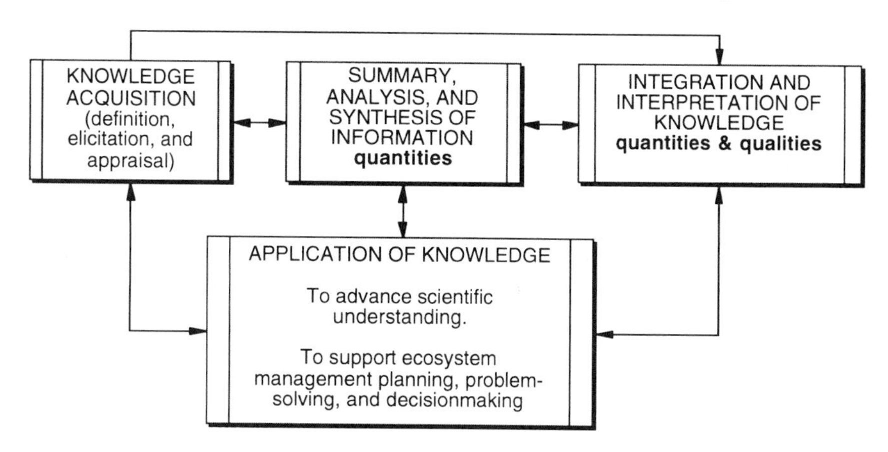

FIGURE 5.1. Basic elements of the knowledge engineering concept and method for ecosystem management (From Coulson et al. 1996, with permission from ERIM International).

synthesis of data and information (quantities), (3) integration and interpretation of knowledge (quantities and qualities), and (4) application of knowledge (Fig. 5.1) (Coulson et al. 1996). The goal of this activity, in the context of ecosystem management, is to facilitate the use of the full extent of knowledge available for the purpose of solving a problem, supporting decision making, or developing a plan of action. Computer-based tools and technologies have been created to formalize and automate the knowledge engineering process, thus greatly expanding human capabilities. Knowledge engineering can be viewed as a computational approach to ecosystem management. It is a direct consequence of the development of digital computers and the infusion of this technology into society. Each of the elements of knowledge engineering is briefly described below. The elements include acquisition of knowledge, analysis and synthesis of data and information, and integration and interpretation of knowledge.

Acquisition of Knowledge

Every program in ecosystem management requires an evaluation of the extant data and information that form the knowledge base for a specific problem. There are three basic activities associated with the knowledge acquisition process: definition, elicitation, and appraisal. Definition is the identification of relevant data and information, and it follows from a systematic evaluation of the problem of interest. Modern computer-based analysis methodologies can assist in this process (e.g., Booch 1991; Rumbaugh et al. 1991). Facilitation tools (e.g., Objectives-Oriented Program Planning™ and various CASE programs) can guide the formulation of specific project objectives. Solution pathways can then be created. Elicitation means acquiring information directly from experts and stakeholders. The goal is to develop a formal knowledge base for a particular topic or problem. A variety of techniques has been devised to guide the process, including focused and structured interviews, probes, goal decomposition, etc. (Muggleton 1990; Scott et al. 1991). Computer-based tools are available to assist in knowledge base construction, maintenance, and documentation (e.g., NETWEAVER™). Recently, computer networks have greatly enhanced our ability to access information. INTERNET and the World Wide Web (WWW) are bidirectional media that can connect planners with the various stakeholders in land-use management. Their opinions can be collected using the WWW and draft plans can be publicly reviewed (Burk and Lime 1996). Appraisal is the evaluation of the data and information that form the knowledge base for a particular problem. All scientists perform this task when they seek to place the results of their research into the existing corpus of knowledge. For complicated problems, where a variety of sources and types of data and information are involved (e.g., evaluating the impact of global warming on biodiversity), computer-based systems (e.g., CLIPS™, KAPPA™, NETWEAVER™, NEXPERT™) are extremely useful for ordering and organizing extant data and information. These systems are also useful in identifying deficiencies in knowledge.

Analysis and Synthesis of Data and Information

Quantitative ecology is a science in which the objects of interest or study (plants, animals, the elements of the environment) can be described by units and dimensions (e.g., biomass in gm^{-2}). Interpretation of results from ecological research usually involves the analysis or synthesis of data and information represented as scaled quantities, that is, as objects defined by units and dimensions (Schneider 1994).

Although the subjects of analysis and synthesis are often discussed in the same context, they are fundamentally different activities. Furthermore, the tools and techniques used are different. Analysis is the process of separating or breaking up of any whole into its parts so as to find out their nature, proportion, function, or relationship (e.g., analysis of variance). There are three common approaches to analysis: graphical, numerical, and statistical. Each of these approaches can be further subdivided. For example, elements of statistical analysis include environmental design and sampling, spatial statistics, statistical ecology, environmental regulatory statistics, environmental monitoring, and environmetrics. Synthesis means putting together of parts or elements so as to form a whole, i.e., it is the antithesis of analysis. The three common approaches are simulation, optimization, and visualization. Each approach represents a substantial and well-developed discipline that can be further compartmentalized. For example, simulation can be viewed as continuous (systems of differential equations) or discrete (object-oriented simulation). Optimization (in relation to restrictions) includes linear programming, non-linear programming, dynamic programming, and control theory. Visualization includes graphics, animation, and four-dimensional representations.

Integration and Interpretation

Although scientists prefer to deal with quantities, managers rely also on qualitative judgments based on their experience (or the experiences of others), that is, they use heuristic knowledge as well as quantitative data and information to solve problems, make decisions, and develop plans. Scientific understanding of a specific ecosystem management problem will rarely be so complete as to eliminate the need for qualitative assessment and human judgment. Therefore, our emphasis herein centers on integrative systems that are useful for blending quantitative as well as quantitative information.

The integrative systems are based on technologies adapted from a variety of subject domains, such as computer science, engineering, mathematics, cognitive psychology, management science. Several types of integrative systems have been used to address ecosystem management problems (Fig. 5.2): expert systems, intelligent geographic information systems, intelligent database management systems, object-oriented simulation, and knowledge-based systems [see AI Applications, (vol. 9, no. 3), for representative examples of these different types of systems].

Note that the increasing complexity of the computer-based tools parallels the levels of human comprehension. The advanced levels (experience processing, shared visions, and epiphanies) (Fig. 5.2) can only be achieved through the use of the full measure of knowledge available on a subject. The current end point (in the continuing evolution of computer-based systems that address this challenge) is the knowledge system environment (KSE) (Coulson et al. 1989, 1995). The KSE is described below in detail, because it represents the facilitating technology for the application of ecosystem management practices.

The Knowledge System Environment

The KSE represents an advanced instance of the integrative computer-based systems that are suitable for addressing complex ecosystem management problems. The KSE is a concept and methodology for ordering and organizing a knowledge base to facilitate planning, problem solving, and decision support. The KSE is a generic approach that can be tailored for different ecosystem management scenarios. In the following sections we examine the general features of the KSE concept and method and examine applications of the approach.

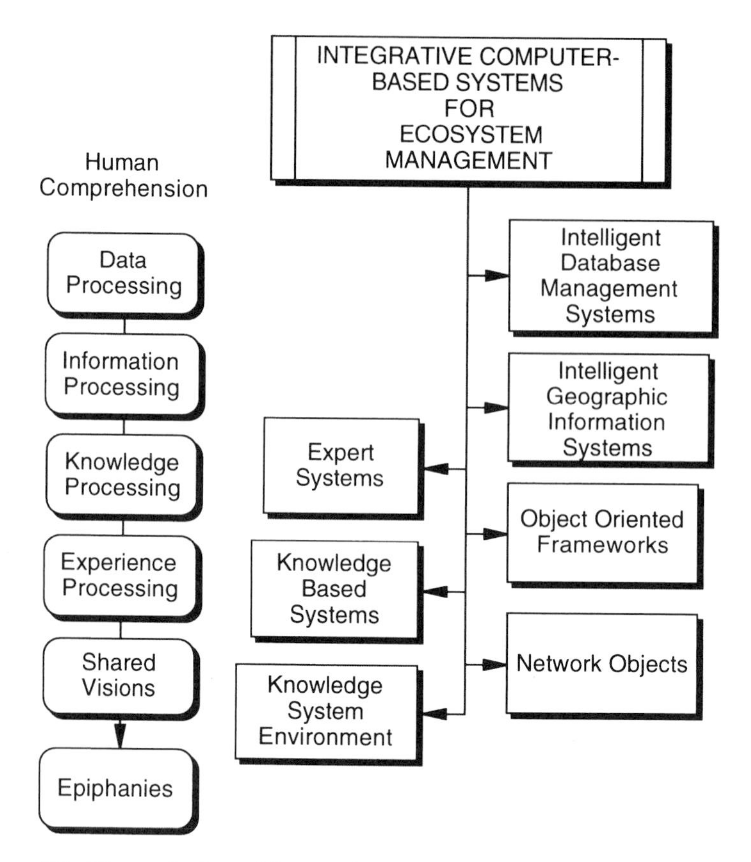

FIGURE 5.2. The evolution of integrative computer-based systems for ecosystem management planning, problem solving, or decision support (From Coulson et al. 1996, with permission from ERIM International).

Concept

The KSE concept centers on a software system consisting of elements to define, order, organize, analyze, integrate, interpret, visualize, model, deploy, and disseminate data, information, and knowledge. The initial challenge in developing the concept was to create a software system that could use qualitative domain information about a specific ecosystem management problem, accommodate problems where algorithmic solutions were not feasible, use metaknowledge (knowledge about knowledge) to create novel problem solution strategies, and integrate the different representations of knowledge available for ecosystem management, that is, tabular and spatially referenced data, simulation output, heuristic judgment of experts, and so on. A KSE consists of six basic elements: presentation management system (user interface), knowledge-based interpreter, management infor-

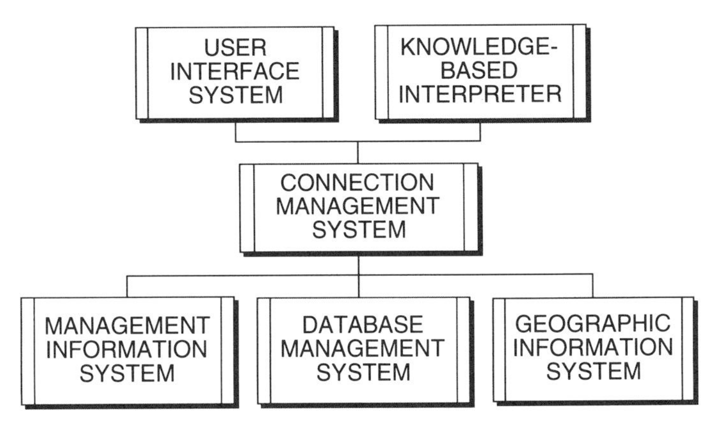

FIGURE 5.3. Elements of a Knowledge System Environment for ecosystem management (Modified from Coulson et al. 1989).

mation system, geographic information system, database management system, and connection management system (Fig. 5.3) (Coulson et al. 1989).

Method

The KSE concept considers what needs to be included in a software system to accommodate the tasks associated with ecosystem management. However, the issue of how to execute the concept requires specification of both a computer hardware platform and software system. Emphasis herein is focused on the software system. A variety of options are available and, therefore, our approach here is to provide an overview of the design strategy and describe an application of KSE.

Object-Oriented Design for a KSE

In 1989, when the KSE concept was described, memory capacity of computer hardware and abstraction capabilities of software tools were only beginning to reach levels that allowed realistic modeling of natural phenomena. The critical problem, from a computing perspective, is how to accommodate the immense complexity associated with ecosystem management problems. This complexity is a consequence of several factors. First, natural systems consist of hierarchies of objects. For example, a forest ecosystem includes the soil, humus, ground vegetation, trees, and microclimate. A tree consists of classes of modules: stem, roots, branches, buds, and foliage (Salminen et al. 1994). Foliage consists of needles or leaves which in turn consist of various classes of tissues. There are large numbers of interactions among objects in ecosystems. Objects in close proximity generally have the greatest reciprocal interaction. Second, ecosystems, and the data and information used to describe them, change in space and time. Third, knowledge aobut the ecosystems is scattered among a variety of domain specialties. Finally, ecosystem management is only a component of the larger issue of environmental management.

In the past, modeling of complex systems was approached using mathematical methods. This technique required excessive simplification which resulted in a significant compromise to reality. The software engineering technology developed to address the need for both abstraction and reality, was objected-oriented programming (OOP) and the modeling methodologies associated with it (object-oriented design and object-oriented analysis).

Object-oriented programming is defined as "a method of implementation in which (computer) programs are organized as cooperative collections of objects, each of which represents an instance of some class, and whose classes are members of a hierarchy of classes united via inheritance relations" (Booch 1991). An object is "something you can do things to." As such objects can be tasks, plans, activities, agents, biological data, and so on. Application of OOP involves development of an object model, which is a software engineering paradigm that emphasizes principles of abstraction, encapsulation, modularity, hierarchy, typing, concurrency, and persistence (Booch 1991). The object model is generally represented as an object diagram (see Saarenmaa et al. 1995, Fig. 1), which defines the objects and their relationships. Saarenmaa et al. (1994, 1995) describe the utility of object-oriented modeling as an approach for investigating problems in forest health management and illustrate how OOP can be used in decision support systems dealing with natural resource management.

Application

The decision support system, ISPBEX-II, developed to assist foresters in their efforts at integrated pest management (IPM) of the southern pine beetle (*Dendroctonus frontalis*) represents an application of the knowledge engineering approach, KSE concept and method, and OOP software engineering technology (Coulson et al. 1995). *Dendroctonus frontalis* is a significant mortality agent of yellow pines (*Pinus* spp.) in the Southern United States, and this insect has been the focus of a considerable research effort for ca. three decades.

IPM of *D. frontalis* is a complicated ecosystem management problem which provides a robust test for computer-aided decision support. The knowledge base for the insect/forest interaction is immense and consists of spatial databases for National Forests, tabular databases maintained for IPM purposes, published scientific studies and reports, a variety of simulation models and economic evaluation functions, and heuristic experiences of forest managers. Aside from the economic damage resulting from the activities of the insect, IPM has to be considered in the context of legal statutes governing endangered species protection and wilderness area management on National Forests.

ISPBEX-II is a computer application that provides expert advice on the selection of proper management actions related to suppression and prevention of *D. frontalis*. An object-oriented message passing architecture was used in the development of ISPBEX-II. A schematic of the approach taken is illustrated in Figure 5.4. The basic objects of ISPBEX-II include the following: (1) a user interface, (2) geographic information system, (3) database manager, (4) services, (5) inference engine, (6) simulation models, and (7) connection manager (Fig. 5.4). Because ISPBEX-II was developed specifically for the USDA Forest Service, care and maintenance of the

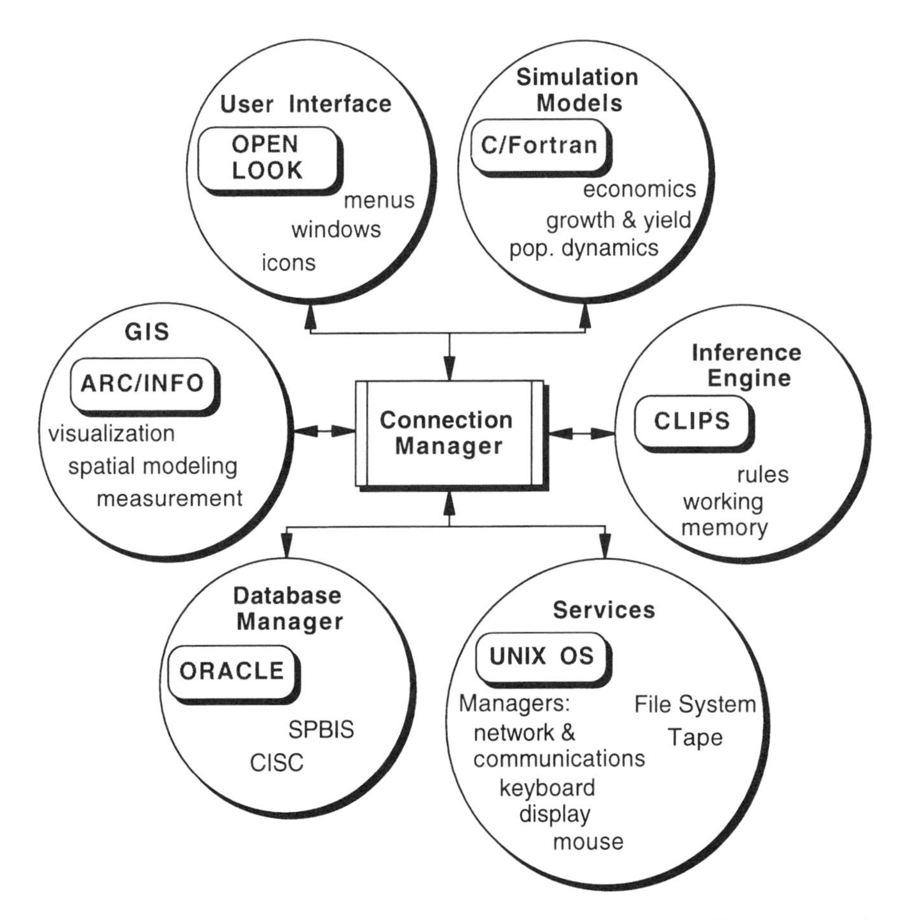

FIGURE 5.4. The object-oriented design for ISPBEX-II, the knowledge system environment for integrated pest management of the southern pine beetle (Modified from Coulson et al. 1995).

system was of paramount concern. For this reason the component software systems used in ISPBEX-II were selected from commercially available products whenever possible. The advantage of using this approach is that their suitability and quality have been evaluated through natural selection in the marketplace. Also, these products are under continuous development and refinement and generally user support is included. The following systems were used in ISPBEX-II: The user interface features a windows approach. OPENLOOK® was the product used. The GIS ARC/INFO® was used for analysis and display of spatially referenced data. The database management system ORACLE® was used to organize and access tabular information. Various types of services (e.g., network and communication, display manager, mouse manager, etc.) were included as part of the UNIX® operating system. The expert system component was CLIPS (C Language Production System), a forward-chaining rule-based language developed by NASA. The model component consisted of a suite of elements developed from previous research on population dynamics, infestation growth and spread, stand growth and yield, economics, etc. These models were C or FORTRAN programs. The connection manager, which served to lace ISPBEX-II together, was a C code (Coulson et al. 1995).

One of the strengths of the object-oriented approach used in ISPBEX-II centers on the fact that the individual elements forming the system can be exchanged with ease. The general concept is expressed in the "rights and responsibility" analogy. For example, the GIS has certain specified functions that it is called on to perform (e.g., production of maps defining the location of *D. frontalis* infestation relative to sensitive forest management areas). The GIS object is "responsible" for providing this information. The object has the "right" to generate the information by any appropriate method. Although the approach used in ARC/INFO® might be quite different than that employed by GRASS®, ISPBEX-II is unaffected as long as the necessary information is provided upon request. This feature allows ISPBEX-II to be updated easily, as the component software systems evolve. It also permits the substitution of one component system with another (e.g., GRASS® for ARC/INFO®). Furthermore, new domain-specific knowledge can be easily incorporated. Only communication protocols between the connection manager and the specific object would need to be changed.

Efficacy, Deployment, and Implementation of Computer-Based Technologies

In the early and mid-1980s computer-based decision support systems were projected for use across a spectrum of applications ranging from business to natural resource management (Coulson et al. 1987). There was particular interest in expert systems (computer programs designed to emulate human problem-solving capabilities), as this artificial intelligence technology was expected to have a significant impact in the commercial marketplace. The impact was not as great as anticipated, and this fact led to an investigation of the fate of the first wave of commercial expert systems (Gill 1995; Yoon et al. 1995). In an examination of approximately 100 such expert systems built before 1987, Gill (1995) found that during the five-year period between 1987 and 1992 approximately two thirds of the applications fell into disuse or were abandoned (Fig. 5.5). Furthermore, the brief longevity of the systems was not attributable to failure to meet technical performance or economic objectives. Prominent factors associated with the demise of the systems included lack of acceptance by users, inability to retain developers, problems in transitioning from development to maintenance, and shifts in organizational priorities (Fig. 5.6) (Gill 1995).

At present, there are not sufficient examples of the integrative systems (Fig. 5.2) to permit an evaluation of their fate. What has been demonstrated to date is that it is possible to develop and deliver complex computer-based systems suitable for addressing ecosystem management problems (Coulson et al. 1995). What has not been demonstrated is that these systems can be used in an operational setting to assist managers in planning, problem solving, and decision-making activities. If the systems are to be used in an operational setting, three additional issues must be addressed: (1) efficacy, (2) deployment, and (3) implementation. These three subjects are considered below.

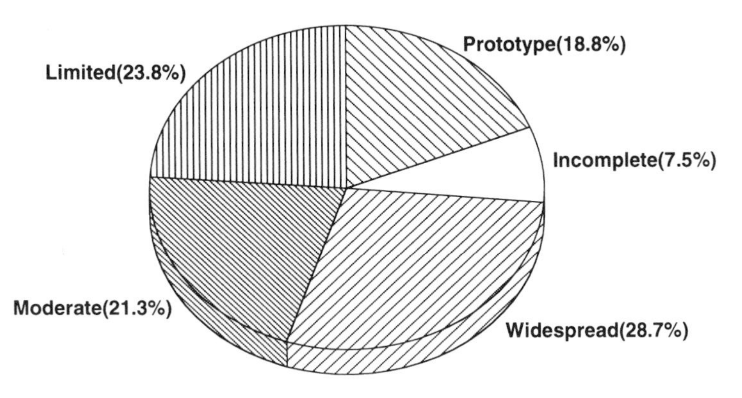

FIGURE 5.5. The fate of approximately 100 expert systems developed before 1987 for commerical purposes. Most of the systems were abandoned or not used operationally (Modified from Gill 1995).

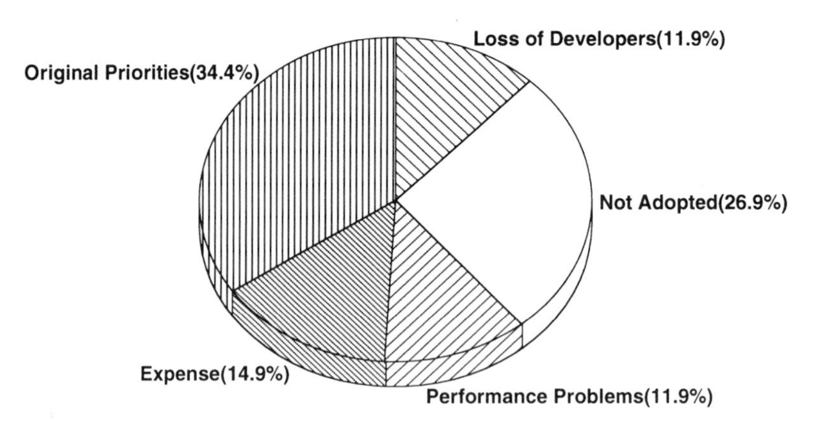

FIGURE 5.6. Reasons why the commercial expert systems were not used. Note that performance problems were not a significant reason for abandoning the systems (Modified from Gill 1995).

Efficacy

Efficacy (i.e., do the computer-based systems actually provide correct and useful ecosystem management plans and enhanced decision support capability) is a subject that has not been addressed in a critical way. This circumstance may simply be a reflection of the fact that few systems are actually in use at the present time. One goal in developing the systems is to utilize extant knowledge. However, rarely will the knowledge base available for a specific management problem be complete. Furthermore, the magnitude of error propagation through the component technologies employed by these systems is unknown. These errors influence the accuracy and precision of advice provided by the system. There are no good recipes for evaluating the efficacy of complex integrative computer systems that use different representations of knowledge. The task appears to involve use of a combination of quantitative (e.g., summary of cost/benefit analyses) and qualitative (evaluating utility of operational maps) criteria. Following are several measures that could provide insight into efficacy: (1) records of system use, (2) records of system performance, (3) analyses of reports generated by the system, (4) cost/benefit analyses of management decisions, (5) time and motion evaluation of system users, and (6) quality control evaluation for management decisions.

Deployment

Deployment refers to the operational use of the computer-based system by an agency or organization. Activities associated with system input and output are critical to deployment. In the design phase of system development, both input requirements and uses of output are considered in detail. Input to the system consists of data and information. A set of procedures for gathering needed input must be defined. If the system is to be used across an organization or agency, a formal protocol for collecting and processing data and information must be established. The task of developing the protocol is not trivial and generally requires a change in normal routine of employees. Output from the system can be used in various ways. Different echelons in an organization or agency will have specialized uses for the system. ISPBEX-II provides a good example of different uses for system output relative to IPM of *D. frontalis*. This system has the capability for dealing with decisionmaking, problem solving, and planning. In general, pest management specialists on National Forests have the responsibility for decisionmaking and problem solving. However, strategic planning uses of the system will generally involve supervisory personnel as well.

Implementation

Activities associated with implementation (establishment of the computer-based technology into the infrastructure of an agency or organization) include (1) enterprise integration, (2) maintenance and evolution, and (3) systems documentation (Fig. 5.7). Enterprise integration refers to the mechanism for establishing a software system into the infrastructure of an agency. Although most government agencies have rigorous procedures for selecting commercial software products, in general, there is no mechanism for integrating custom software products developed through contract research. A protocol that addresses this issue must be developed if the system is to receive agency-wide distribution and use. Maintenance and evolution of the software includes tasks associated with (1) adapting the computer code, (2) perfecting the computer code, and (3) correcting the computer code. Perfective alterations will follow from rigorous use of the system by practitioners. Corrective alterations involve "bug" fixes and changed requirements. Adaptive alterations include changes in hardware or operating system requirements, component software products, and documentation. Without constant vigil to maintenance and evolution, complex integrative software systems will decompose and quickly lose their functionality. Technology development refers to the initial contract for custom software. At that stage it is possible to define in detail the type and extent of the technical and user documentation required for the software. Such documentation will greatly influence the ease of support and maintenance following completion of system development.

Epilogue to Efficacy, Deployment, and Implementation

INTERNET technology can be used to address some of the problems associated with efficacy appraisal, deployment, and implementation of the KSE. Following are specific examples. The user interface for the KSE could

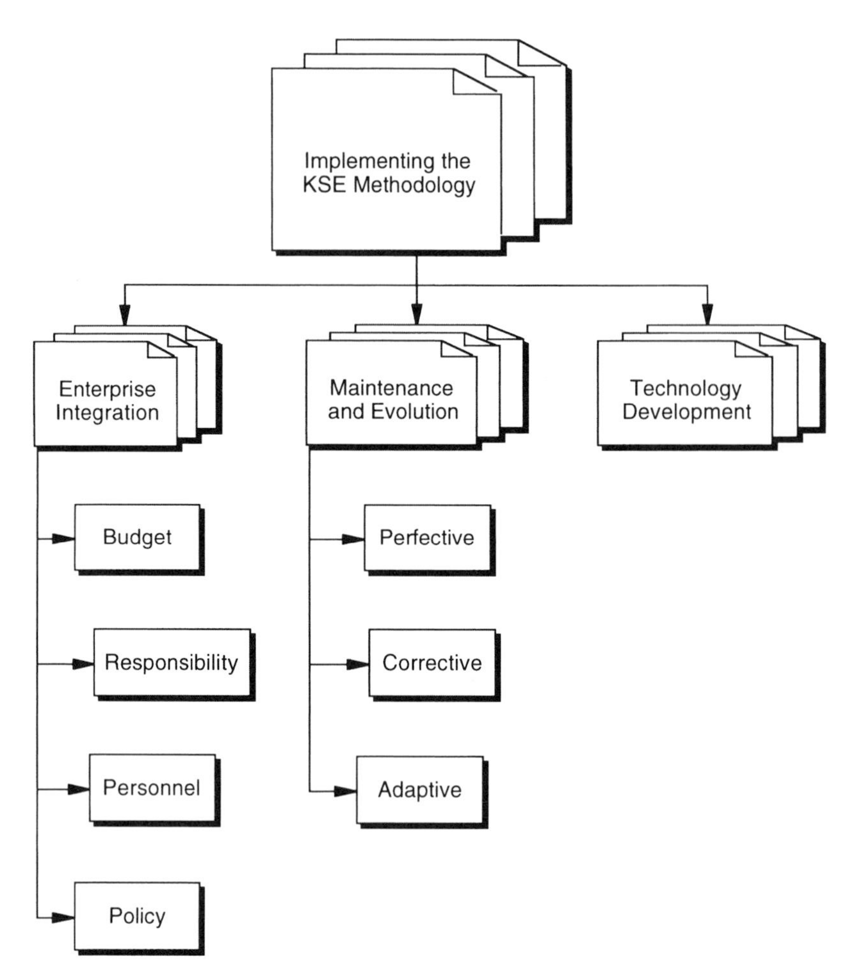

FIGURE 5.7. Issues that must be addressed in order to implement a knowledge system environment for ecosystem management.

be simplified through use of a single WWW front end. Net browsers, such as NETSCAPE NAVIGATOR™, provide access to the various types of subsystems that form the KSE, for example, databases, expert systems, simulation models, GIS, and so on (e.g., Steinke et al. 1996; Väkevä et al. 1996). The connection manager, which is the unique element of each KSE, can be dramatically simplified by replacing current software engineering approaches with the HyperText Transport Protocol (HTTP) and its future successors on the WWW. Component applications of the KSE can reside in remote servers (i.e., they do not have to be housed and maintained at field offices). Furthermore, applications can be built with network objects using the JAVA language, which gives the advantage of hardware platform independence. The knowledge-based interpreter of

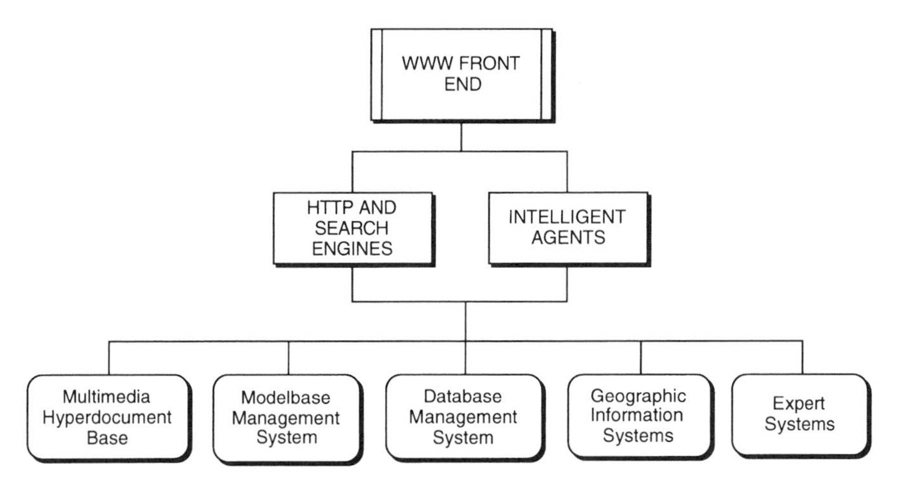

FIGURE 5.8. Elements of a next-generation network-aware Knowledge System Environment.

the KSE can also be replaced with intelligent agents (Saarenmaa et al. 1995) that can retrieve pertinent information from the networks (Fig. 5.8). In summary, INTERNET access will greatly facilitate downsizing and simplification of the KSE and will likely attract a broader clientele of users.

Conclusion

Application of ecosystem management requires new tools and technologies to define, order, organize, analyze, integrate, interpret, model, visualize, deploy, and disseminate existing data and information. Interest in technologies for ecosystem management has reached the level of the U.S. Congress (e.g., Morrissey 1995) and, indeed, the world community of nations. While new tools for field data collection are certainly needed, computer based technologies that facilitate planning, problem solving, and decision support are vital to the implementation of ecosystem management concepts. The KSE approach described herein represents the current end point in the evolution of computer-based integrative systems. The ISPBEX-II system illustrates a specific application of the KSE to a practical and complicated problem of land-use management. Although the methodologies for constructing a KSE are reasonably well developed, past experience with expert systems indicates that other considerations are crucial to actual use of these systems in operational settings. These considerations relate to the pragmatic constraints associated with deployment, implementation, and efficacy of the computer-based systems. Deployment of a KSE on the INTERNET can provide an important test of its usefulness and can be used to collect stakeholder feedback on both utility and improvements.

This chapter has represented computer-based systems as an enabling technology for ecosystem management and emphasized the importance of using existing knowledge. However, scientific understanding of a specific ecosystem management problem will rarely be so complete as to eliminate the need for seeking further knowledge through directed research. We suggest that extant knowledge is the foundation upon which informed judgments are made. The task of addressing ecosystem management problems begins with using the full measure of knowledge currently available.

References

Alpert, P. 1995. Integrated conversation and development. Ecological Applications 4:857–860.

Booch, G. 1991. Object Oriented Design. Benjamin/Commings, Menlo Park, CA.

Bormann, B.T., M.H. Brookes, E.D. Ford, A.R. Kiester, C.D. Oliver, and J.F. Weigand. 1994a. A framework for sustainable-ecosystem management. Vol. 5. USDA Forest Service PNW-GTR 331.

Bormann, B.T., P.G. Cunningham, M.H. Brookes, V.W. Manning, and M.W. Collopy. 1994b. Adaptive ecosystem management in the Pacific Northwest. USDA Forest Service PNW-GTR-341.

Bormann, F.H., and G.E. Likens. 1979. Pattern and Process in a Forested Ecosystem. Springer-Verlag, New York.

Boyce, G.G. 1995. Landscape Forestry. John Wiley and Sons, New York.

Burk, T.E., and S.D. Lime. 1996. Distributing tactical forest planning information via the INTERNET. In: INTERNET Applications And Electronic Information Resources in Forestry and Environmental Sciences. H. Saarenmaa and A. Kempf (eds.). European Forest Institute Proceedings 10:79–90.

Caldwell, L.E., C.F. Wilkinson, and M.A. Shannon. 1994. Making ecosystem policy: three decades of change. Journal of Forestry 92:7–10.

Callicott, J.B. 1994. A brief history of American conservation philosophy. In: Sustainable Ecological Systems: Implementing an Ecological Approach to Land Management. W.W. Covington and L.F. DeBano (eds.). USDA Forest Service General Technical Report, RM-247.

Christensen, N.L., A.M. Bartuska, J.H. Brown, S. Carpenter, C. D'Antonio, R. Francis, J.F. Franklin, J.A. MacMahon, R.F. Noss, D.J. Parsons, C.H. Peterson, M.G. Turner, and R.G. Woodmansee. 1996. The report of the Ecological Society of America committee on the scientific basis for ecosystem management. Ecological Applications 6:665–691.

Costanza, R., B.G. Norton, and B.D. Haskell (eds.). 1993. Ecosystem Health. Island Press, Washington, DC.

Coulson, R.N., L.J. Folse, and D.K. Loh. 1987. Artificial intelligence and natural resource management. Science 237:262–267.

Coulson, R.N., M.C. Saunders, D.K. Loh, F.L. Oliveria, D.B. Drummond, P.A. Barry, and K.M. Swain. 1989. Knowledge system environment for integrated

pest management in forest landscapes: the southern pine beetle. Bulletin of the Entomological Society of America 34:26–33.
Coulson, R.N., W.C. Daugherity, M.D. Vidlak, J.W. Fitzgerald, S.H. Teh, F.L. Oliveria, D.B. Drummond, and W.A. Nettleton. 1995. Computer-based planning, problem solving, and decision making in forest health management: An implementation of the knowledge system environment for the southern pine beetle, ISPBEX-II. In: Proceedings of the IUFRO Symposium on Current Topics in Forest Entomology. F.P. Hain, S.M. Salom, W.F. Ravlin, T.L. Payne, and K.F. Raffa (eds.). Maui, HI.
Coulson, R.N., W.C. Daugherity, E.J. Rykiel, H. Saarenmaa, and M.C. Saunders. 1996. The pragmatism of ecosystem management: planning, problem-solving, and decisionmaking with knowledge based systems. Proceedings of Econforma '96 Global Networks for Environmental Information. 10:342–350.
Covington, W.W., and L.F. DeBano (eds.). 1993. Sustainable Ecological Systems: Implementing an Ecological Approach to Land Management. USDA Forest Service General Technical Report, RM-247.
Diaz, N., and D. Apostol. 1993. Forest Landscape Analysis and Design: A Process for Developing and Implementing Land Management Objectives for Landscape Patterns. USDA Forest Service PNR, R6 ECO-TP-043-92.
Forest Ecosystem Management Assessment Team. 1993. Forest ecosystem management: an ecological, economic, and social assessment. Report of the Forest Ecosystem Management Assessment Team. U.S. Government Printing Office, Washington, DC.
Forman, R.T.T. 1995a. Some general principles of landscape and regional ecology. Landscape Ecology 10:133–142.
Forman, R.T.T. 1995b. Land Mosaics. Cambridge University Press, New York.
Franklin, J.F. 1994. Ecological science: a conceptual basis for FEMAT. Journal of Forestry 92:21–23.
General Accounting Office. 1994. Ecosystem management: additional actions needed to adequately test a promising approach. GAO/RCED-94-111.
Gill, T.G. 1995. Early expert systems: where are they now? MIS Quarterly 19:51–80.
Golley, F. 1993. A History of the Ecosystem Concept: More Than the Sum of the Parts. Yale University Press, New Haven.
Groombridge, B. (ed.). 1995. Global Biodiversity. World Conservation Monitoring Center and Chapman and Hall, London.
Grumbine, R.E. 1994. What is ecosystem management? Conservation Biology 8: 27–38.
Heywood, V.H., and R.T. Watson (eds.). 1995. Global Biodiversity Assessment. Cambridge University Press, New York.
Holling, C.S. (ed.). 1978. Adaptive Environmental Assessment and Management. John Wiley and Sons, New York.
Holling, C.S. 1992. Cross-scale morphology, geometry, and dynamics of ecosystems. Ecological Monographs 63:447–502.
Kaufmann, M.R., R.T. Graham, D.A. Boyce Jr., W.H. Moir, L. Perry, R.T. Reynolds, R.L. Bassett, P. Mehlhop, C.B. Edminister, W.M. Bock, and P.S. Corn. 1995. An ecological basis for ecosystem management. USDA Forest Service General Technical Report, RM 246.
Knopp, T.B., and E.S. Caldbeck. 1990. The role of participatory democracy in forest management. Journal of Forestry 88:13–18.

Lackey, R.T. 1995. Ecosystem health, biological diversity, and sustainable development: research that makes a difference. Renewable Resources Journal 13:8–13.
Lackey, R.T. 1996. Seven pillars of ecosystem management. Landscape and Urban Planning May 9, 1996:1–14.
Levin, S.A. 1993. Forum, science and sustainability. Ecological Applications 3: 445–489.
Loikkanen, T.T. 1995. Public participation in natural resource management. European Forest Institute Proceedings 4:19–25.
Lubchenco, J., A.M. Olson, L.B. Brubaker, S.R. Carpenter, M.M. Holland, S.P. Hubbell, S.A. Levin, J.A. MacMahon, P.A. Matson, J.M. Melillo, H.A. Mooney, C.H. Peterson, H.R. Pulliam, L.A. Real, P.J. Regal, and P.G. Risser. 1991. The sustainable biosphere initiative: An ecological research agenda. Ecology 72:371–412.
Lucas, O.W.R. 1991. The Design of Forest Landscapes. Oxford University Press, New York.
McNab, W.H., and P.E. Avers. 1994. Ecological subregions of the United States: section descriptions. USDA Forest Service, WO-WSA-5, Washington, DC.
Ministerial Conference on the Protection of Forests in Europe. Helsinki 16–17 June 1993. European list of criteria and most suitable quantitative indicators. Ministry of Agriculture and Forestry in Finland, 1994.
Monnig, E., and J. Byler. 1992. Forest health and ecological integrity in the Northern Rockies. USDA Forest Service Forest Pest Management Report 92.7.
Morrissey, W.A. 1995. Ecosystem management tools and techniques. Proceedings of the Congressional Research Service Workshop. Library of Congress, 95,430 SRP, Washington, DC.
Muggleton, S. 1990. Inductive Acquisition of Knowledge. Addison-Wesley, New York.
Noss, R.F., E.T. LaRoe III, and J.M. Scott. 1995. Endangered ecosystems of the United States: a preliminary assessment of loss and degradation. USDI National Biological Service. Biological Report 28.
Overbay, J.C. 1992. Ecosystem management. In: Taking an Ecological Approach to Management, Proceedings of the National Workshop, USDA Forest Service WO-WSA-3, April 27–30, 1992 (original not examined, definition quoted from Lackey 1995).
Risser, P.G., J. Lubchenco, and S.A. Levin. 1991. Biological research priorities: a sustainable biosphere. BioScience 41:625–627.
Robertson, F.D. 1992. Ecosystem management on the National Forests and Grasslands. USDA Forest Service, Washington, DC. 1330-31 policy letter, June 4, 1992.
Rumbaugh, J., M. Blaha, W. Pramerlani, F. Eddy, and W. Lorensen. 1991. Object-Oriented Modeling and Design. Prentice-Hall, Englewood Cliffs, NJ.
Saarenmaa, H., J. Perttunen, J. Vakeva, and A. Nikula. 1994. Object-oriented modeling of tasks and agents in integrated forest health management. AI Applications 8:43–59.
Saarenmaa, H., J. Perttunen, J. Vakeva, and J.M. Power. 1995. Multiagent problem solving and object-oriented decision support systems for natural resource management. AI Applications 9:99–111.

Salminen, H., H. Saarenmaa, J. Perttunen, R. Sievänen, J. Väkevä, and E. Nikinmaa. 1994. Modeling trees using an object-oriented scheme. Mathematics and Computer Modelling 20:49–64.
Salwasser, H. 1994. Ecosystem management: can it sustain diversity and productivity. Journal of Forestry 92:6–11.
Sample, V.A. 1991. Land Stewardship in the Next Era of Conservation. Grey Towers Press, Milford, PA.
Schneider, D.C. 1994. Quantitative Ecology. Academic Press, San Diego, CA.
Scott, A.C., J.E. Clayton, and E.L. Gibson. 1991. A Practical Guide to Knowledge Acquistition. Addison-Wesley, New York.
Sedjo, R.A. 1995. Toward an operational definition of ecosystem management (Keynote). In Analysis in support of ecosystem management. J.E. Thompson (ed.). USDA Forest Service, Ecosystem Management Analysis Center, Washington, DC.
Slocombe, D.S. 1993. Implementing ecosystem-based management. BioScience 43:612–622.
Steinke, A., D.G. Green, and D.J. Peters. 1996. On-line environmental geographic information systems. In: INTERNET Applications And Electronic Information Resources in Forestry and Environmental Sciences. H. Saarenmaa and A. Kempf (eds.). European Forest Institute Proceedings 10:91–100.
Thomas, J.W. 1994. Forest ecosystem management assessment team: objectives, processes, and options. Journal of Forestry 92:12–19.
Thompson, J.E. 1995. Analysis in support of ecosystem management. USDA Forest Service, Ecosystem Management Analysis Center, Washington, DC.
Urban, D.L. 1993. Landscape ecology and ecosystem management. In: Sustainable Ecological Systems: Implementing an Ecological Approach to Land Management. W.W. Covington and L.F. DeBano (eds.). USDA Forest Service General Technical Report, RM-247.
USDA Forest Service. 1996. Forest health for the Nation's future: A policy and strategic direction. Washington, DC.
Van Dyne, G.M. 1969. The Ecosystem Concept in Natural Resource Management. Academic Press, New York.
Väkevä, J.J. Perttunen, H., Saarenmaa, and J. Saarikko. 1996. A diagnostic information service about damaging agents of trees in the World Wide Web. In: INTERNET Applications And Electronic Information Resources in Forestry and Environmental Sciences. H. Saarenmaa and A. Kempf (eds.). European Forest Institute Proceedings 10:131–139.
Vogt, K.A., J.C. Gordon, J.P. Wargo, D.J. Vogt, H. Asbjornsen, P.A. Palmiotto, H.J. Clark, J.L. O'Hara, W.S. Keaton, T. Patel-Weynand, and E. Witten. 1997. Ecosystems: Balancing Science with Management. Springer-Verlag, New York.
Watt, K.F.E. 1968. Ecology and Resource Management: A Quantitative Approach. McGraw-Hill, New York.
Wilson, E.O. (ed.). 1992. Biodiversity, National Academy Press. Washington, DC.
Wondolleck, J. 1988. Public Lands: Conflict and Resolution. Plenum Press, New York.
Woodley, S., J. Kay, and G. Francis (eds.). 1993. Ecological Integrity and the Management of Ecosystems. St. Lucie Press.

Yoon, Y. 1995. Exploring the factors associated with expert systems success. MIS Quarterly 19:83–106.
Zonneveld, I.S. 1990. Scope and concepts of landscape ecology as an emerging science. In: Changing Landscapes: An Ecological Perspective. I.S. Zonneveld and R.T.T. Forman (eds.). Springer-Verlag, New York.

Appendix: Definitions of Ecosystem Management

As a consequence of the wide-spread interest in ecosystem management as a paradigm for land-use management, several definitions of the concept have been offered. Following is a sample. These definitions are taken out of context, and the original publications should be examined for elaboration and clarification.

1. "By ecosystem management, we mean an ecological approach will be used to achieve the multiple use management of our National Forests and Grasslands. It means that we must blend the needs of people and environmental values in such a way that the National Forests and Grasslands represent diverse, healthy, productive, and sustainable ecosystems" (Robertson 1992, in Kaufmann et al. 1995).
2. "The careful and skillful use of ecological, economic, social, and managerial principles in managing ecosystems to produce, restore, or sustain ecosystem integrity and desired conditions, uses, products, values, and services over the long term" (Overbay 1992).
3. "A strategy or plan to manage ecosystems to provide for all associated organisms, as opposed to a strategy or plan for managing individual species" (FEMAT 1993).
4. "The use of an ecological approach that blends social, physical, economic, and biological needs and values to assure productive, healthy ecosystems" (from Interim Directive, USDA Forest Service, Washington Office, 1994).
5. "Ecosystem management integrates scientific knowledge of ecological relationships within a complex sociopolitical and values framework toward the general goal of protecting native ecosystem integrity over the long term" (Grumbine 1994).
6. "A system of making, implementing, and evaluating decisions based on the ecosystem approach, which recognizes that ecosystems and society are always changing" (Bormann et al. 1994a; see also Bormann et al. 1994b).
7. "To restore and maintain the health, sustainability, and biological diversity of ecosystems while supporting sustainable economies and communities" (EPA 1994, quoted in Lackey 1995).
8. "The application of ecological and social information, options, and constraints to achieve desired social benefits within a defined geographic area and over a specified period" (Lackey 1995).
9. "Ecosystem management is management driven by explicit goals, executed by policies, protocols, and practices, and made adaptable by moni-

toring and research based on our best understanding of the ecological interactions and processes necessary to sustain ecosystem composition, structure, and function" (Christensen et al. 1996).
10. The orchestrated modification or manipulation of primary production, consumption, decomposition, and abiotic storage (this paper).

6
A Multiscale Ecological Model for Allocation of Training Activities on U.S. Army Installations

W. Michael Childress, Terry McLendon, and David L. Price

An understanding of ecological processes at multiple scales is important in making sound evaluations of the effects of management practices for natural resources. Hierarchical approaches to these processes should be able to link these processes across multiple scales into simulation models of community and landscape dynamics (Urban et al. 1987; Pickett et al. 1987). In recent years, considerable progress has been made in landscape ecological analysis of spatial patterns in order to make inferences on spatial processes (Cale et al. 1989; Turner 1990; Turner and Gardner 1991), and on hierarchically structured landscapes in particular (Lavorel et al. 1993). A complementary approach is to examine processes to determine their effects on resulting spatial patterns; this is the approach taken in simulation modeling of landscapes (Gardner et al. 1987; Gardner and O'Neill 1991). "Because landscape ecology explicitly considers spatial heterogeneity, it follows that models of population dynamics in a landscape context, even those that include only simple dynamics, are analytically intractable. A possible alternative is the use of simulation models" (Fahrig 1991). Both approaches are valuable in establishing the connections between pattern and process at all spatial scales.

However, responsible ecologists cannot wait until these tools have been "perfected" before presenting them to natural resource managers. Ecologists have been justly criticized for being reluctant to provide this best-available technology in a world where environmental problems loom so large (Erhlich and Daily 1993; Hillborn and Ludwig 1993; Underwood 1995). Instead best technology and knowledge can be used to implement a scheme of "adaptive management" (Walters 1986) in which managers and ecologist learn to manage by doing.

However, ecologists cannot adequately understand and manage large-scale systems while ignoring the dynamics of their component smaller-scale systems. "Heterogeneities far smaller than anything that could be defined as a patch in plant community ecology can control ecosystem-level nutrient flows" (Vitousek 1985). Species-level dynamics are also important in understanding dynamics at higher hierarchical levels. "Changes in the abundance of species—especially those that influence water and nutrient dynamics, trophic interactions, or disturbance regimes—affect the structure and functioning of ecosytems" (Chapin et al. 1997). The underlying mechanisms are increasingly recognized as important components of models used to adequately predict ecosystem dynamics (DeAngelis 1988). The extent to which these mechanisms, rather than abstract, empirical, or phenomenological processes, are implemented in models may largely determine our confidence in their application in conventional or novel management situations.

Models also must operate at the scale at which resources are actually managed, such as training areas in military installations, park units in National Parks, and watersheds in National Forests. This means that community-level processes and dynamics must be extrapolated up to these management units by aggregating the dynamics of different communities into spatially explicit models.

Objectives

We present a description of a modeling approach taken in developing a central component in a management scenario for the U.S. Army: allocation of training activities on Army installations. Many of these lands contain significant amounts of relatively undisturbed areas, sensitive habitats, and endangered species. However, concentration of training activities on remaining operational installations, as other installations are closed, may result in significant deterioration of these natural resources.

Our approach is to develop a general hierarchical model EDYS to simulate small- to large-scale ecological dynamics. Work to date has resulted in "first-pass" completion of modules at the quadrat (1×1 m) and community (1 ha) levels, with an initial design for and linkages to a landscape-level module that represents the basic management unit on installations, the training area, typically of several thousand hectares in area. Here the design for each module is presented, along with results from preliminary simulation runs and an outline of field studies now underway or planned for model calibration, testing, and validation. Several additional applications of this modeling approach to special management applications in several National Parks and other areas in the United States are also described.

Environmental Management

The Problem: Allocation of Training Activities

The specification of management problems and scenarios is crucial to the development of satisfactory and successful approaches to their resolution. This section presents a background for the modeling approach that is similar to those for other government agencies managing public lands throughout the United States.

In 1984, the U.S. Army convened an independent panel of experts in range, wildlife, and forest sciences to evaluate and recommend changes in Army environmental management programs (Jahn et al. 1984). As a result of their recommendations, the U.S. Army began development of an Integrated Training Area Management (ITAM) program by the U.S. Army Construction Engineering Research Laboratories (USACERL). This program has four major components: (1) Environmental Awareness, (2) Land Condition-Trend Analysis (LCTA), (3) Land Rehabilitation and Maintenance, and (4) Training Requirements Integration. The development of the ITAM program has been driven by four major factors (Tazik et al. 1992a): (1) the Army's unique land management challenge, (2) the need for sufficient training land, (3) recommendations of natural resource experts, and (4) environmental compliance requirements (e.g., National Environmental Policy Act, Endangered Species Act, and Clean Water Act).

A significant need of the Army is the development of management strategies to meet general and special environmental requirements and constraints for individual installations. In particular, allocation of training activities among different areas within an installation must appropriately balance training objectives and costs with environmental stewardship and constraints, that is, the training activities should not exceed the "carrying capacity" of the installation at which unacceptable environmental degradation begins to occur. A formal allocation strategy would also assist in compliance with environmental regulations that require documentation of the status of natural resources both before and after training use.

Planning of training activities at the largest Army installations is truly environmental management at the landscape level. These installations cover large areas and encompass diverse and often unique habitats. Because significant portions of these large tracts are relatively undisturbed, some of the potential training areas have remained in near pristine condition. Threatened and endangered species of plants and animals are found on many installations because of this lack of disturbance (e.g., golden-cheeked warbler and black-capped vireo at Fort Hood, Texas; Tazik et al. 1992b). Managing military activities and natural resources on the installations therefore requires complex strategies to meet all of these concerns.

Overall Management Approach: A Decision Support System

The overall management objective is to develop a computer-based approach for assisting training officers in allocating activities in a cost-effective, environmentally sound, and accountable manner, while still meeting training needs for military preparedness (McLendon et al. 1996). The conceptual approach is a "decision support system" (DSS, also known as "computer-based decision aids"; Coulson and Saunders 1988). This system would have a high degree of commonality among installations and ecoregions, but would be specially tailored to meet each installation's special environmental concerns. The system would include a number of modules, each of which would address one of the cost and/or environmental factors described above and which would be integrated by a prioritization system for weighting the different factors. However, this system would be designed as an interactive tool to be used in an iterative and ongoing fashion to assist the responsible officers in allocating activities, rather than as a "black-box" expert system from which unequivocal decisions are issued (Coulson et al. 1987). Decision support systems and related expert systems have been developed for natural resource management in a variety of areas, such as forest management, pest management, and agricultural systems (reviews by Stock 1987; Moninger and Dyer 1988; Davis and Clark 1989).

The system performs two primary tasks: (1) allocation of training activities among different areas in an installation in a cost-effective manner while meeting training needs for military readiness and minimizing ecological impacts, and (2) serving as a tool for natural resource managers to evaluate outcomes over time of various stressors, natural and anthropogenic, on the ecosystems within the installation. The system is one that can be applied to any installation in any ecoregion with a minimum of modifications, while maximizing site-specific output. The system is PC-based and user friendly, with many users expected to be personnel with high technical expertise in military operations, but with limited ecological/environmental knowledge. To meet these challenges, the conceptual design and initial core model for this system will be incrementally developed and reviewed and tested by U.S. Army staff, along with other user groups, over the next several years.

The current conceptual design for this system, termed here the "Management Model," is comprised of a number of linked modules (Fig. 6.1). These are organized by the types of information that are held in databases, entered by the system user, computed by the system, or saved in databases for later use and reference. The Management Module would specifically address environmental impacts and costs. Within this Module, a Community Model would be used to simulate ecosystem and landscape dynamics. Ordinarily, the total system would be used in the allocation of a single activity, but would hold information about other past and planned activities across the installation so that planning for all future activities could be coordinated.

Most of the work currently underway focuses on development of the Community Model (EDYS), described in detail later in this chapter. An important requirement of this model is that it be generally applicable to terrestrial ecosystems so that it can be applied not only on military lands, but in other management scenarios on public as well as private lands.

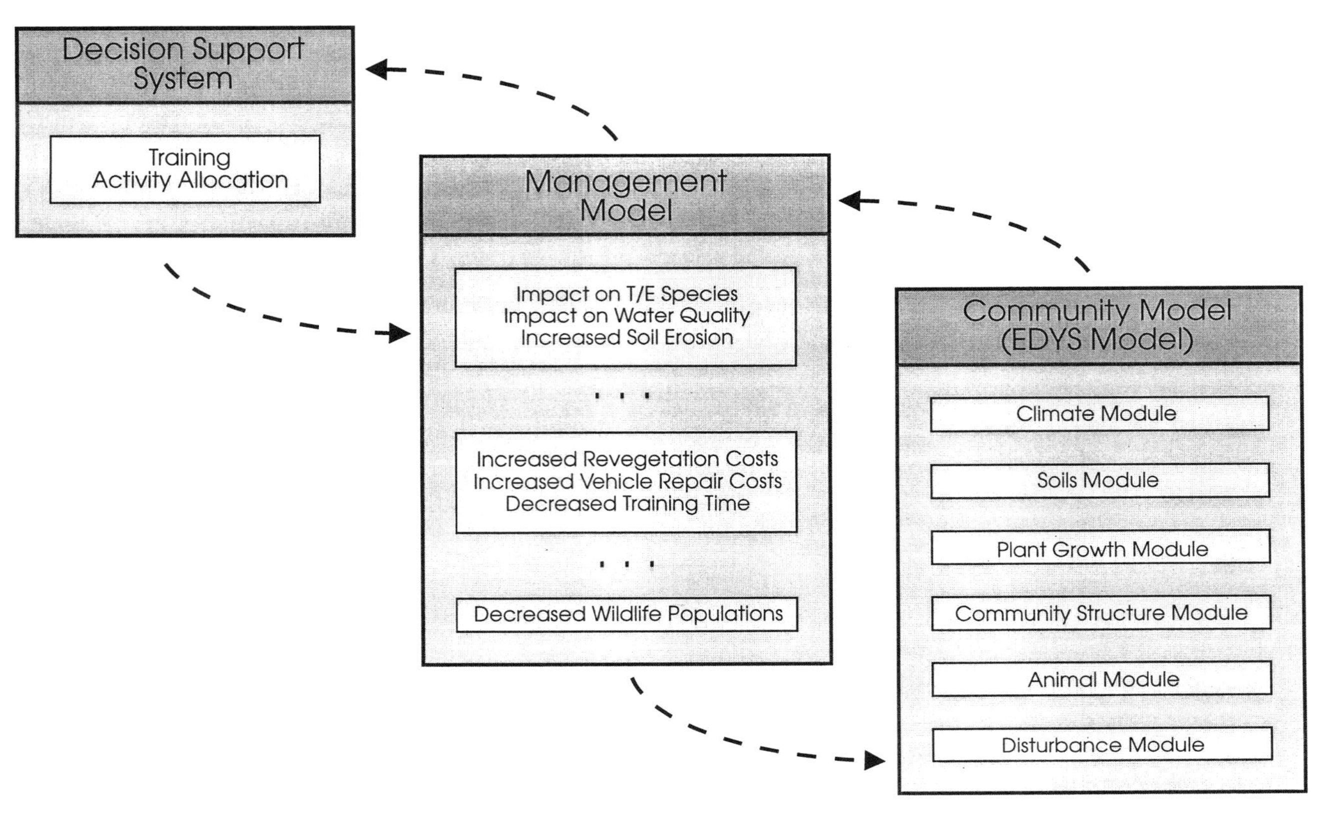

FIGURE 6.1. Conceptual organization of Decision Support System for management of training activities on U.S. Army installations.

Rather than attempt to overdesign this model so that it is applicable to only a single ecosystem or management scenario, the objective is to develop a model that is sufficiently general in design, with all relevant ecosystem components, so that it can be easily adapted to any environmental scenario.

Modeling Approach: Ecological Dynamics System (EDYS)

Overall Model Design

The overall model design for EDYS implements a hierarchical system of modules that incorporates different ecological processes at different spatial scales. Hierarchical approaches to ecological processes are widely advocated in ecology (Urban et al. 1987; Pickett et al. 1987, 1989; Wiens et al. 1993). Processes and mechanisms should be explicitly incorporated into models not only to increase their realism and predictive accuracy, but to allow consideration of all kinds of internal and external stressors on the system (DeAngelis 1988; Hilborn and Ludwig 1993). A spatially explicit model is necessary because individual plant growth is often crucial to understanding of ecosystem dynamics (Clark 1992), there is considerable heterogeneity in plant species and community types at all spatial scales (Milne 1991; O'Neill et al. 1991), there is considerable heterogeneity in the physical environment within communities, especially nutrients, soils, slope, and aspect (Robertson et al. 1988), and there should be considerable heterogeneity in the effects of training activities and impacts within a training area.

A large spatial scale is needed so that ecological effects of large-scale training activities can be evaluated and integrated across community types in training areas that include more than one community type (Davis et al. 1988). We are interested in entire training areas because these are the primary management units on Army installations. These are on the order of thousands of hectares, each encompassing a number of community types. Each installation has GIS map representations of vegetation, soils, slope, and aspect data, and typically extensive satellite and aerial photographic imagery. Following the tenets of hierarchy theory, linking these different models is not overly difficult if the separate processes at each scale are kept separate (Allen and Starr 1982; O'Neill et al. 1986; King 1991).

Initial multivariate statistical analyses of vegetation surveys conducted in the LCTA program at five installations in Texas, Kansas, Colorado, and Washington indicate that although there are anywhere from 15 to 30 statistically unique plant communities present at each installation, four general community associations cover 80% to 95% of the land surface at each installation (Table 6.1). This indicates that community-level models would be an appropriate scale for simulating and projecting functional responses

TABLE 6.1. Major plant associations in five U.S. Army Installations.

Association	Type	Transects (%)
Fort Bliss, TX		
Mesquite-snakeweed	Shrubland	45.6
Creosotebush-snakeweed	Shrubland	23.3
Black grama-blue grama	Grassland	11.9
Blue grama-black grama	Grassland	6.7
Others		12.5
Fort Carson, CO		
Blue grama	Grassland	49.0
Blue grama-snakeweed	Grassland	4.5
Kochia	Shrubland	20.1
Juniper-pinyon pine	Woodland	21.6
Others		4.0
Fort Hood, TX		
Juniper	Woodland	17.9
Oak-juniper	Woodland	20.1
Little bluestem-dropseed	Grassland	16.7
Wintergrass-buffalograss	Grassland	34.7
Others		10.6
Fort Riley, KS		
Big bluestem-indiangrass	Grassland	39.3
Indiangrass-dropseed	Grassland	27.9
Indiangrass-little bluestem	Grassland	9.3
Oak-elm	Woodland	5.0
Others		18.5
Yakima TC, WA		
Sandberg bluegrass	Grassland	35.6
Bluebunch wheatgrass-Sandberg bluegrass	Grassland	34.7
Big sagebrush-bluebunch wheatgrass	Shrubland	5.9
Big sagebrush-Sandberg bluegrass	Shrubland	13.8
Others		9.0

Each association is an aggregation of similar community classes identified by multivariate analyses of LCTA transect data from each installation. The multivariate methods are described in McLendon and Dahl (1982). The number of transects at each installation ranges from 140 to 202.

of these ecosystems to disturbance. Further, only these four associations need to be incorporated into large-scale models to represent 80% to 95% of the landscapes at each installation.

Our approach is to build separate simulation modules at three spatial scales: quadrats (1×1 m), communities (1 ha), and landscapes (training areas: 1000s of hectares) (Fig. 6.2). Separate units in each lower-scale module will be imbedded in a spatially explicit grid at the next higher level. In addition, we build in modularity for all processes within each unit. This facilitates updates to model algorithms and special adaptations (either simplifications or complications) for special ecological conditions at certain sites, as well as allows teams of modelers to design, implement,

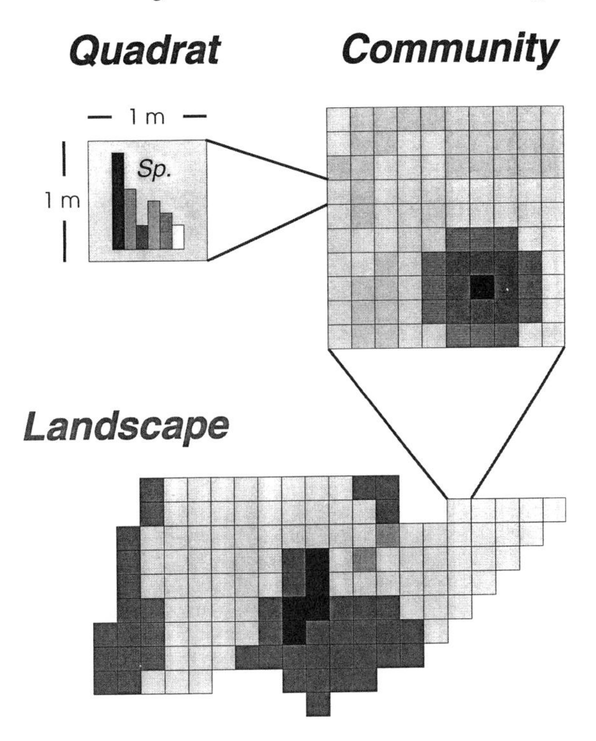

FIGURE 6.2. Scaling of three modules in EDYS: Quadrat, Community, and Landscape Modules. The Quadrat Module is the basic grid unit in the Community Module, and the Community Module is the basic grid unit in the Landscape Module.

and test computer code for each process separately (Reynolds and Acock 1997).

Quadrat Module

The Quadrat Module in EDYS holds the major mechanistic elements of the overall approach: plant growth, soil substrate, and hydrology. The approach implements small-scale dynamics of plant species so that their interactions and contributions to total system functions and dynamics can be explicitly incorporated into the overall model. This will ensure that the complete EDYS model can adequately represent changes in ecosystem function and dynamics as species composition changes with succession, disturbance, and invasion (Chapin et al. 1997). The modeling format is a compartment model like that used in many forest gap models (e.g., Shugart and Seagle 1985).

A schematic diagram of the components in the Quadrat Module is presented in Figure 6.3. Above- and belowground plant components, soil horizon structure, soil and nutrient availability, and interactions among plant

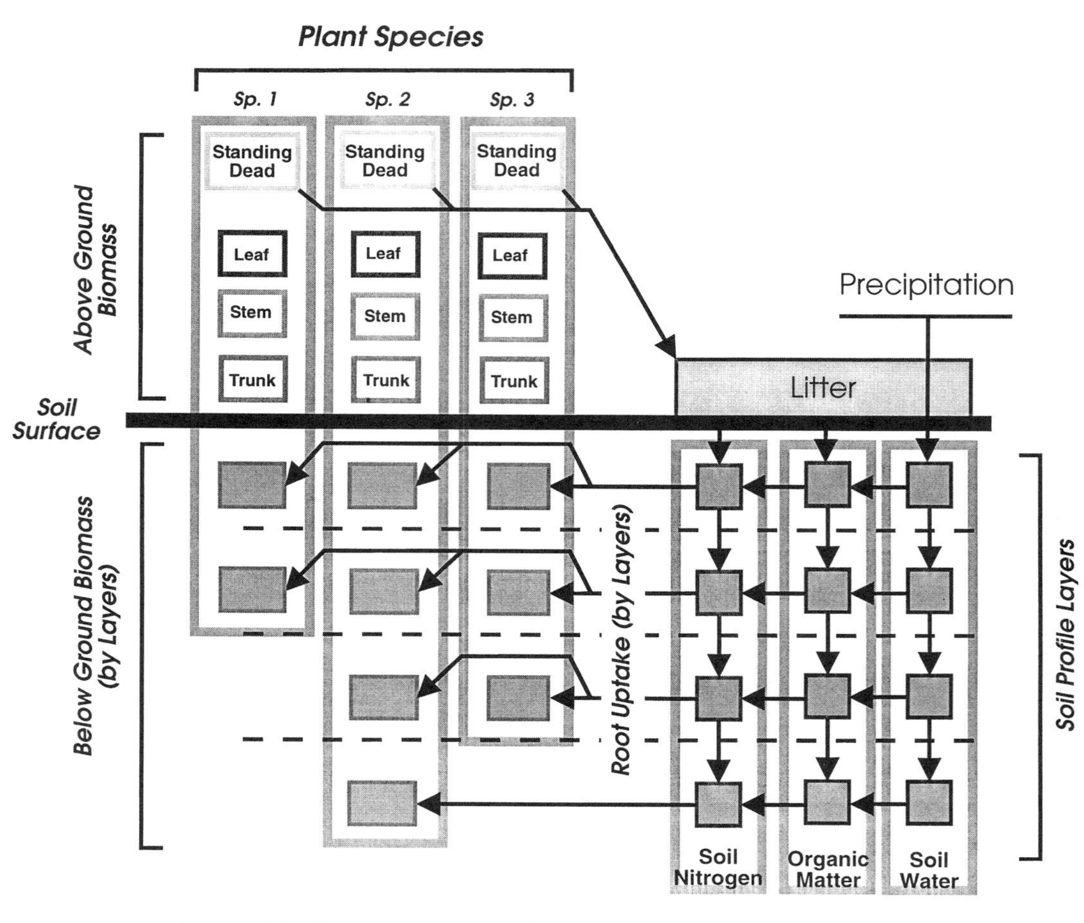

FIGURE 6.3. Components and major linkages of the Quadrat Module.

species are explicitly represented because they are crucial to understanding and simulating plant community dynamics (Van Auken and Bush 1997). Water dynamics are based on precipitation, soil moisture, evaporation, and plant uptake for maintenance and growth. Nitrogen and water availabilities are explicitly represented as pools in each layer of the soil profile. Precipitation percolates through the profile, transporting nitrogen and organic matter and recharging soil moisture. Plant species biomasses are represented by several aboveground components (trunk, stem, leaves, seeds, and standing dead stems and leaves), and by root biomasses in each soil layer. Each species can take up water and nitrogen from those layers in which it has roots, depending on its seasonal demand for these resources, the availability of these resources in each layer, the amount of its roots in that layer, and competition from other species with roots in that layer. Primary production is simulated as growth under optimal conditions, constrained by season and nitrogen and water uptake.

Nitrogen and water are usually limiting resources in plant communities (DeAngelis 1992), and their dynamics are crucial to rates and patterns of successional recovery (Vitousek 1985; McLendon and Redente 1991, 1992). The Quadrat Module specifically implements a complete nitrogen cycle to address limitations of plant production at the species level as a result of nitrogen and water constraints (Fig. 6.4). Rates for the various transfer functions in this Module are taken from the ecological literature. Derivation of site-specific rate data is a major focus of field studies now underway in experimental plots.

Herbivory plays an important role in determining community structure and function in many grassland, savanna, and woodland systems like those at Forts Bliss and Hood in Texas, not just by large ungulate grazers, but for small animals as well (e.g., Weltzin et al. 1997). Herbivory rates in the Quadrat Module are based on literature data for consumption rates and preferences for different plant species by different herbivores. A preference matrix for different plant parts by species and by different herbivores (e.g., grasshoppers, rabbits, and cattle) is used to calculate specific vegetation losses to herbivory each day in the simulation runs. Complete animal components are currently under development, but will separately simulate population dynamics of major herbivores.

Fire is an important factor to ecosystem function because of its general and selective effects on plant growth and standing crop and on water and nutrient dynamics (Christensen 1985). Effects of fire on the plant community are implemented in the Quadrat Module as losses of biomass from particular plant components and from the litter, depending on the intensity of the fire and climatic conditions. Fire initiation and propagation processes are implemented in the Community Module.

The initial version of the Quadrat Module has been calibrated in detail for a little bluestem community at Fort Hood, Texas, using LCTA survey data to derive initial biomasses for the six major plant species, and literature

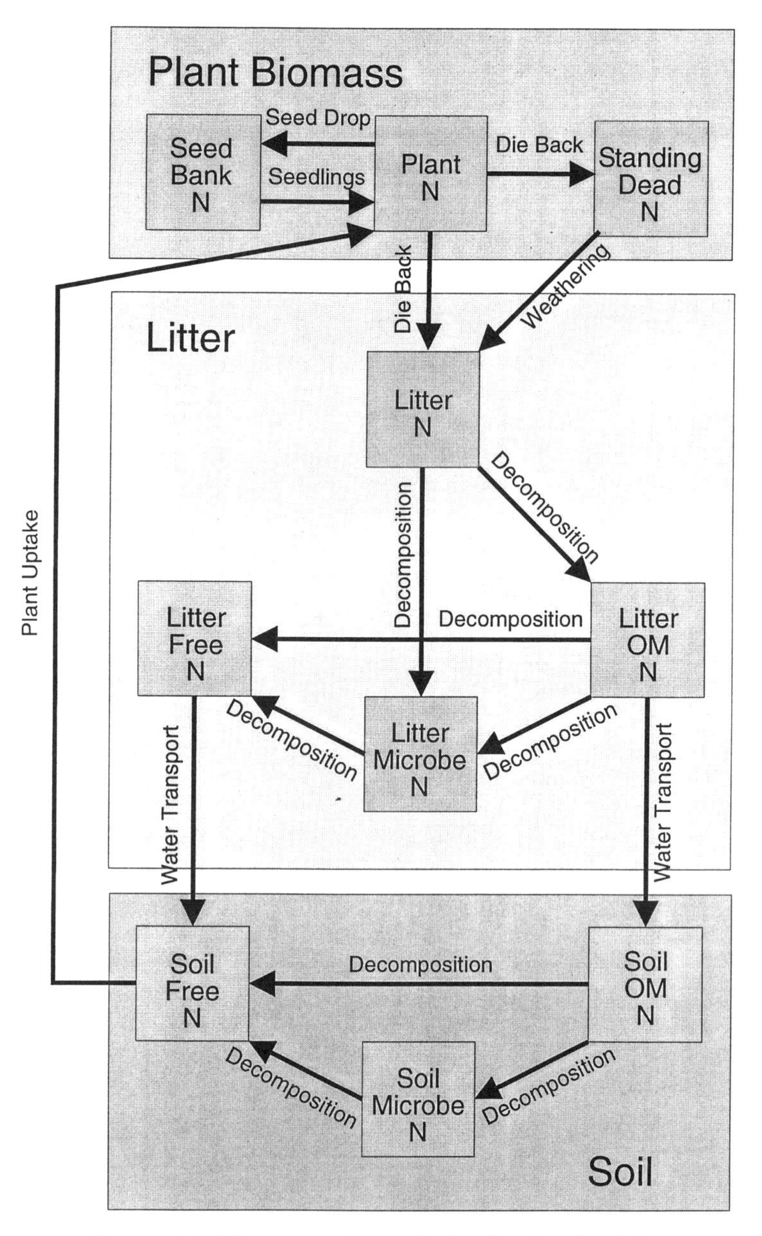

FIGURE 6.4. Nitrogen pathways in the Quadrat Module.

values for all other data requirements of the Module, in particular, plant growth rates and information about the soil profile. Ashe juniper typically is suppressed in this system by periodic cool fires, but with fire suppression, invasion by juniper and several oak species converts this grassland into a nearly closed canopy woodland with limited understory.

Twenty-year simulation runs of the Quadrat Module for this community indicate that the community composition varies greatly under different climatic and disturbance scenarios. Runs with fire and herbivory excluded and the initial juniper biomass set to 0 result in a community dominated by little bluestem with the other major species largely eliminated (Fig. 6.5). When the average juniper biomass per square meter across this community is added to the simulations, the result is a quadrat completely dominated by juniper as the seedlings grow into mature trees over the 20-year run (Fig. 6.6). Simulation runs with juniper excluded have also been made for this community in which climatic, fire, and herbivory are varied (Fig. 6.7). Although an increase in precipitation results merely in an increase in the dominant bluestem biomass after 20 years, a 75% reduction in precipitation results in a shift from a community dominated by bluestem to one comprised of four grass species. Fire added in regular three-year intervals

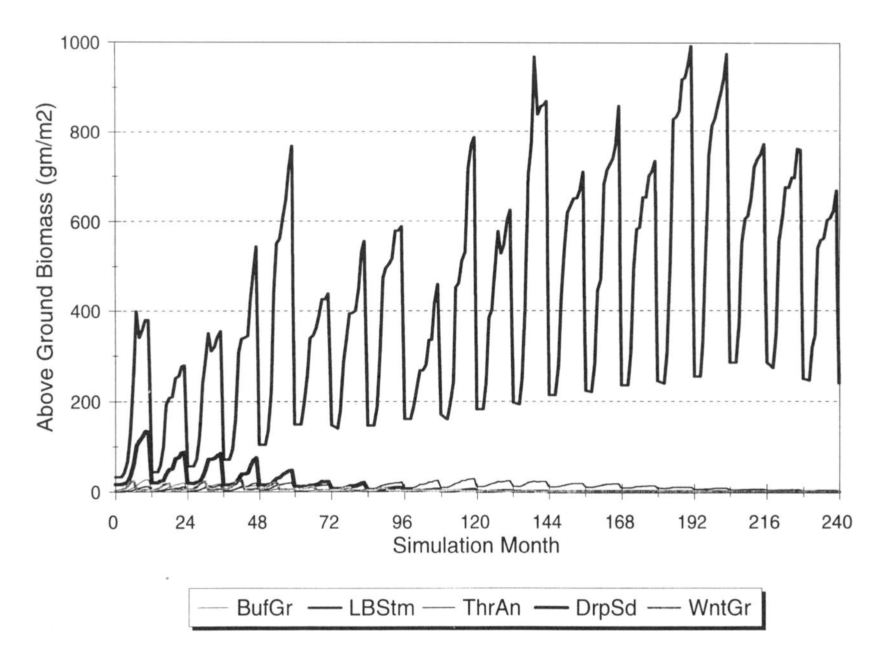

FIGURE 6.5. Dynamics of above ground biomass for major plant species in a Quadrat Module simulation of a little bluestem community at Fort Hood, Texas. Ashe juniper was excluded from this simulation.

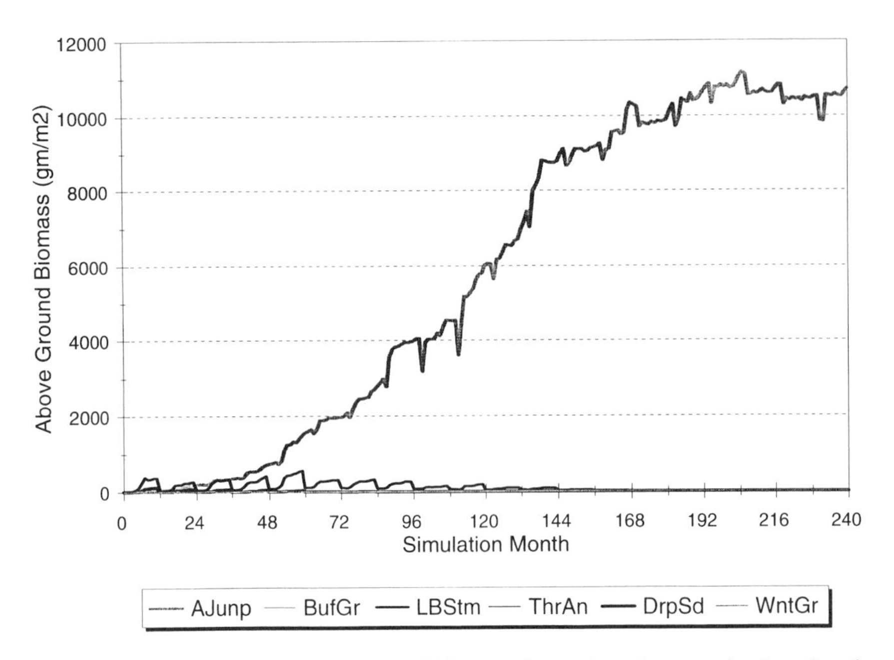

FIGURE 6.6. Dynamics of above ground biomass for major plant species in a Quadrat Module simulation of a little bluestem community at Fort Hood, Texas. All species, including Ashe juniper, were included in this simulation.

results in a community dominated by only two grass species. An intense grazing simulation results in a community dominated again by a single grass species, little bluestem.

The initial community composition is also important in determining composition in 20-year simulations. Variations in the initial biomass of little bluestem in runs with juniper excluded result in dramatic shifts in species composition (Fig. 6.8). This indicates the importance of implementing a variety of community vegetation subtypes in community-level simulations specifically to represent the initial and ongoing heterogeneity in the community.

Extensive field work is now underway for model calibration and testing at Forts Hood and Bliss in Texas. These include field plots in unburned and recently burned areas to:

1. Characterize community structure at different spatial scales using multivariate statistical analyses,
2. Characterize heterogeneity among square meters within communities,
3. Monitor changes in community structure over time, and
4. Conduct experiments using nitrogen addition/depletion and water addition to examine species and quadrat-level responses.

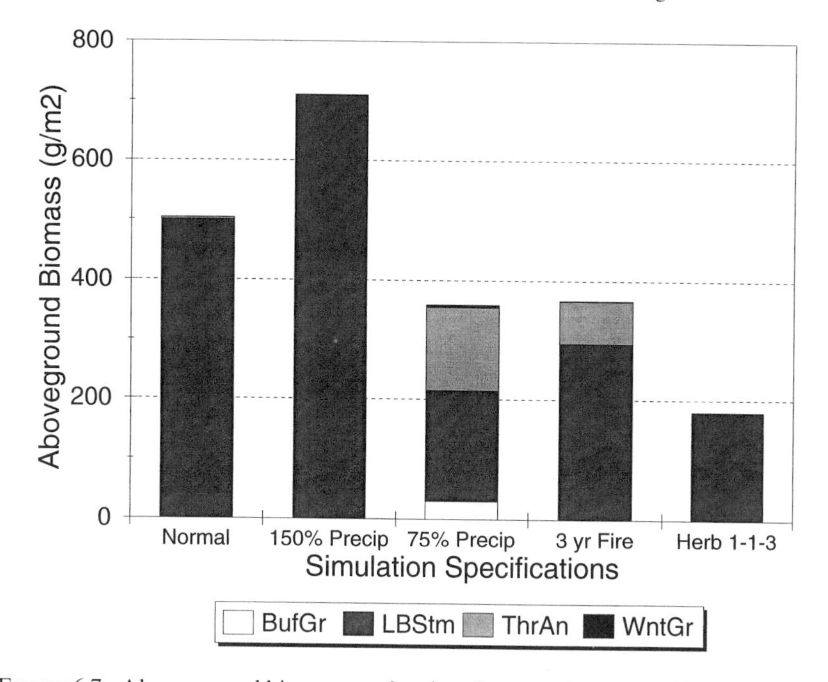

FIGURE 6.7. Aboveground biomasses of major plant species at year 20 of alternative Quadrat Module simulations. Ashe juniper was excluded from these simulations. "Normal" refers to a simulation with normal precipitation and fire and herbivory excluded; "150%" and "75%" refer to simulations with the indicated adjustment to daily precipitation; "3 yr Fire" refers to a simulation in which there is a fire at the end of August every third year; "Herb 1-1-3" refers to a simulation in which there was light herbivory from grasshoppers and cottontail rabbits, but heavy grazing by cattle.

Community Module

The Community Module in the EDYS model is a 1 ha, spatially explicit representation of 10,000 1×1 m quadrats, plus a number of ecological processes that operate at ecological scales larger than that of the single quadrat. These processes include a fire regime (initiation and spatial propagation among quadrats), movement of surface and subsurface water among quadrats, erosion, and dispersal of seeds. The initial spatial pattern of quadrats is either taken directly from spatial patterns at particular locations in the field, or devised to be typical or representative of LCTA transect data.

The Community Module does not implement separate Quadrat Modules for each quadrat in the grid, but instead has Quadrat Modules for each of the different vegetation subtypes that are found within the community (Fig. 6.9). Each Quadrat Module has different soil profiles, soil water content,

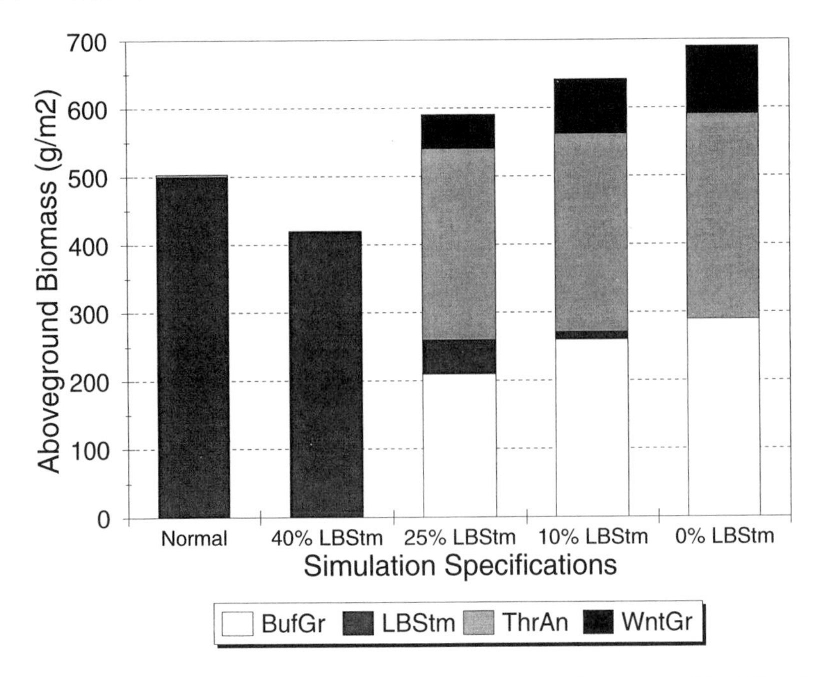

FIGURE 6.8. Aboveground biomasses of major plant species at year 20 of Quadrat Module simulations with decreasing initial biomass for little bluestem.

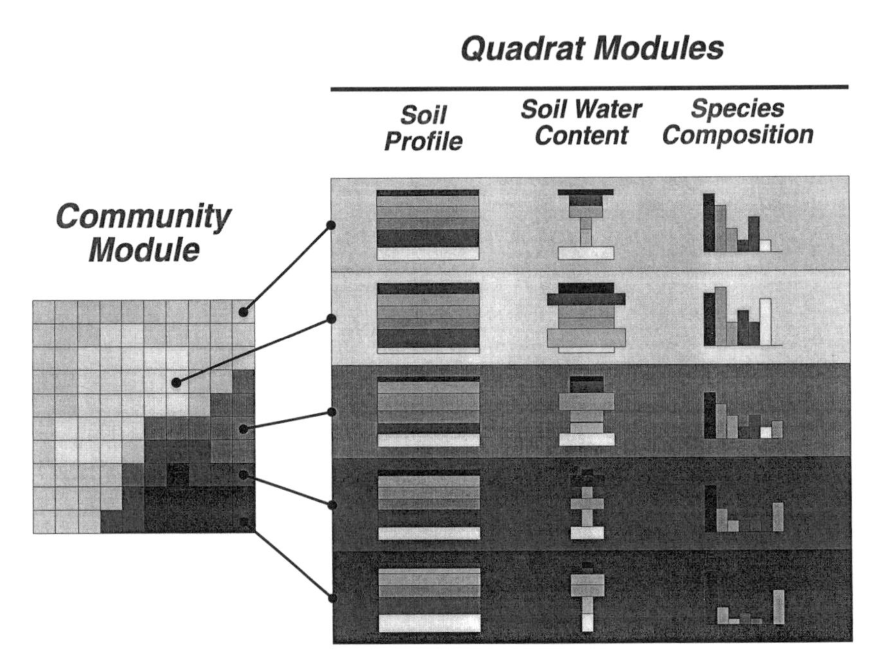

FIGURE 6.9. Different soil profiles, soil water content, and species composition in different quadrat types for the Community Module.

and plant composition. The results of independent but simultaneous runs of the Quadrat Module are then applied to the cells in which each of the corresponding vegetation types occurs within the community. A complication in this scheme is the representation of large woody plants which extend over several adjacent quadrats (Fig. 6.10). The quadrat types influenced by these shrubs and trees are modified to account for shading and root presence by these plants.

Water movement among quadrats is a major emphasis of the Community Module. Linkages among Quadrat Modules in the Community Module grid allow simulation of surface and subsurface water movement down slopes (Fig. 6.11). The Quadrat Modules implement only up and down movement of water throught the soil profile, but the Community Module simulates movement of water and dissolved and suspended materials among quadrats. This feature of the Community Module has the particular aim of addressing effects of plant cover and physical disturbance on surface flow and erosion.

The Community Module displays the vegetation type of each quadrat in the 1 ha grid as well as the canopy extent of large woody plants. The initial version implements a stochastic change matrix for quadrat-level dynamics

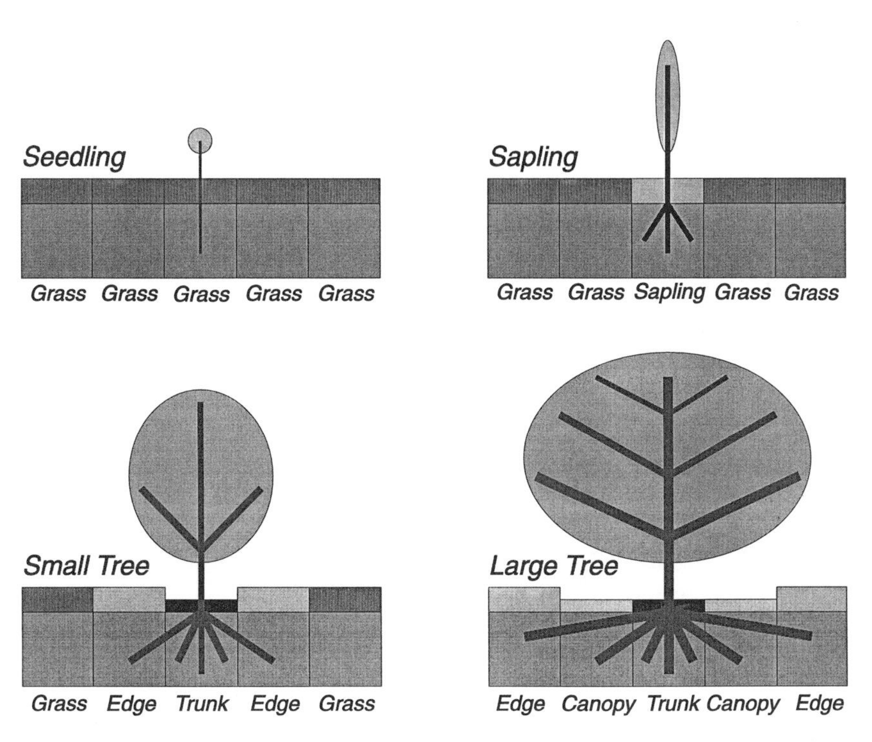

FIGURE 6.10. Changes in quadrat types around woody plant quadrats in the Community Module. As the plant grows over time, its canopy and roots extend into adjacent quadrats, causing the quadrat types to change.

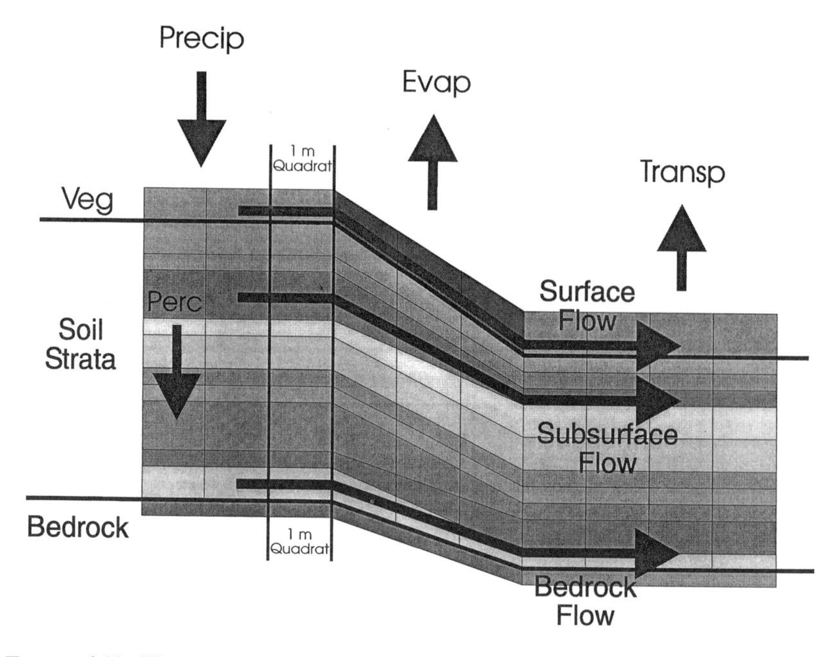

FIGURE 6.11. Water transport among quadrats in the Community Module. Water moves between quadrats as surface flow without entering the soil profile, or as lateral movement between the same strata in adjacent quadrats. "Precip" indicates precipitation, "Perc" indicates percolation through soil strata, "Evap" indicates evaporation, and "Transp" indicates plant transpiration from the roots to aboveground components.

rather than complete Quadrat Modules. We present here simulation results for a little bluestem grassland community at Fort Hood in central Texas. This community is ordinarily dominated by little bluestem with several other subdominant grasses. However, suppression of fire results in rapid invasion by Ashe juniper and, to a lesser extend, by various species of oaks. The initial community in the simulation runs in a uniform grass-dominated state in all quadrats. The dynamics of the simulation involve stochastic invasion by either juniper or oak seedlings, and subsequent growth and extension of the canopy into adjacent cells in a series of steps as indicated in Figure 6.10. The annual probabilities for these changes, including that of death of the woody plant, are:

	Seedling probability	Growth probability	Death probability
Oak	0.005	0.10	0.01
Juniper	0.005	0.10	0.01

When fire is suppressed in 50-year simulation runs of the Community Module for this system, the grassland is converted to a mixed oak-juniper woodland in which more than 70% of the quadrats are covered by juniper or oak trees or shrubs (Fig. 6.12). The most current version of the Community Module with preliminary Quadrat Modules for each quadrat-level vegetation type incorporated yields similar results.

Spatially explicit fire models are common in the ecological literature (e.g., Christensen 1985; Turner and Dale 1991; Karafyllidis and Thanailakis 1997). A similar approach is used in the Community Module by implementing a stochastic model for fire initiation (0.20 annual probability of a burn initiation) and propagation among quadrats, using as modifiers annual climatic condition and fuel loads (i.e., vegetation type) in each quadrat. With fire added to the 50-year simulations, the woody vegetation is greatly reduced, but fire propagation across the community does not result in complete burns of all quadrats (Fig. 6.13).

Disturbance is implemented in this Module as linear patterns (roads, vehicle tracks, and staging areas) of vegetation state change in quadrats. The default frequency of disturbance in simulation runs is every three years.

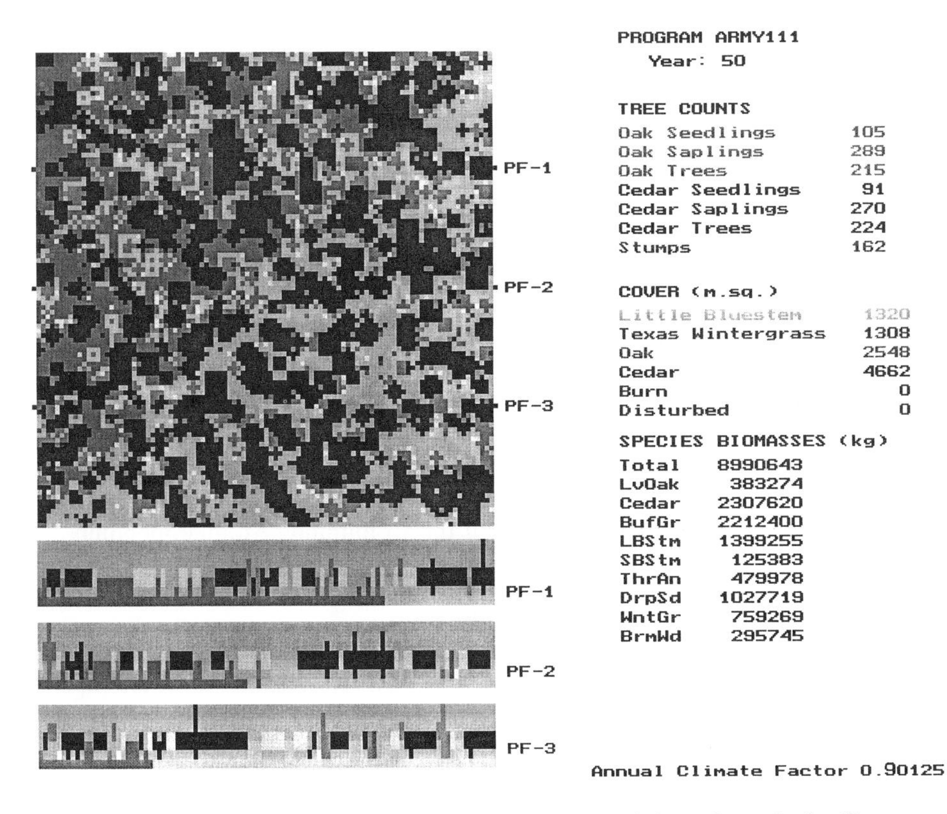

FIGURE 6.12. Graphical display of the Community Module at the end of a 50-year simulation. This image shows spatial pattern and statistics for aboveground vegetation for a simulation which began with no woody plants and excluded fire and training disturbance.

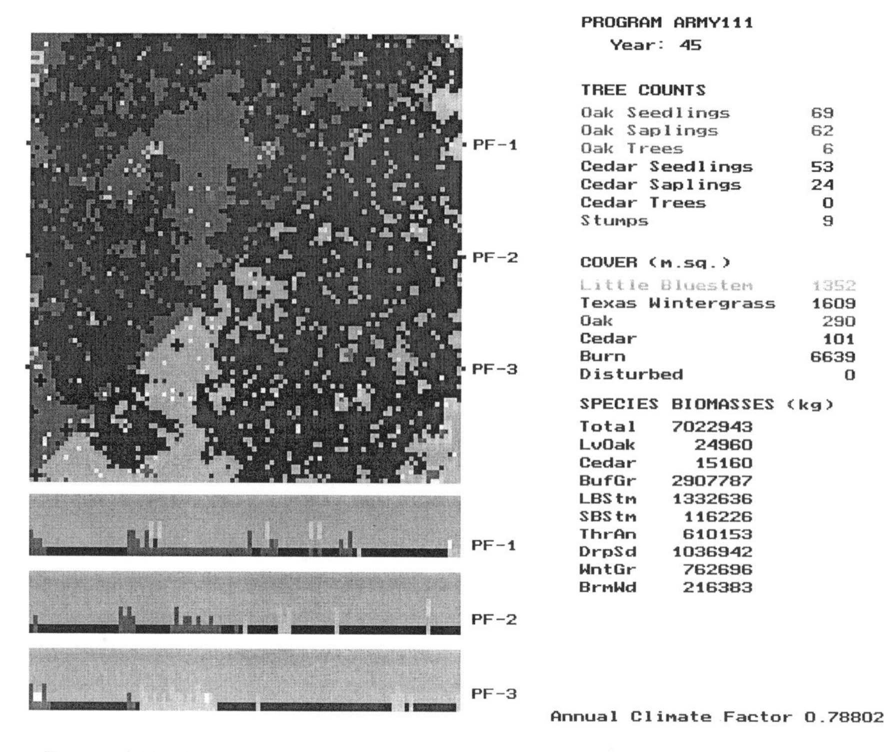

FIGURE 6.13. Graphical display of the Community Module at year 45 during a burn event. Unburned areas just to the left of the vertical center of the image are areas in which training has occurred previously, removing much of the woody vegetation.

Community dynamics in terms of vegetation types in the Community Module change considerably depending on whether fire and disturbance are implemented in the 50-year simulation runs (Fig. 6.14). With both fire and physical disturbance the percent cover of woody plants decreases from more than 70% to approximately 20%.

Herbivory is not yet implemented in the Community Module, pending final incorporation of Quadrat Modules explicitly in the model. Herbivory will simulate patchiness in grazing and quadrat and community scales because of inherent heterogeneity in these communities. Herbivores both create and respond to spatial heterogeneity in the plant community (Wiens 1985; Hyman et al. 1991; Moen et al. 1997). Grazing by cattle, for example, is not uniform in grasslands, but is typically more intense on previously burned areas, along cattle paths, and near watering and shelter areas. The intensity of herbivory in each quadrat will be determined by the Community Module both stochastically and as a function of the vegetation composition in the quadrat and spatial patterns of vegetation in nearby quadrats.

Field work underway will also support development of the Community Model. Replicated 10 × 10 m study plots have been established for long-term monitoring in mixed vegetation communities in recently burned and unburned areas at Forts Hood and Bliss, Texas. Field surveys in these plots will determine the amount and persistence of spatial heterogeneity within each community to guide in determining number of quadrat types that are required for adequate simulations.

The U.S. Army Contruction Engineering Laboratories and other Army units are actively engaged in characterizing actual ecological disturbance caused by training activities (e.g., Tazik et al. 1992b). This involves two separate efforts. First, characterizing the actual activities typical of training exercises (e.g., how many miles driven or walked for different unit types, whether these are on roads or dispersed across the terrain, and in what seasons and climatic conditions does training occur). Second, characterization of the ecological effects of each type of activity (e.g., effects of tank treads on different types of vegetation). The results of these studies will be used to more realistically simulate spatial patterns and intensity of disturbance typical of training activities in each community type (e.g., grasslands, shrublands, and forest) at each installation.

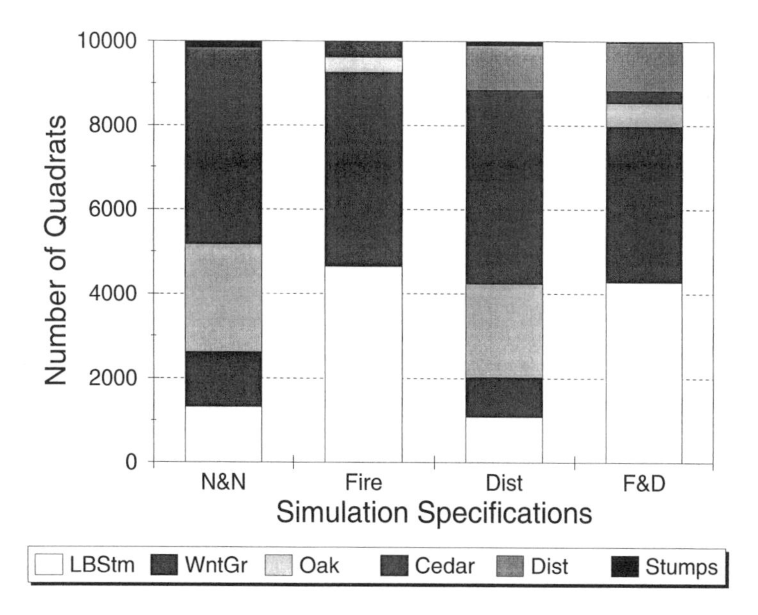

FIGURE 6.14. Number of quadrats of each type at year 50 for alternative simulation runs of the Community Module. "N&N" indicates a simulation without fire or training disturbance; "Fire" indicates a simulation with fire but no training; "Dist" indicates training disturbance but no fire; "F&D" indicates both fire and training disturbance implemented.

Landscape Module

The Landscape Module of EDYS imbeds the Community Module into a grid in a manner similar to that in which the Quadrat Modules are imbedded in the Community Module. This scale is important for implementing processes such as fire initiation and propagation, grazing, and brush control efforts. Development of the Landscape Module has just begun, using a sample training area from Fort Hood broken up into 1 ha cells, the same scale as the extent of the Community module. Fort Hood is located in central Texas on the Edwards Plateau vegetation zone, which is dominated by mixed grasslands and oak/juniper woodlands. The spatial representation of a selected training area (Area 52) at Fort Hood, Texas, as implemented in the Landscape Module is displayed in Figure 6.15. This map was adapted from a vegetation cover map of the installation prepared by the U.S. Army at about 30 m resolution. In the Army vegetation map and in the Landscape Module, the current cell state is indicated as type and percent cover of woody species in the hectare. The different vegetation types are indicated

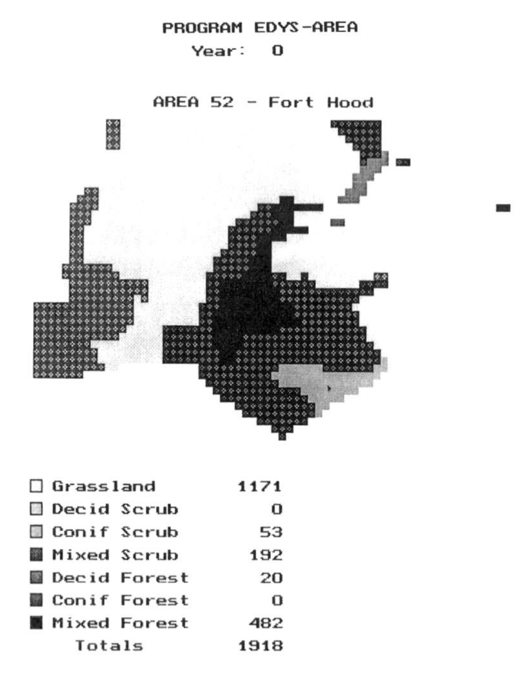

FIGURE 6.15. Graphical display of the Landscape Module adapted to Training Area 52 at Fort Hood. This image indicates the initial community types across the training area.

by different shadings and are defined primarily by the type of dominant woody cover: shrubs, trees, or no woody plants, and deciduous, coniferous, or mixed species.

Successional dynamics are based on hypothetical annual transition probabilites for percent cover of shrubs and trees. In essence, there are fixed probabilities of invasion of each cell by deciduous or conifereous shrubs and of growth of these shrubs into mature trees. These probabilities are modified by the climatic conditions in the year, with better years resulting in higher invasion and growth probabilities. If a successful invasion occurs, then the state of the cell changes to reflect the presence of the new species composition. Fire is implemented as stochastic propagation of burns from cell to cell in a manner very similar to that of fire dynamics in the Community Module. After field testing of the Quadrat and Community Modules at Fort Hood, these will be imbedded into the Landscape Module for further development and testing of the full EDYS Model.

After a 10-year simulation run of the current stochastic Landscape Module with fire and disturbance excluded, there are significant shifts towards woody vegetation types (Fig. 6.16). Both deciduous and coniferous woody

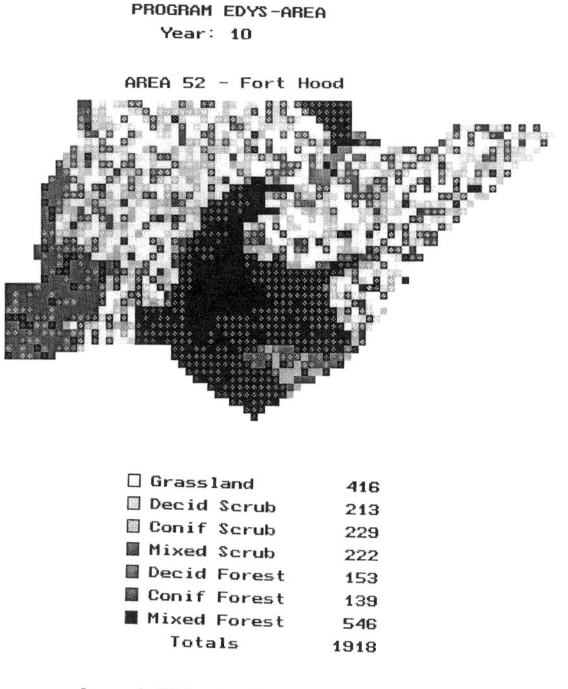

FIGURE 6.16. Graphical display for Training Area 52 at simulation year 10 in the Landscape Module. This simulation implements successional dynamics but not fire.

species have encroached on the grassland areas, and many of the shrubland areas have been converted to tree-type vegetation. The shift in major cover types over the 10-year simulation is as follows:

	Initial	10 year
Grassland	61%	22%
Shrubland	13%	35%
Woodland	26%	44%

The Texas Hill Country was originally dominated by grasslands with isolated pockets of woodland in protected canyons and drainageways. With active suppression of fire for more than 100 years, the dominant vegetation in this region is now oak/juniper woodland. The simulation results reasonably reflect the rapid invasion and growth of these woody species.

Additional Applications

The current interest in developing the Quadrat, Community, and Landscape Modules on Army installations focuses on training activities, endangered species, fire processes and control, and grazing. The approach is to derive a model that is mechanism oriented, not stochastically or empirically based. The objective is to develop a management tool that is sufficiently robust to be adapted to most terrestrial ecosystems, at any spatial scale, and under a variety of management scenarios. It therefore has potential for applications far beyond that of the military. A list of sites and communities for which the EDYS model will be calibrated and tested is presented in Table 6.2. These locations include military installations, National Parks, and research sites across the United States; the ecosystems of interest include deserts, grasslands, shrublands, woodlands, and forests. Specific potential applications of the EDYS modeling approach for public lands management at these sites include:

- Evaluating revegetation options in disturbed areas.
- Predicting natural successional dynamics.
- Evaluating effects of the population dynamics of natural herbivores and their control.
- Evaluating impacts of visitor use along trails and other concentration areas.
- Predicting dynamics associated with re-introduction of native animals.
- Predicting impacts of alien plant species on native vegetation.

The initial Community Module has been adapted to simulate effects of elk grazing on grasses and aspen in mountain forests typical of those in National Parks in the Rocky Mountains. Preliminary results demonstrate tight linkages among grazing, available fuels, and aspen dynamics. It is typically held that increased elk populations result in overgrazing and suppression of aspen populations. However, Community Module simulations indicate that increased elk grazing on both grasses and aspen results in lower fuel loads for fire, altering the fire regime in favor of aspen growth and to the detriment of competing conifers, thereby causing a net increase in aspen production.

The Community Module has been adapted for another National Park scenario: invasion of Canada thistle along horse trails in Rocky Mountain National Park. Preliminary results show that this invasive plant is eventually excluded by native vegetation, but this process takes decades, even with revegetation efforts. The management application of the Community Module is in assessing whether restoration is cost effective in accelerating recovery of vegetation and exclusion of the thistle along abandoned horse trails.

TABLE 6.2. Current research sites and communities for adaptation, testing, and validation of the EDYS Model.

Research site	State	Community	Scheduled
Air Force Academy	CO	Blue grama grassland	Spring 1988
"		Early-seral grassland	Spring 1988
"		Gambel oak shrubland	Spring 1988
"		Little bluestem grassland	Spring 1988
"		Pine-oak woodland	Spring 1988
"		Ponderosa pine woodland	Spring 1988
"		Smooth brome grassland	Spring 1988
"		Willow riparian shrubland	Spring 1988
Central Plains Experimental Range	CO	Blue grama grassland	Established
Fort Carson	CO	Blue grama grassland	None
Piceance Basin	CO	Big sagebrush shrubland	Established
Rocky Mountain National Park	CO	Spruce-fir forest	Fall 1998
Fort Riley	CO	Big bluestem prairie	None
Acadia National Park	ME	Freshwater marsh	None
Glacier National Park	MT	Fescue-aspen ecotone	Summer 1998
		Subalpine spruce forest	Summer 1998
Lake Mead National Rec. Area	NV	Creosotebush shrubland	Established
		Gypsum desert	Spring 1998
Big Bend National Park	TX	Creosotebush shrubland	Established
		Oak-juniper woodland	Established
Fort Bliss	TX	Black grama grassland	Established
		Creosotebush shrubland	Established
Fort Hood	TX	Little bluestem prairie	Established
		Wintergrass-buffalograss	Spring 1998
		Oak-juniper woodland	Spring 1998
Yakima Training Center	WA	Big sagebrush shrubland	None
Grand Tetons National Park	WY	Big sagebrush shrubland	Summer 1998
Yellowstone National Park	WY	Big sagebrush shrubland	Summer 1998
		Lodgepole pine forest	Summer 1998
		Riparian grassland	Summer 1998

A number of environmental concerns can be addressed using this approach as well, including:

- Evaluation of ecological effects of contaminants, from effects on plant growth to transport through the soil profile.
- Design of vegetation covers for landfills.
- Incorporation as the ecological component for advanced decision-making systems.

Discussion

This approach is to develop a model specifically for management use. There will never be a perfect final answer to any perceived problems because of the uncertainties in all ecological systems (Hilborn and Ludwig 1993). We have not developed any new paradigms either in ecological modeling or natural resource management, but rather had the opportunity to put a number of existing approaches together into a single package to meet the complex needs of environmental stewardship of public lands on U.S. Army installations. First, an explicitly hierarchical series of model units, built on fine-scale ecosystem dynamics up to the extent of entire training areas and beyond, is implemented. Second, the model is only one component in a larger system designed specifically for direct assistance in management planning; purely ecological considerations are necessarily integrated with training needs to meet U.S. Army readiness standards, and with a variety of environmental concerns and regulations that are now being stringently applied to U.S. Army lands. Third, this is an iterative system in which the model will be continuously improved based on ecological scientific advances and feedback from actual management experiences. Modeling technologies, including new methods and criteria for testing model effectiveness and utility (e.g., Fong et al. 1997; Loehle 1997) are advancing rapidly, and these can be adapted and implemented as they become available.

That ongoing revisions and improvements in both modeling tools and management are needed as part of adaptive management is widely recognized and advocated (Walters and Holling 1990). Just because models are new and under development does not mean that they cannot be applied in management situations, as long as it is recognized that these are part of an ongoing management development scenario. Waiting until the models are well tested and widely applied does not help meet current ecological management needs such as those for U.S. Army installations (Ludwig et al. 1993). Modern management of environmental processes is itself an important area of research (Underwood 1995), and it is important that application of the best available tools be included in its development. Immediate application of these models allows learning for both modelers and managers in how to use these tools in real management situations (Johnson 1995).

Acknowledgments. We acknowledge financial support from the U.S. Army Construction Engineering Research Laboratories, Champaign, Illinois.

References

Allen, T.F.H., and T.B. Starr. 1982. Hierarchy: Perspectives for Ecological Complexity. University of Chicago Press, Chicago.

Cale, W.G., G.M. Henebry, and Y.A. Yeakley. 1989. Inferring process from pattern in natural communities. BioScience 39:600–605.

Chapin, F.S. III, B.H. Walker, R.J. Hobbs, D.U. Hopper, J.H. Lawton, O.E. Sala, and D. Tilman. 1997. Biotic control over the functioning of ecosystems. Science 277:500–504.

Christensen, N.L. 1985. Shrubland fire regimes and their evolutionary consequences. In: The Ecology of Natural Disturbance and Patch Dynamics, pp. 85–100. S.T.A. Pickett and P.S. White (eds.). Academic Press, San Diego.

Clark, J.S. 1992. Relationships among individual plant growth and the dynamics of populations and ecosystems. In: Individual-Based Models and Approaches in Ecology, pp. 421–454. D.L. DeAngelis and L.J. Gross (eds.). Chapman & Hall, New York.

Coulson, R.N., L.J. Folse, and D.K. Loh. 1987. Artificial intelligence and natural resource management. Science 237:262–267.

Coulson, R.N., and M.C. Saunders. 1988. Types and characteristics of computer-based decision aids. AI Applications in Natural Resource Management 2:39–43.

Davis, J.R., and J.L. Clark. 1989. A selective bibliography of expert systems in natural resource management. AI Applications in Natural Resource Management 3:1–18.

Davis, J.R., P. Whigham, and I.W. Grant. 1988. Representing and applying knowledge about spatial processes in environmental management. AI Applications in Natural Resource Management 2:17–25.

DeAngelis, D.L. 1988. Strategies and difficulties of applying models to aquatic populations and food webs. Ecological Modelling 43:57–73.

DeAngelis, D.L. 1992. Dynamics of Nutrient Cycling and Food Webs. Chapman & Hall, London.

Ehrlich, P.R., and G.C. Daily. 1993. Science and the management of natural resources. Ecological Applications 3:558–560.

Fahrig, L. 1991. Simulation methods for developing general landscape-level hypotheses of single-species dynamics. In: Quantitative Methods in Landscape Ecology, pp. 417–442. M.G. Turner and R.H. Gardner (eds.). Springer-Verlag, New York.

Fong, P., M.E. Jacobson, M.C. Mescher, D. Lirman, and M.C. Harwell. 1997. Investigating the management potential of a seagrass model through sensitivity anslysis and experiments. Ecological Applications 7:300–305.

Gardner, R.H., B.T. Milne, M.G. Turner, and R.V. O'Neill. 1987. Neutral models for the analysis of broad-scale landscape pattern. Landscape Ecology 1:19–28.

Gardner, R.H., and R.V. O'Neill. 1991. Pattern, process and predictability: the use of neutral models for landscape analysis. In: Quantitative Methods in Landscape Ecology, pp. 289–307. M.G. Turner and R.H. Gardner (eds.). Springer-Verlag, New York.

Hilborn, R., and D. Ludwig. 1993. The limits of applied ecological research. Ecological Applications 3:550–552.

Hyman, J.B., J.B. McAninch, and D.L. DeAngelis. 1991. An individual-based simulation model of herbivory in a heterogeneous landscape. In: Quantitative Methods in Landscape Ecology, pp. 443–475. M.G. Turner and R.H. Gardner (eds.). Springer-Verlag, New York.

Jahn, L.R., C.W. Cook, and J.D. Hughes. 1984. An evaluation of U.S. Army natural resource management programs on selected military installations and civil works projects. Report to the Secretary of the Army, U.S. Department of the Army. Washington, DC.

Johnson, B.L. 1995. Applying computer simulation models as learning tools in fishery management. North American Journal of Fisheries Management 15:736–747.

Karafyllidis, I., and A. Thanailakis. 1997. A model for predicting forest fire spreading using cellular automata. Ecological Modelling 99:87–97.

King, A.W. 1991. Translating models across scales in the landscape. In: Quantitative Methods in Landscape Ecology, pp. 479–517. M.G. Turner and R.H. Gardner (eds.). Springer-Verlag, New York.

Lavorel, S., R.H. Gardner, and R.V. O'Neill. 1993. Analysis of patterns in hierarchically structured landscapes. Oikos 67:521–528.

Loehle, Craig. 1997. A hypothesis testing framework for evaluating ecosystem model performance. Ecological Modelling 97:153–165.

Ludwig, D., R. Hilborn, and C. Walters. 1993. Uncertainty, resource exploitation, and conservation: lessons from history. Ecological Applications 3:547–549.

McLendon, T., W.M. Childress, and D. Price. 1996. Use of Land Condition-Trend Analysis (LCTA) to develop a community dynamics simulation model as a factor for determination of training carrying capacity of military lands. Proceedings of the 1996 Integrated Training Area Management Workshop, LaCrosse, WI. University of Wisconsin-Stevens Point, Stevens Point, WI.

McLendon, T., and B.E. Dahl. 1983. A method for mapping vegetation utilizing multivariate statistical techniques. Journal of Range Management 36:457–462.

McLendon, T., and E.F. Redente. 1991. Nitrogen and phosphorus effects on secondary succession dynamics on a semi-arid sagebrush site. Ecology 72:2016–2024.

McLendon, T., and E.F. Redente. 1992. Effects of nitrogen limitation on species replacement dynamics during early secondary succession on a semiarid sagebrush site. Oecologia 91:312–317.

Milne, B.T. 1991. Heterogeneity as a multiscale characteristic of landscapes. In: Ecological Heterogeneity, pp. 69–84. J. Kolasa and S.T.A. Pickett (eds.). Springer-Verlag, Berlin.

Moen, R., J. Pastor, and Y. Cohen. 1997. A spatially explicit model of moose foraging and energetics. Ecology 78:505–521.

Moninger, W.R., and R.M. Dyer. 1988. Survey of past and current AI work in the environmental sciences. AI Applications in Natural Resource Management 2:48–52.

O'Neill, R.V., D.L. DeAngelis, J.B. Waide, and T.F.H. Allen. 1986. A Hierarchical Concept of Ecosystems. Princeton University Press, Princeton, NJ.

O'Neill, R.V., R.H. Gardner, B.T. Milne, M.G. Turner, and B. Jackson. 1991. Heterogeneity and spatial hierarchies. In: Ecological Heterogeneity, pp. 85–96. J. Kolasa and S.T.A. Pickett (eds.). Springer-Verlag, Berlin.

Pickett, S.T.A., S.L. Collins, and J.J. Armesto. 1987. A hierarchical consideration of causes and mechanisms of succession. Vegetatio 69:109–114.

Pickett, S.T.A., J. Kolasa, J.J. Armesto, and S.L. Collins. 1989. The ecological concept of disturbance and its expression at various hiearchical levels. OIKOS 54:129–136.

Reynolds, J.F., and B. Acock. 1997. Modularity and genericness in plant and ecosystem models. Ecological Modelling 94:7–16.

Robertson, G.P., M.A. Huston, F.C. Evans, and J.M. Tiedje. 1988. Spatial variability in a successional plant community: patterns of nitrogen availability. Ecology 69:1519–1524.

Shugart, H.H., and S.W. Seagle. 1985. Modelling forest landscapes and the role of disturbance in ecosystems and communities. In: The Ecology of Natural Disturbance and Patch Dynamics, pp. 353–368. S.T.A. Pickett and P.S. White (eds.). Academic Press, San Diego.

Stock, M. 1987. AI and expert systems: an overview. AI Applications in Natural Resource Management 1:9–17.

Tazik, D.J., S.D. Warren, V.E. Diersing, R.B. Shaw, B.J. Brozka, C.F. Bagley, and W.R. Whitworth. 1992a. U.S. Army Land Condition-Trend Analysis (LCTA) plot inventory field methods. Technical Report N-92/03, U.S. Army Corps of Engineers Construction Engineering Research Laboratories, Champaign, IL.

Tazik, D.J., J.D. Cornelius, D.M. Herbert, T.J. Hayden, and B.R. Jones. 1992b. Biological assessment of the effects of military associated activities on endangered species at Fort Hood, Texas. Special Report EN-93/01, U.S. Army Corps of Engineers Construction Engineering Research Laboratories, Champaign, IL.

Turner, M.G. 1990. Spatial and temporal analysis of landscape patterns. Landscape Ecology 4:21–30.

Turner, M.G., and R.H. Gardner (eds.). 1991. Quantitative Methods in Landscape Ecology. Springer-Verlag, New York.

Turner, M.G., and V.H. Dale. 1991. Modeling landscape disturbance. In: Quantitative methods in landscape ecology, pp. 323–351. M.G. Turner and R.H. Gardner (eds.). Springer-Verlag, New York.

Underwood, A.J. 1995. Ecological research and (and research into) environmental management. Ecological Applications 5:232–247.

Urban, D.L., R.V. O'Neill, and H.H. Shugart, Jr. 1987. Landscape ecology. BioScience 37:119–127.

Van Auken, O.W., and J.K. Bush. 1997. Growth of Prosopis glandulosa in response to changes in aboveground and belowground interference. Ecology 78:1222–1229.

Vitousek, P.M. 1985. Community turnover and ecosystem nutrient dynamics. In: The Ecology of Natural Disturbance and Patch Dynamics, pp. 325–333. S.T.A. Pickett and P.S. White (eds.). Academic Press, San Diego.

Walters, C. 1986. Adaptive Management of Renewable Resources. Macmillan, New York.

Walters, C.J., and C.S. Holling. 1990. Large-scale management experiments and learning by doing. Ecology 71:2060–2068.

Weltzin, J.F., S. Archer, and R.K. Heitschmidt. 1997. Small-mammal regulation of vegetation structure in a temperate savanna. Ecology 78:751–763.

Wiens, J.A. 1985. Vertebrate responses to environmental patchiness in arid and semiarid ecosystems. In: The Ecology of Natural Disturbance and Patch Dynamics, pp. 160–193. S.T.A. Pickett and P.S. White (eds.). Academic Press, San Diego.

Wiens, J.A., N.C. Stenseth, B. Van Horne, and R.A. Ims. 1993. Ecological mechanisms and landscape ecology. OIKOS 66:369–380.

7
HARVEST: A Timber Harvest Allocation Model for Simulating Management Alternatives

ERIC J. GUSTAFSON

Landscapes primarily comprised of forests have become highly valued by society because of the multiple benefits derived from forested habitats. These benefits include stable quantities and quality of water, forest products, resources to support the ecological and human communities that are associated with forests, and a relatively natural setting for recreation and aesthetic enjoyment. A large proportion of the North American landscape is directly and indirectly structured by forestry practices. Traditional forest management has largely focused on the management of individual stands (Tang and Gustafson 1997) and only recently has the need to take a landscape perspective been embraced. As a result, forest management has become increasingly complex because of the multiple values, resources, and scales that must be considered in the development of a forest management plan. Timber harvesting strategies are especially scrutinized because logging introduces disturbance that can have effects at multiple scales and on many attributes of ecological systems. Both time (when an activity will occur) and space (where an activity will occur) need to be considered simultaneously when evaluating the cumulative impacts of forest management activities. However, this is a complex task. Because there are ecological consequences related to landscape patterns in space and time, quantitative, spatial tools are needed to assess the long-term spatial consequences of alternative management strategies (Franklin and Forman 1987; Hemstrom and Cissel 1991; Li et al. 1992; Thompson 1993). Simulation models that incorporate both temporal and spatial dynamics (e.g., Czaran and Bartha 1992; Li et al. 1993; Liu 1993; Wallin et al. 1994; Gustafson and Crow 1996; Gustafson 1996) can be helpful in these assessments.

Landscape ecology has emphasized the relationship between the spatial pattern of landscape elements and ecological processes. Much progress has been made in determining details of this relationship, particularly in forested landscapes (e.g., forest area, fragmentation, and isolation effects on community dynamics [Askins et al. 1987; Blake and Karr 1987]; predation [Small and Hunter 1988; Robinson et al. 1995]; dispersal patterns

[Pulliam 1988]). Because we now have some basis for predicting the ecological response to spatial pattern, it is reasonable to inform management decisions by predicting the spatial pattern that management options can be expected to produce. To this end, a timber harvest allocation model (HARVEST) has been constructed that provides a visual and quantitative means to predict the spatial pattern of forest openings produced under alternative harvest strategies (Gustafson and Crow 1994). The model allows simulation of differences in the size of timber harvest units, the total area harvested, rotation lengths (minimum interval between harvest treatments), and the spatial distribution of harvested areas. This chapter presents a detailed description of the use of HARVEST to simulate specific management alternatives. The alternatives simulated were developed as part of the ongoing forest planning efforts of the Hoosier National Forest (HNF) in southern Indiana. They provide a realistic set of alternatives to illustrate the use of HARVEST and the information that HARVEST can provide. HARVEST has general relevance as a spatial model of a human activity that functions as an ecological disturbance agent. By predicting the spatial nature of the disturbance, it is possible to predict the ecological conditions that may result (e.g., Gustafson and Crow 1994). These predicted conditions may provide feedback to the managers as they plan the disturbance to achieve the desired results.

Background and Description

HARVEST is a timber harvest allocation model that allows the input of specific rules to allocate forest stands for even-age harvest (clearcuts and shelterwood) and group selection, using parameters commonly found in National Forest Plan standards and guidelines. The model produces landscape patterns that have spatial attributes resulting from initial landscape conditions and potential timber management activities. HARVEST was designed to require minimal input data, and the model makes a number of simplifying assumptions that enable its use for strategic planning on large areas over long time frames. The first assumption is that the location of stands harvested typically take a spatially random distribution within timber production zones over a 10-year period. However, harvest allocations are spatially constrained by the locations of stands that are old enough to be harvested and by the boundaries of the management zones where timber harvest is allowed. The spatially random assumption is based on an analysis of stands reaching rotation age and past harvest allocations. Using nearest neighbor analysis (Davis 1986) on 10 subsets of HNF stand maps (mean size of subsets = 3366 ha, SD = 1062 ha), the observed mean nearest-neighbor distance between stands of similar age was compared to the distance expected if stands were randomly distributed, and a z-statistic was computed. The null hypothesis that stands are randomly distributed could not be rejected at the 95% confidence level for 8 of the 10 subsets (see Gustafson and Crow 1996).

HARVEST also ignores forest types, with the exception of a single, secondary class (e.g., conifer), for which the user can (optionally) define a different size distribution for harvests. This feature was incorporated into HARVEST to allow larger harvest units on conifer plantations. If some forest types will not be harvested in proportion to their abundance within the timber land base, then harvest of various forest types would need to be simulated independently, and the results combined using a GIS. HARVEST uses age as a surrogate for merchantability and ignores stocking density and size class. Access and operability are assumed to be uniform across the land base. Significantly large areas known to be inoperable should be excluded from the timber land base for the simulations. It is important to remember that HARVEST was designed to allow comparison of the impacts of broad management strategies on forest spatial pattern over large spatial areas, and not for more detailed, stand-specific decisions.

HARVEST includes the ability to modify a number of parameters that are commonly specified in the standards and guidelines of management plans. These include harvest size distributions, total area harvested, rotation length (understood by HARVEST to mean the minimum age of stands that can be harvested), silvicultural method (even-age or group selection), and the width of buffers that must be left around harvests. An important capability of HARVEST is the ability to allocate harvests only in portions of the landscape that are designated for harvest. Equally important is the ability to apply different management strategies to different portions of the landscape (management areas). Modeling the process of allocating timber harvests in space and time allows experimentation to link variation in management strategies with the resulting pattern of forest openings and the distribution of forest age classes.

HARVEST was developed using ERDAS Toolkit routines to run within the ERDAS v. 7.4^{+} GIS environment. A version has also been produced that is independent of ERDAS, allowing use of files exported from other raster GIS systems. Timber harvest allocations are made by HARVEST using a digital map of the forest, where grid-cell values reflect the age (in years) of the forest in that cell. HARVEST takes this GIS age map as input and produces a new age map incorporating harvest allocations, where harvested cells take a value of 1 and unharvested cells increase in age by the length of the time-step of the simulation.

An ASCII file is produced by HARVEST in which is recorded the harvest history of stands. Each record represents one stand, and the user specifies an integer (treatment code) that is written to the records representing stands that are harvested during the model run. The meaning of these integer "treatment codes" is left to the user, but they allow a flexible means of controlling the way HARVEST allocates in stands treated during previous runs. The user can also use this file to force HARVEST to revisit group selection stands at the appropriate time. This file is the link between successive model runs, and represents institutional memory of previous management activity.

The two algorithms built into the model for determining the spatial dispersion of allocations are a random dispersion and a group selection dispersion. HARVEST selects harvest locations randomly within the timber production zones, checking first to ensure that the forest is old enough to meet rotation length requirements. Group selection is implemented by HARVEST such that a user-specified proportion of a group-selected stand is cut during each entry. The number of cells (n) harvested in a stand during each entry is calculated by HARVEST as a user-specified proportion (p) of the size of the stand (A):

$$n = \mathrm{NINT}(A * p)$$

Selection of new stands for group selection is achieved with the Generate option of HARVEST, in which stands are randomly selected from those stands with an age greater than the prescribed rotation length, and small openings (groups of trees) within those stands are then randomly placed, with at least the user-specified distance between openings. Reentry into previously group selected stands is achieved with the Lookup option, in which HARVEST allocates groups in all stands that have a specific "treatment code" value stored in the ASCII "treatment file." The stands are allocated in numerical order by stand ID. The user must ensure that reentries occur by invoking the Lookup option and specifying the proper "treatment code" during model runs representing the reentry dates (see example below).

Portions of the land base can be excluded from timber harvest by presenting HARVEST with an age map of only areas where harvest is allowed. Independent runs of HARVEST on different portions of a larger map may be used to simulate different management strategies on different portions (management areas) of the landscape. The maps produced for each management area can later be combined with the rest of the land base (using a GIS) to produce a map that characterizes the entire land base.

HARVEST generates a normal distribution of harvest sizes around a user-specified mean. The user may truncate either (or both) tail(s) of the distribution. HARVEST allows the user to (optionally) specify a different size distribution of harvests for up to one additional specific land cover type (for example, conifer).

Example: Simulation of Alternatives Using HARVEST

To illustrate the use of HARVEST, the simulation of 3 management alternatives proposed at various times for the Hoosier National Forest (Indiana) will be described. Two of the alternatives were proposed in the Final Environmental Impact Statement (FEIS, USDA Forest Service 1985)

TABLE 7.1. Area harvested (ha) under each prescription set (denoted by decimal labels) for the 3 simulated alternatives. The alternatives vary in the amount (and location) of land dedicated to the various prescriptions.

	Prescription set					
Management alternatives	2.1	2.8	3.1	3.2	6.1	Total[1]
Alternative 1	—	—	—	11956.4	2610.3	14566.7
Alternative 2	319.5	—	4960.7	6421.6	2610.8	14312.6
Alternative 3	—	8041.2	—	—	—	8041.2

[1] Total area of all timber harvest M.A.s on the study area. Differences among the alternatives reflect variation in the area set aside from timber harvest.

for the original HNF Forest Plan: (1) continuation of current management (primarily clearcutting over 80% of the forest), and (2) emphasis on a variety of dispersed recreation, visual, wildlife habitat, and timber products. The third is an alternative that was eventually adopted in the Land and Resource Management Plan Amendment (USDA Forest Service 1991), with an emphasis on a more natural appearing forest. The parameters used by HARVEST to simulate these alternatives are given in Table 7.1.

These alternatives were simulated on a 49,515 ha subset of the Tell City Unit of the HNF, using a forest age map of National Forest land within the study area, with ages calculated as of 1988 and gridded to 30 m cells. Land use on non-Forest Service land was derived from Landsat Thematic Mapper (TM) imagery collected in 1988, as described in Gustafson and Crow (1994).

The Forest Service has developed a series of vegetation management prescriptions that are used to achieve specific goals, and these are given decimal names (e.g., 2.8, 3.1). Each alternative included a map in which the land base was divided into various management areas with specific timber harvest prescriptions for each (Fig. 7.1). Thus, the details of prescriptions did not vary among alternatives, but the location, extent, boundary locations, and number of management areas where the prescriptions were applied did vary (Fig. 7.1; Table 7.1). Forest planning documents do not necessarily provide precise values for many of the parameters required by HARVEST. For example, a range of acceptable harvest sizes may be specified, while HARVEST requires a mean size and a standard deviation to produce a normal distribution of harvest sizes. The user must specify a mean and standard deviation that will produce the desired range. It is useful to recall that for a normal distribution, approximately 95% of the values will fall within two standard deviations of the mean. The specification of minimum and maximum allowable sizes allows the user to truncate either tail of the distribution. The parameters derived from the FEIS and the 1991 Plan Amendment are given in Table 7.2. Because we are using a subset of the entire forest as a study area, we must convert the percentages specified in the Forest Plan to number of pixels to be harvested by tabulating the

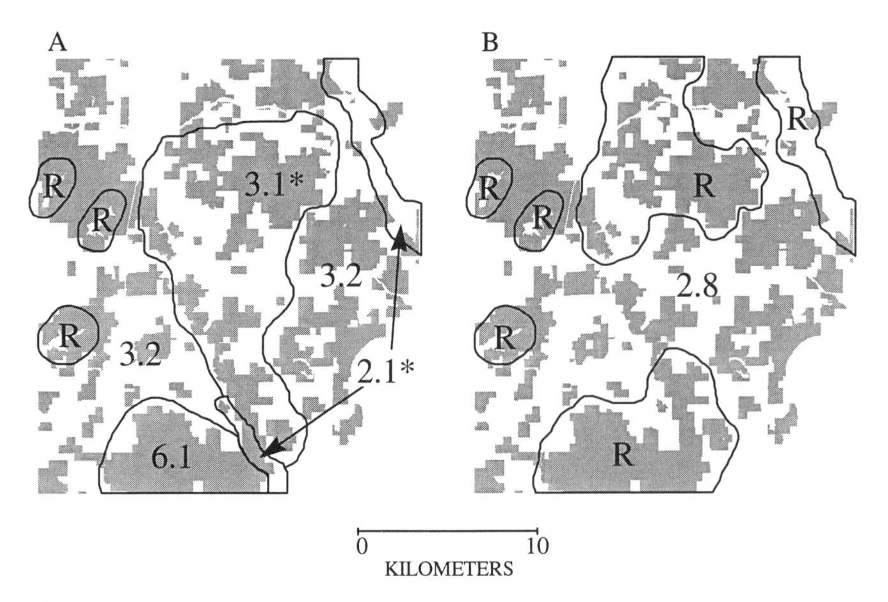

FIGURE 7.1. Map of the study area on the Hoosier National Forest showing ownership (HNF land is shaded) and management area boundaries (outlined) for the three alternatives. Each management area is labeled according to the prescription set applied to it under each alternative (see Tables 7.1 and 7.2). Alternatives 1 and 2 are shown in (A), and management areas denoted by asterisks were delineated as such under Alternative 2, and were designated as prescription set 3.2 under Alternative 1. Alternative 3 is shown in (B). R designates areas reserved from timber harvest.

total number of forest cells in the management area and taking the percentage specified. Using HARVEST, five decades of implementation for each alternative were simulated. Fifty years represents the planning horizon of the FEIS. A detailed description of data preparation and the procedures used to simulate these alternatives is provided in the Appendix.

Use of HARVEST Simulation Results

The analysis of simulation results can take several forms. The output maps are best suited for analyses that focus on the spatial pattern of contiguous forest, forest openings, and forest edge (Gustafson and Crow 1994, 1996; Gustafson 1996). HARVEST produces a forest age map that typically does not cover the entire area of interest (i.e., there are gaps within or beyond the timber management area due to the ownership pattern or presence of non-forested land uses.) To evaluate the pattern of forest openings across a large landscape, it is necessary to have digital land use data for those parts of the landscape not simulated by HARVEST. Using a GIS it is possible to overlay the pattern of forest and openings produced by HARVEST onto land cover data generated from remotely sensed data. Using this composite map it is relatively straightforward to conduct spatial pattern analyses using GIS modeling functions or other analysis software (such as FRAGSTATS [McGarigal and Marks 1995]). For example, the amount and distribution of forest interior and edge produced by various management alternatives using the output from HARVEST has been analysed (Gustafson and Crow 1994, 1996; Gustafson 1996). The age maps produced by HARVEST are also useful for assessing the age class distribution of the managed forest (Gustafson and Crow 1996). The results of these analyses are especially revealing when plotted as a function of simulated time, showing trends and cumulative impacts of management over time. Simulations can easily be

TABLE 7.2. Harvest intensities as derived from the 1985 Hoosier National Forest Final Environmental Impact Statement and the 1991 Forest Plan Amendment. Clearcut and shelterwood treatments use the same parameters in Prescription Sets 3.1, 3.2, and 6.1 and are simulated as one treatment. Prescription set 2.8 has clearcut, shelterwood, and group selection treatments simultaneously.

	Prescription set				
Model parameter	2.1	2.8	3.1	3.2	6.1
Mean clearcut opening size (ha)					
deciduous	—	1.80	7.02	4.95	3.96
conifer	—	3.60	7.02	4.95	3.96
Standard deviation clearcut size (ha)					
deciduous	—	0.09	2.34	1.44	0.90
conifer	—	0.18	2.34	1.44	0.90
Mean shelterwood opening size (ha)	—	3.15	7.02	4.95	3.96
Standard deviation shelterwood size (ha)	—	0.45	2.34	1.44	0.90
Mean group opening size (ha)					
deciduous	0.36	0.22	—	—	—
Standard deviation group size (ha)					
deciduous	0.08	0.05	—	—	—
Minimum opening size (ha)					
deciduous	—	0.72	1.08	0.72	1.08
conifer	—	0.72	1.08	0.72	1.08
group	0.09	0.09	—	—	—
Maximum opening size (ha)					
deciduous	—	3.96	10.80	7.20	5.40
conifer	—	3.96	10.80	7.20	5.40
Harvest rate/decade[1] (%)	5.4	10.1[2]	11.5	7.8	7.5
Rotation length (years)	90	90	80	120	120
Buffer width[3] (m)	30	30	30	30	30

[1] Represents percent of forest within a Management Area (i.e., Area using this prescription set) that is harvested each decade (see Table 7.1).

[2] Distributed among the following harvest treatments: 35% clearcuts, 15% shelterwood, and 50% group selection.

[3] The width of buffers left between harvest allocations and other harvests, streams, and openings.

replicated to provide a statistical estimate of the variability in the spatial patterns produced.

Results

Alternatives 1 and 2 differ primarily in that Alternative 2 divides the large area designated as management area 3.2 in Alternative 1 into smaller blocks designated management areas 2.1, 3.1, and 3.2 (Fig. 7.1A). Management area 6.1 is essentially the same in both alternatives. It is not surprising then that these two alternatives produced very similar patterns (Figs. 7.2–7.4) because they designate approximately the same total area to timber production under similar prescriptions (Table 7.1). Alternative 3 produces markedly different patterns than the other two (Figs. 7.2–7.4) by both limiting the amount of land harvested, and clustering land into relatively large reserves (Fig. 7.1B).

Using HARVEST it has been shown that the spatial (and temporal) configuration of the timber land base has more of an effect on the amount of forest interior and edge than does the intensity of harvest within that land base (Gustafson and Crow 1996). We have also shown that heavy use of group selection reduces forest interior and increases edge more than harvest methods that produce larger, but fewer openings (Gustafson and Crow 1994, 1996).

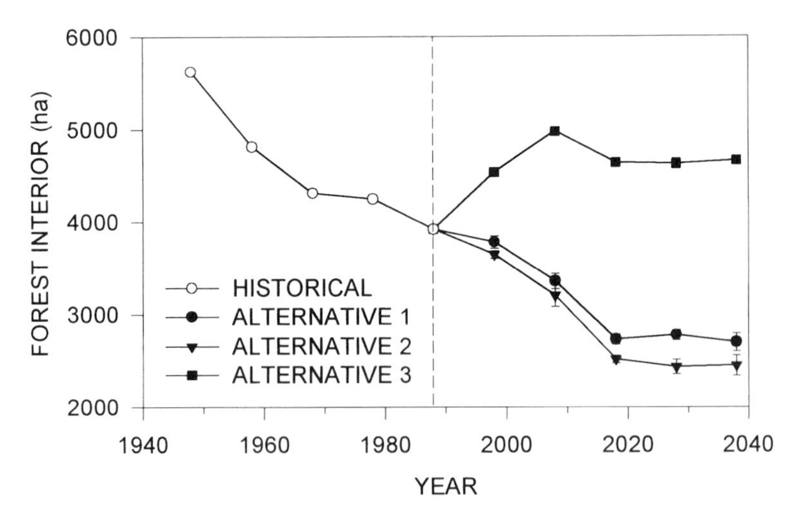

FIGURE 7.2. Changes in the amount of forest interior habitat (forest >210 m from an edge) over time resulting from simulation of three management alternatives on the Hoosier National Forest, beginning with 1988 forest conditions. Error bars show one standard deviation from the mean of the three replicates.

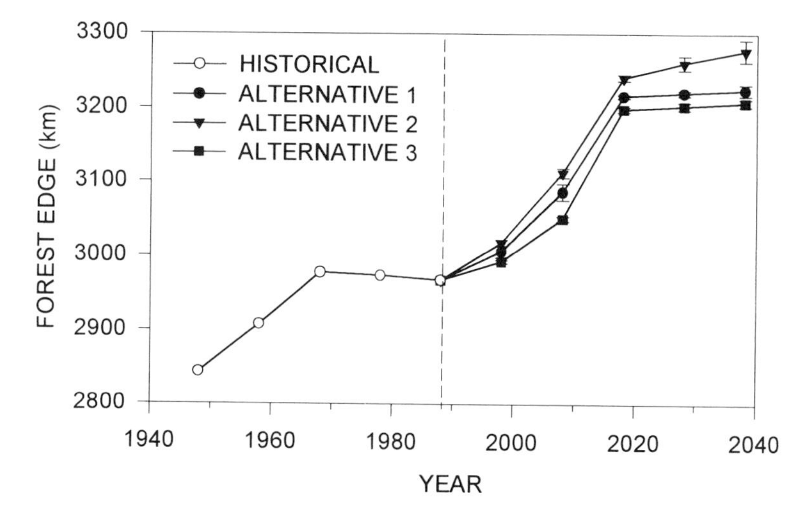

FIGURE 7.3. Changes in the amount of linear forest edge over time resulting from simulation of three management alternatives on the Hoosier National Forest, beginning with 1988 forest conditions. Error bars show one standard deviation from the mean of the three replicates.

Developments in landscape ecology have suggested that explicit design of the spatial and temporal application of vegetation management activities can be used to produce desired ecological conditions. Furthermore, the large spatial and temporal scales commonly considered by landscape ecologists open the possibility of novel forest management strategies. In a recent study using the entire HNF as a study area, HARVEST was used to investigate one such novel management strategy (Gustafson 1996). The experimental treatments consisted of alternative designations of timber harvesting areas on the HNF that varied as to where timber harvest was allowed during each decade and for how many decades it was allowed there. For the static zoning alternative, harvest was allowed throughout the timber harvest land base (65% of the HNF) during all 15 decades. The three dynamic zoning alternatives varied in the degree to which harvests were clustered in time and space. Harvest intensity was held constant among all four zoning alternatives. For the 50-year hiatus alternative, the timber land base was divided into three subsets; timber harvest was allowed on only two of these subsets at a time, and the third was temporarily set aside from timber harvest for 50 years. The treatments were rotated every five decades, so that each subset was harvested for ten successive decades and then set aside from timber harvest for five decades. For the 100-year hiatus alternative, the same three subsets were used, but timber harvest was allowed on only one of these subsets at a time, and the other two were temporarily set aside from timber harvest. Again, the treatments were rotated every five decades, and each subset was harvested for five successive decades and then

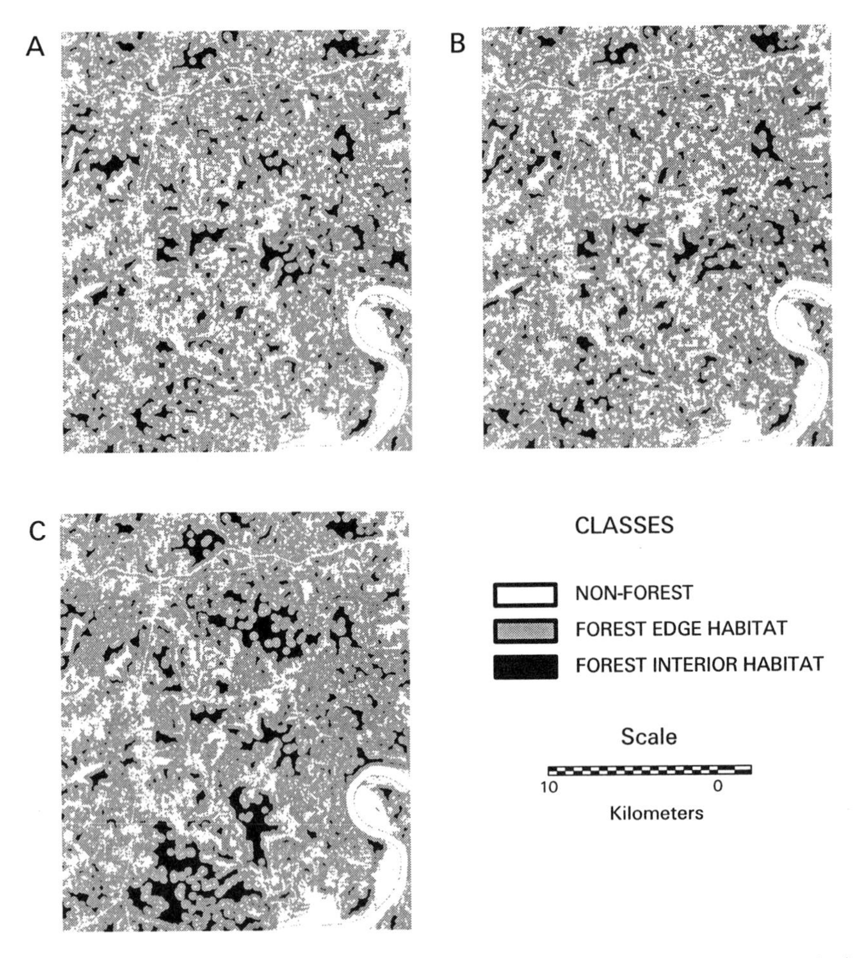

FIGURE 7.4. Spatial distribution of forest interior (forest ≥210 m from a forest edge) at the end of simulation of five decades of forest management on the Hoosier National Forest under (A) Alternative 1; (B) Alternative 2; (C) Alternative 3.

set aside for ten decades. Finally, for the 120-year hiatus alternative, the timber harvest land base was divided into five subsets, and each subset was harvested for 3 decades and then was set aside for 12 decades. Total area harvested, size of harvest openings, and rotation interval (minimum age for cells to be harvested) were held constant across all treatments and decades, so that timber production was the same for all four scenarios. The dynamic zoning strategies resulted in substantial increases in the amount of forest interior and reductions in the amount of forest edge across the landscape, as well as an increase in the average age of stands when harvested. The greatest reduction in fragmentation was produced by the alternative that most tightly clustered harvests in time and space (i.e., intensive harvesting of small blocks in a relatively short time). When harvest intensity was high, this alternative produced amounts of forest interior and edge comparable to those of the dispersed alternative with half the rate of harvest.

The previous example illustrates of the potential of HARVEST as a strategic planning tool. It provides a link between the forested landscape that logistically must be managed by relatively small, site-specific manipulations, and the broad spatial and temporal domains shown to be important by developments in landscape ecology. Because there is a strong relationship between spatial pattern and ecological process, HARVEST allows managers to predict the spatial patterns to be expected by management activity, to allow prediction of the ecological conditions that may result. Such a capability is critical for the development of new management strategies to meet the changing demands of society in the next century.

HARVEST Software Availability

HARVEST may be obtained free of charge via the Internet. The software, updates, source code, and detailed user documentation are available through the North Central Research Station World Wide Web site (http://www.ncfes.umn.edu) under "Research Products."

Acknowledgments. I thank an anonymous reviewer for constructive comments that helped improve the manuscript. The development of HARVEST was funded by the North Central Forest Experiment Station and the Hoosier National Forest.

References

Askins, R.A., M.J. Philbrick, and D.S. Sugeno. 1987. Relationship between the abundance of forest and the composition of forest bird communities. Biological Conservation 39:129–152.

Blake, J.G., and J.R. Karr. 1987. Breeding birds of isolated woodlots: area and habitat relationships. Ecology 68:1724–1734.

Czaran, T., and S. Bartha. 1992. Spatiotemporal dynamic models of plant populations and communities. TREE 7:38–42.

Davis, J.C. 1986. Statistics and Data Analysis in Geology, pp. 308–312. John Wiley & Sons, New York.

Franklin, J.F., and R.T.T. Forman. 1987. Creating landscape patterns by forest cutting: ecological consequences and principles. Landscape Ecology 1:5–18.

Gustafson, E.J. 1996. Expanding the scale of forest management: allocating timber harvests in time and space. Forest Ecology and Management 87:27–39.

Gustafson, E.J., and J.R. Crow. 1994. Modeling the effects of forest harvesting on landscape structure and the spatial distribution of cowbird brood parasitism. Landscape Ecology 9:237–248.

Gustafson, E.J., and T.R. Crow. 1996. Simulating the effects of alternative forest management strategies on landscape structure. Journal of Environmental Management 46:77–94.
Hemstrom, M., and J. Cissel. 1991. Evaluating alternative landscape patterns over time—an example from the Willamette National Forest in Oregon. In: Proceedings of the 1991 Society of American Foresters National Convention, SAF Publication 91-05, pp. 231–235. Society of American Foresters, Bethesda, MD.
Li, H., J.F. Franklin, F.J. Swanson, and T.A. Spies. 1992. Developing alternative forest cutting patterns: a simulation approach. Landscape Ecology 8:63–75.
Liu, J. 1993. ECOLECON: an ECOLogical-ECONomic model for species conservation in complex forest landscapes. Ecological Modelling 70:63–87.
McGarigal, K., and Marks, B.J. 1995. FRAGSTATS: spatial pattern analysis program for quantifying landscape structure. General Technical Report PNW-GTR-351. USDA Forest Service, Pacific Northwest Research Station, Portland, OR.
Pulliam, H.R. 1988. Sources, sinks, and population regulation. American Naturalist 132:652–661.
Robinson, S.K., F.R. Thompson, T.M. Donovan, D.R. Whitehead, and J. Faaborg. 1995. Regional forest fragmentation and the nesting success of migratory birds. Science 267:1987–1990.
Small, M.F., and M.L. Hunter. 1988. Forest fragmentation and avian nest predation in forested landscapes. Oecologia 76:62–64.
Tang, S.M., and E.J. Gustafson. 1997. Perception of scale in forest management planning: challenges and implications. Landscape and Urban Planning 39:1–9.
Thompson, F.R. III. 1993. Simulated responses of a forest-interior bird population to forest management options in central hardwood forests of the United States. Conservation Biology 7:325–335.
USDA Forest Service. 1985. Final Environmental Impact Statement, Land and Resource Management Plan, Hoosier National Forest. USDA Forest Service, Eastern Region, Hoosier National Forest, Bedford, IN.
USDA Forest Service. 1991. Plan Amendment, Land and Resource Management Plan, Hoosier National Forest, pp. 1–220. USDA Forest Service, Eastern Region, Hoosier National Forest, Bedford, IN.
Wallin, D.O., F.J. Swanson, and B. Marks. 1994. Landscape pattern response to changes in pattern generation rules: land-use legacies in forestry. Ecological Applications 4:569–580.

Appendix

HARVEST requires up to three input GIS map layers: a forest age map, a map of stand ID numbers that is derived from the forest age map by clustering pixels of the same age, and an optional land use/land cover map layer. The input forest age GIS map can be imported from an existing digital stand age map or by digitizing from a hard copy map. This stand age map may be further modified to exclude certain portions of the map (e.g., exclusion of land in other ownerships, exclusion of areas in different management units, or exclusion of inoperable areas.) Separate GIS layers containing these boundaries can be used to generate a forest age map that excludes stands outside the area to be harvested (by setting the GIS class to zero) using the map algebra capabilities of the GIS. Independent runs of HARVEST on different portions of a larger map may be used to simulate different management strategies on different portions of the landscape. These separate output maps may later be combined using the GIS.

A map of stand ID numbers must be produced from the forest age map by using the clumping function of the GIS. Each group of contiguous pixels of the same age at the start of a time series of simulations are thus considered a stand throughout the series of simulations. An existing stand ID map may be used if it has the same extent and resolution as the forest age map, and each stand ID value is a unique integer ≤3000. This map of stand IDs is used unchanged throughout a simulation of multiple time periods.

Recall that HARVEST allows specification of a different harvest size distribution for a second forest type. If this capability is needed, a land cover map that (minimally) includes the pixels of the secondary forest type must be prepared, so that HARVEST can identify stands of that type. This map must have the same extent and resolution as the input age map. An extensive land cover map including land outside the timber land base is useful to combine with the output maps produced by HARVEST, to represent the pattern of forest openings across the larger landscape.

To simulate a specific time period of management activity, a copy of the input map is made by HARVEST, adding a user-specified offset (value added to each non-zero pixel value) to simulate the increase in stand ages during the time period represented by the model run. For example, if one run of HARVEST represents 10 years of management activity, then HARVEST adds 10 to each non-zero pixel. This process is completed during initialization of the output map.

Pixels harvested by HARVEST are assigned a GIS class value of one in the output file. Should an offset of zero be specified, it may be desirable to recode any 1's that do not represent harvested cells in the input map to some other unique value before running HARVEST, so they can be distinguished from allocations produced by HARVEST. They can later be recoded back to their original value, or some other appropriate value.

As an example of the use of HARVEST to simulate alternatives, consider the simulation of Alternative 3 (Fig. 7.1B). There is only one management area where timber harvest is prescribed (using prescription set 2.8; Table 7.1), so one will have to prepare only a single input age map consisting of HNF land within management area 2.8. However, there are three distinct silvicultural treatments (group selection, shelterwood, and clearcut; Table 7.2) that will have to be simulated with three model runs. In each decade, group selected stands are allocated first, followed by shelterwood cuts and finally clearcuts. Although this order is arbitrary, it may have consequences on the results. For example, group selection requires more stands to achieve a given level (number of pixels) of harvest, and these stands are effectively removed from the timber base for other harvest treatments. On the HNF, group selection typically removes trees in small patches that total about 1/6 (16.7%) of a stand, at 20 to 30 year intervals.

Simulations would be implemented in one decade intervals, using parameters found in Table 7.2, as follows:

1. Beginning with the forest age map of beginning conditions, first decade of group selection is simulated by using an offset of 10 years, Stand locations are generated and a proportion of 0.167 is specified to be cut in each group-selected stand. For this first decade of treatment, a "treatment code" of 1 is used, which will represent group stands initiated in the first decade of the simulation. We will plan to reenter group-selected stands at 30-year intervals. Minimum harvest size should be one pixel (0.09 ha), minimum GIS value (rotation length) is 100 years (since an offset of 10 years was specified).
2. A number of stands have been committed to group selection and have resulted in a new forest age map that has these new openings incorporated into it. Before the next decade of group selection is simulated the allocations to shelterwood and clearcuts for the first decade must be made. The new forest age map just produced will be the input for the shelterwood allocations. Since 10 is already added to the cell ages in the previous run, it will not be done again (offset = 0). Remember, we are still working on the first decade. Do not select group selection, and enter the shelterwood parameters given in Table 7.2. An arbitrary "treatment code" of 4 will be used for shelterwood. (A "treatment code" of 5 for clearcuts will be used later.) Because shelterwood treatments cannot be made in group selected stands, but can be made where shelterwood and clearcuts have been implemented in the past, we will use a minimum "treatment code" of 4 and a maximum "treatment code" of 5 for both shelterwood and clearcuts throughout the simulation. (Note that if HARVEST attempts to allocate in a stand previously treated by clearcut or shelterwood, recently cut pixels will be too young to be harvested, but any pixels within the stand that were not harvested in the previous treatment would still be eligible for allocation.)
3. To complete the first decade of simulation, clearcut treatments are allocated. The map just produced by allocating shelterwood is the input map. Again, an offset of zero and a "treatment code" of 5, and the other parameters as in Table 7.2 are used. After this step, we will have produced three maps. The first two can be considered intermediate and deleted, since the last map also contains the allocations made during the

first two steps. All cells allocated during these three runs will have a value of 1 in this third map, and the "treatment file" will have the appropriate "treatment code" recorded for each stand treated during these three runs.

4. To simulate the second decade, the process is essentially repeated, keeping in mind that in this step it is too early to reenter the group stands allocated in the first decade. So a new group stands is generated (choosing the Generate option of group selection), using a "treatment code" of 2 to represent group stands initiated in the second decade of the simulation. As a new decade is now being initiated, an offset of 10 for this run is used. Other parameters may or may not be the same as the first decade of group selection, but in this case fixed management scenario is simulated, so they will remain the same.

5. Shelterwood and clearcut treatments repeat steps 2 and 3 exactly, although management parameters could be varied without creating complications.

6. Decade 3 will be a repeat of the steps for decade 2, except the "treatment code" for group selection is 3, to represent group stands initiated in the third decade of the simulation.

7. In decade 4, it is time to reenter the group-selection stands initiated in decade 1. After specifying group selection, choose the Lookup option, and specify a "treatment code" of 1. All stands with a "treatment code" of 1 (recorded in the "treatment file") were those group-selected in decade 1. This causes HARVEST to read the "treatment file" to identify the ID number of stands having a "treatment code" of 1, and then to allocate additional group openings on previously unharvested cells in those stands.

8. Shelterwood and clearcuts repeat steps 2 and 3. These are not forced reentries to previously cut stands—only group selection uses forced reentry. However, stands only partially cut by shelterwood or clearcutting may incidentally be reentered on subsequent model runs.

9. Decade 5 is similar to decade 4. To reenter group selection stands initiated in decade 2, use a "treatment code" of 2. Simulating subsequent decades would be a straightforward iteration of this process, again reentering stands initiated in decade 1 during decade 7, and so on. One could continue to "look up" the stands indefinitely (group selection in perpetuity), or, alternatively, Generate new stands at some point, allowing shelterwood and clearcuts in stands formerly group selected by using a "minimum treatment code" of 1 when allocating shelterwood and clearcuts.

Simulation of Alternatives 1 and 2 is somewhat different. Here multiple management areas are within the study area (Fig. 7.1A), but only a single harvest treatment is prescribed in each. (Shelterwood and clearcuts are simultaneously prescribed in prescription sets 3.1, 3.2, and 6.1, but because their parameters are the same and the final shelterwood reentry is within 5 years of initiation, we can simulate these treatments as a single treatment.)

The simulation approach here requires preparation of separate forest age maps for each management area within the study area. Each management area is simulated independently, and the output maps are subsequently combined to produce an integrated map for the entire study area at each decade.

Part III
Planning Strategies

8
A Hierarchical Framework for Conserving Biodiversity

Denis White, Eric M. Preston, Kathryn E. Freemark, and A. Ross Kiester

Society recognizes a large variety of values associated with biodiversity including aesthetic, economic, conservation, and educational (McNeely et al. 1990; Heywood and Watson 1995). These values are all ultimately related to the definition of biodiversity as a manifestation of genetic diversity, the primary raw material that is filtered by natural selection, resulting in evolutionary and ecological adaptation of biota to environmental conditions. Minimizing additional loss of biodiversity will provide the best assurance that biota will adapt to the increasing rate and spatial extent of environmental change (Pratt and Cairns 1992), and that societal values will be sustained.

Traditionally, the management of biodiversity has focused on rescuing rare, threatened, or endangered species. Huge sums have been spent on recovery programs for a small number of species. Although strong conservation arguments exist for preserving these species, the effort expended can easily become out of proportion to the contribution that these species make to genetic diversity, and therefore to the fitness of the biota to adapt to environmental change. In a time when resources for environmental management are decreasing, prioritizing effort so that resources are allocated in proportion to risk and value can optimize conservation effectiveness (Pulliam and Babbitt 1997).

Multiple-scale, hierarchical approaches are needed for conserving biodiversity (Freemark 1995; Freemark et al. 1995; Davis and Stoms 1996). Such approaches should be interdisciplinary, including contributions not only from biology and ecology but also from other applied sciences such as hydrology, agriculture, and forest science, and from the social sciences and arts as well. With collaboration from many perspectives, richer databases and analytic approaches can be formulated. More significantly, a multifaceted approach promises better linkage between scientific perspectives and the spatial, temporal, and political structure of decision making (Lubchenco 1995). Clarifying the scientific status of biodiversity can set the stage for moving the biodiversity debate from one primarily about the facts of the issue to one about values.

This chapter presents a hierarchical framework for assessment and management of biodiversity. The framework advocates (1) understanding associations of biodiversity with environmental factors over large regions, (2) identifying those areas within large regions having species assemblages that contribute the greatest diversity to the biota, and (3) evaluating alternative approaches for managing those important areas in order to explicitly include conservation of biodiversity in land use decisions.

Interaction of Political and Ecological Hierarchies

Hierarchical Structure of Human Decision Making

In Western societies, the hierarchy of sociopolitical entities often follows a common structure, with five to seven levels: villages, towns, or city precincts at the lowest level (smallest human populations); townships or cities at the next level; then counties; states or provinces; multiple state or interstate regions; nations; and finally continents. The boundaries of these entities bear no direct relationship to the spatial boundaries of ecological units, though those at larger extents (e.g., continents) may approximate each other.

Value-based policy about natural resources is usually located at one or more levels in the political hierarchy. For example, an assessment of the implications of alternative biodiversity conservation policies for a state or province must be ultimately constrained to the area within their jurisdiction. However, biodiversity policy for a state or province is more likely to be effective if considered in the context of ecological or biotic regions, of which the state or provincial biota is only a subset for many taxa. Once adopted, a policy applies and is generally implemented by smaller administrative units, such as counties. Therefore the policy must also consider the roles and relative intensity of effort required by smaller units in allocating appropriate resources to get the job done. As a general rule of thumb, ecological policy assessment needs to look both upward for the context at a larger extent and downward for implications of implementation.

The intent of ecological policy assessments varies in emphasis somewhat as a function of scale. Larger government units (nation, state, or province) need to efficiently allocate scarce resources to a complex array of ecological issues that they face. Although political considerations will always be important, objective policy analysis that suggests priorities in the allocation of resources is particularly useful. To do this one needs an objective way to compare a variety of different issues in common terms. Comparability is a prime criterion.

Smaller political units have fewer resources to allocate but greater responsibility for management and implementation of policy. Land-use decisions that are likely to affect biodiversity are traditionally made at the county and municipal level, particularly in the United States. These administrative units may be asked and funded by the nation or state or province to cooperate in a coordinated effort to achieve a larger goal within a relative set of priorities. Ecological management activities at this scale in the hierarchy are likely to be directed much more towards effective, on the ground, conservation activities. For these activities, understanding the particular history and nature of human institutions is important in achieving effective results.

Hierarchical Structure in Ecology

Biodiversity is usually measured at different levels of biological organization: organisms that are composed of cells that contain genes; species populations (or the set of local populations forming a metapopulation) that are composed of individual organisms; communities that are groups of populations interacting with each other; ecosystems that are communities together with their abiotic environment; landscapes that are spatial groupings of ecosystems, and so on to the biome and biosphere. Spatial and temporal scales are conceived as increasing up the hierarchy from genes to biosphere. However, quantifying spatiotemporal scales can be problematical because levels such as populations, communities, and landscapes are open systems with spatiotemporal domains that vary widely among species and processes (Turner 1989; Wiens 1989).

From a different perspective, levels of organization can be viewed as alternative, conceptual constructs that are not hierarchical per se (Allen and Hoekstra 1992). Ecosystem and community conceptions can be compared across a landscape of a given area as well as at larger and smaller spatial extents. A given landscape can be seen to contain smaller landscapes, while itself being a part of a larger landscape. Lastly, community patterns at a given scale may be related to the landscape context at a larger extent. In practice, spatiotemporal scaling is done by the observer so that, at a particular scale, the biological levels of interest appear most cohesive, explicable, and predictable. For adequate understanding, it is necessary to consider three levels and/or scales at once: the one in question, the one below that gives mechanisms, and the one above that gives context, role, or significance (Pickett et al. 1994).

The implications of the biological levels of biodiversity for assessment and management are profound. At large spatial extents, biodiversity priorities should take into account the properties of sets of species. A component of prioritization through complementarity analysis is spatial comparison of sets of species to determine those subsets of the whole that give the greatest representation of total diversity in the least number of samples (Williams and Humphries 1994, but see Faith and Walker 1996 for an alternative view). At small extents, assessment and management of biodiversity are more concerned with the functional interaction among the local set of species. Understanding the mechanisms causing local problems is crucial to solving them.

Subsets of species with similar ecological capabilities provide a functional redundancy which buffers against changes in capacity of any one species of the subset. Since species must cooccur in space to provide redundancy and functional substitution, spatial patterns in diversity are one important descriptor of biodiversity at any scale. Maps of spatial pattern can be useful in setting priorities for biodiversity conservation and in suggesting management options. Spatial pattern recognition is crucial to assessment of risk to values derived from biodiversity and ultimately to managing those risks. Pattern is used as a surrogate measure of process because process is presumed to produce pattern but is more costly and difficult to observe at the large spatial extents relevant to biodiversity (Brown 1995). Therefore, we describe and evaluate patterns of species diversity and anthropogenic and natural modifying factors, and interpret these in the light of processes that are thought to be important at the spatial scale of the study.

Analysis Methods and Data

Types of Analysis

We have incorporated ideas of biodiversity structure and assessment into a hierarchical framework for analysis. At large spatial extents there are two kinds of analyses studying two different questions. First, what are the possible associations of environmental factors with the spatial distribution of biodiversity? Investigations into this question can help to reveal possible mechanisms, including anthropogenic disturbance, responsible for the patterns in biodiversity. These patterns may help to predict biodiversity patterns for areas for which data do not exist, help to predict biodiversity patterns for other taxonomic groups, or help to guide policy development and management implementation by indicating places at greater risk. The second question is, given the distribution of biodiversity, where are the centers, or most important locations, of biodiversity? In other words, with limited resources to study or conserve biodiversity, where are the best places to start further investigations or conservation activities?

For the first question, one approach is a statistical analysis of a response variable representing some measure of biodiversity against a set of predictor variables representing environmental factors, in an exploratory pattern analysis and hypothesis generating mode, rather than in a confirmatory hypothesis testing mode (Brown 1995). Regression trees using the CART methodology (Breiman et al. 1984) are a powerful method for this exploratory analysis. Regression trees can reveal hierarchical spatial structure in the relationship between the explanatory variables and the response, therefore assisting in formulation of hypotheses about mechanisms of control of the distribution of the response at multiple spatial scales. In regression tree development, the midpoints between all values of all of the predictor variables that are present in the data form the possible splits for the tree. In the first step, sums of squares of differences between the observation response values and their means are computed for all binary divisions of the observations formed by all of the splits. The minimum sum determines the split. The observations are then divided into two sets based on the split and the process recursively repeats on the two descendent sets. Splitting continues until a stopping criterion is reached. We used the cross-validation pruning techniques of Breiman et al. (1984), as implemented by Clark and Pregibon (1992), to determine the optimal size of trees.

The second question can be rephrased as which places in the study area jointly contain the greatest number of species? This is different than the question, which places individually contain the greatest number of species? The answer to the latter is simply those places with the highest total numbers. The most species-rich places will likely have a high overlap in their species lists and may also be concentrated in one part of the study area so that policy targeted there would ignore other less rich but important parts. Places with the greatest joint species richness tend to be located in different parts of a study area, reflecting the contributions of complementary faunas or floras. The complementarity question can be posed with a limit on the number of places, in the manner of an optimization problem. For example, what is the greatest number of species that can be found in any four places? A related question is what is the least number of places to jointly contain all species?

Using optimization methodology, sets of varying numbers of places can be determined as possible answers to questions about important places (Church et al. 1996). We used integer programming optimization techniques (Csuti et al. 1997) to obtain solutions for our case studies. For Oregon the problem size was computing joint species richness for all combinations of 441 hexagons taken 8 at a time (selecting the best eight places), a total of approximately 3.3×10^{16} computations. In Pennsylvania the problem size was all combinations of 211 hexagons taken 6 at a time, a total of 1.1×10^{11} computations. (The number of species is not an important factor in the computational complexity.)

When multiple combinations of places provide optimal coverage of species, this methodology identifies those places that are singularly valuable or irreplaceable because they occur in all combinations, and those places that are optimal but also interchangeable with other optimal places, offering options for conservation activities (Pressey et al. 1993).

Giving all species in an indicator group, such as vertebrates, equal weight in a complementarity analysis invokes the fewest additional assumptions. On the other hand, methods for solving the complementarity problem can be adapted to use species weights based on phylogenetics, ecological function, or conservation ranks (see Chapter 9). If species have different levels of importance because of different roles in different places, then spatially varying weights could be used.

At the level below the analyses described above, assessment focuses on the question of what are the consequences of possible land use changes at the places identified as important at the higher level, larger area, through the complementarity analysis? Given several places of importance, what should be done about them? One answer is to study the impacts of possible change on the biodiversity of these places. This type of study constitutes a change in scale, now focusing on a local (place or places) scale of concern. Biological data are combined with land use and habitat maps for the existing or current conditions and for one or more alternative scenarios about how the region might change. The alternative scenarios are created to represent a range of possible changes in the amount and spatial distribution of land use and habitat (Harms et al. 1993). With these data, risk statistics can be calculated for various measures of biodiversity, showing the proportion of habitat gained or lost in each alternative scenario relative to the current conditions. This approach can also be used to study change between the current conditions and a reconstruction of past conditions.

Spatial Accounting Units

Different spatial frameworks are appropriate for different levels of analysis. For the larger area studies of environmental associations and of complementarity of biodiversity distributions, a structure that provides comparability is most appropriate (see Conroy and Noon 1996 on issues of using habitat patches). We have used a spatial framework that was designed to provide a regular, systematic, hierarchical spatial structure for environmental monitoring and assessment (White et al. 1992). The basic structure of this framework is a tessellation, or grid, of hexagons of approximately 640 square kilometers in size, with a point-to-point (center-to-center) spacing of approximately 27 kilometers. In the eastern part of the conterminous United States, from about the 103rd meridian eastward, where counties are of a moderately uniform size, this density corresponds, on the average, to about two and two-thirds hexagons per county.

The hexagon sampling cells provide an accounting mechanism that serves several purposes. First, a single set of analysis units facilitates comparison of different data sets. Second, some of the uncertainty in species occurrence data obtained from range maps can be accounted for by limiting the precision of location assignment to this grid. But in addition, concerns about the confidentiality of precise locations of occurrence for certain rare species may be alleviated by generalizing the location assignment to the grid. Finally, there is a theoretical argument for generalizing species distributions from the precise data of field observations in order to account for the biases in observation locations and the presumed broader distributions over time.

It is an advantage to use equal area accounting units, other things being the same. The equal area grid provides a common spatial unit for comparison of diverse data types whereas ecoregions, for example, are not comparable, rather they are by definition unique. Equal area units also minimize confounding due to species-area relationships, a potential problem if units such as counties in the United States are used (for example, Dobson et al. 1997). A hexagon tessellation minimizes spatial distortion and provides an equal area sample (if constructed on an equal area map projection). Furthermore, hexagons are generalizable to both larger and smaller spatial scales. This becomes important for extending assessments to continents and the globe.

The size of the accounting units reflects a compromise between the desire for spatial detail, on the one hand, and the constraints of reasonable spatial representation of species life histories, of data collection, of confidentially, and of computational feasibility, on the other hand. Solutions to statistical analyses of associations and to complementarity analyses can depend, of course, on the size selected (Stoms 1994).

For the assessment of impacts of alternative futures in a smaller area, a more appropriate spatial structure is the ecological units that comprise the study area, usually patches of habitat. Patches may be defined as polygons or aggregates of remote sensing pixels, depending on the source of data.

Biodiversity Response Variables

In choosing how to represent biodiversity in analyses, one principle is that it is preferable to base the response variable on well-defined concepts. In this view, genes and species should form the basis for the mapping and monitoring. Because of the practical impossibility of using the gene level we are led to using species. Although there is considerable controversy about the details of the theory of species (biological species concepts versus evolutionary species concepts, for example, see Rojas 1992; Bush 1993), in our applications it is clear in most cases how to decide which species to consider. For a contemporary assessment, where we have a single slice through time, the biological and evolutionary species concepts largely overlap and species are considered to be more or less independent collections of genes (among other things). Hence species have their own identity and are good surrogates for genetic diversity. So our metric of choice to quantify biodiversity for analysis is species richness.

Our objective was to describe the distribution of biological species across the conterminous United States. We have chosen to work with The Nature Conservancy and its cooperating network of state Natural Heritage Programs to begin developing the first comprehensive nationwide database that includes standard range information from published literature and expert sources plus specific location data on plants and animals of conservation concern that has been assembled by TNC (Master 1996). These data sets include all vertebrate species, butterfly and skipper species, tree species, and freshwater mussel species for hexagons covering the states of Washington, Oregon, California, Pennsylvania, Maryland, Delaware, West Virginia, and Virginia. For each species in each hexagon of states initially sampled, the following information was recorded: the occurrence status (confidently assumed or known, probable, possible, or not present); the origin of the species in the hexagon (native, introduced, reintroduced, or unknown); the best source of information for the occurrence information; and residency (year round/seasonal) and breeding (confirmed, probable, nonbreeder) status for migrant species (e.g., birds, bats). The occurrence status is more precisely defined as "confident or certain" (>95% chance of occurrence), "predicted or probable" (80%–95% chance of occurrence), "possible" (10%–80% chance of occurrence; this category may not be used), and "not significant" (<10% chance of occurrence). Taxonomic experts in each state extensively reviewed the assignments of species to hexagons.

Explanatory Variables

Among the processes that have been hypothesized to account for spatial patterns of species diversity are climatic extremes, climatic stability, productivity, and habitat heterogeneity (Brown and Gibson 1983; Wickham et al. 1997). Data sets were assembled from existing sources to represent these processes.

Data were compiled for topographic elevation, January and July temperatures, and annual precipitation on a rectangular grid at a resolution of one kilometer for the conterminous United States. The elevation data were derived from a 15 arc second digital elevation model obtained from Sue Jensen at the USGS EROS Data Center, by projecting and resampling to the coarser resolution. January and July mean temperature data were modeled and compiled using the method of Marks (1990). The initial data values were the means, over the 40-year period from approximately 1948 to 1988, of the means, over the respective month, of the daily mean temperatures at approximately 1,200 stations in the Historical Climate Network database. These values were first corrected to potential temperatures at a reference air pressure of 1,000 mb using the station elevations and assuming a normal adiabatic lapse rate. The potential temperatures were then interpolated to the 1-km grid using a linear model. Finally, the interpolated values were then converted to estimated actual temperatures from the adiabatic lapse rate correction using the corresponding elevation values at each grid point. Annual precipitation data were compiled from the 10-km resolution dataset prepared by Daly et al. (1994). These authors used a locally adaptive regression model to estimate annual precipitation values for unknown locations from known stations and from the elevation structure in the local region. The 10-km data were interpolated to 1 km using a linear model.

Data for stream density were developed from the USGS/EPA River Reach File, version 3, corresponding to, and derived from, in part, the USGS 1:100,000 Digital Line Graphs for hydrography. The USGS AVHRR land classification (Loveland et al. 1991) and the USGS Gap Vegetation Classification Map for Oregon (Kagan and Caicco 1992) provided representations of land cover and vegetation heterogeneity. The AVHRR data also provided a measure that estimates net primary productivity, the Normalized Difference Vegetation Index (NDVI) (Reed et al. 1994). Data for road density and human population density were developed by Wickham et al. (1997).

All data not collected by hexagon cells were aggregated or summarized by cell with several statistics, including the mean, median, minimum, maximum, range, and standard deviation. The AVHRR and Gap Vegetation data were converted to class richness values by cell. Slope statistics were calculated from the elevation values in each cell.

Results of Analyses

We will describe examples of analyses at two scales. The regional analyses of environmental associations and complementarity of biodiversity distributions were done in the states of Oregon and Pennsylvania (Fig. 8.1). Based partly on these analyses, landscape level analyses were conducted within the two states, one in a county of Pennsylvania, and the other in a small watershed in Oregon.

Environmental Associations

This analysis investigated the relationship between bird species richness in Oregon and climatic, topographic, hydrographic, land cover, and anthropogenic variables. Richness values were the sum of native summer resident breeding bird species in each hexagon, from a total of 252 species for the state as whole. Regression tree analysis was used to predict the number of bird species by hexagon grid cell across the state.

The final tree had 6 leaves, or terminal nodes, and used 4 of 19 possible predictor variables to explain 73% of the variation in the response variable (Fig. 8.2). To interpret the tree, a map (Fig. 8.3) of the cases belonging to each leaf is very helpful (O'Connor et al. 1996; White and Sifneos 1997). The most important predictor variable for the data for the state as a whole was minimum elevation which separated a lower richness area in most of western Oregon and the Columbia Plateau from the rest of the state. This split confirmed the strong east–west division in Oregon formed by the

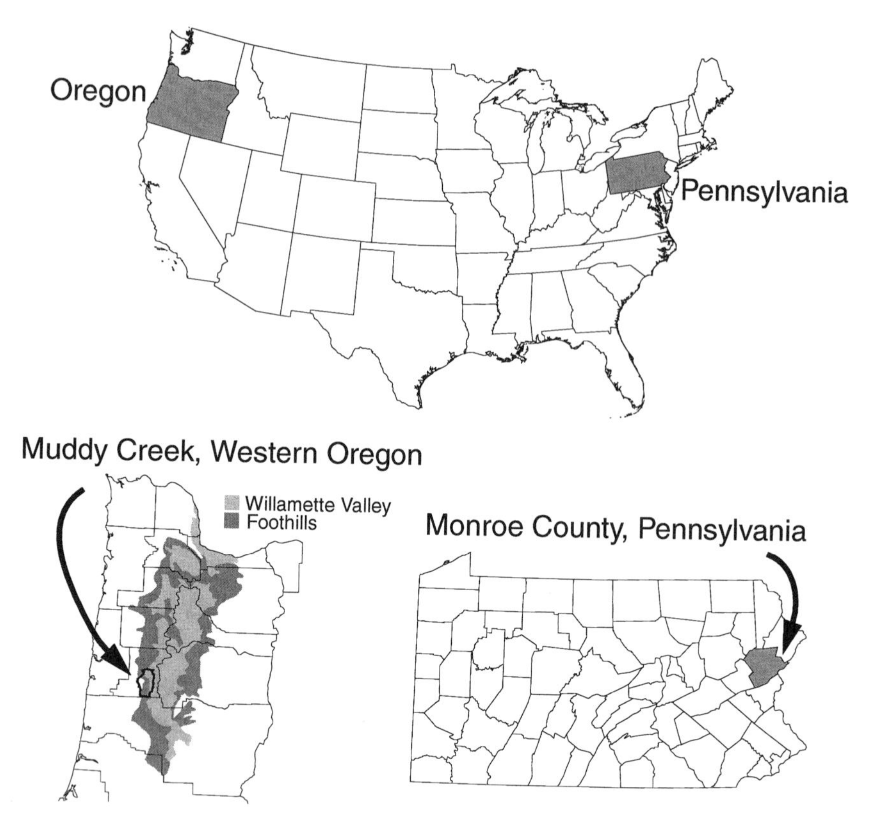

FIGURE 8.1. Locations of study areas for biodiversity analyses are Oregon and Pennsylvania. One landscape level study is set in Monroe County, Pennsylvania and the second in the Muddy Creek watershed in western Oregon.

Cascades Mountains. Lower species richness in the west was not expected, however. The second split was based on AVHRR-derived total NDVI and separated the drier, less-productive parts of eastern Oregon having lower species richness from the remainder of the hexagons that were more forested and had greater species richness. The drier part of the eastside was further split by the number of Gap vegetation classes into a larger group of hexagons that was less rich both in vegetation and in bird species, and into the remaining hexagons associated with playas, permanent lakes, or higher elevation mountain ranges, hexagons that had a higher average species richness. The other two splits in the tree were based on annual precipitation. The first of these separated higher precipitation, higher elevation hexagons in the Cascades and other high mountains having lower species richness from the remainder. Hexagons remaining from this split were separated by the final split into dry, less rich areas at the margins of the forested part of the eastside, and the core of the forested areas of the eastside that, among all the groups identified, had the highest average value species richness.

How does this prediction geography in Oregon correspond to knowledge of patterns in other areas? At a global scale, higher species richness is generally associated with areas that are: (1) warmer rather than colder, (2) wetter rather than drier, (3) less seasonal rather than more, (4) more varied in topography and climate rather than less, and (5) larger rather than smaller (Caldecott et al. 1996, but also see Scheiner and Rey-Benayas 1994). The last condition does not apply because equal area accounting units were used. Three of the other global patterns were contradicted in the study. Western Oregon is distinguished from eastern Oregon by having greater annual precipitation, less seasonality, and slightly greater mean annual temperature, yet bird species richness was lower in western Oregon than in large areas in eastern Oregon. Only association (4) may

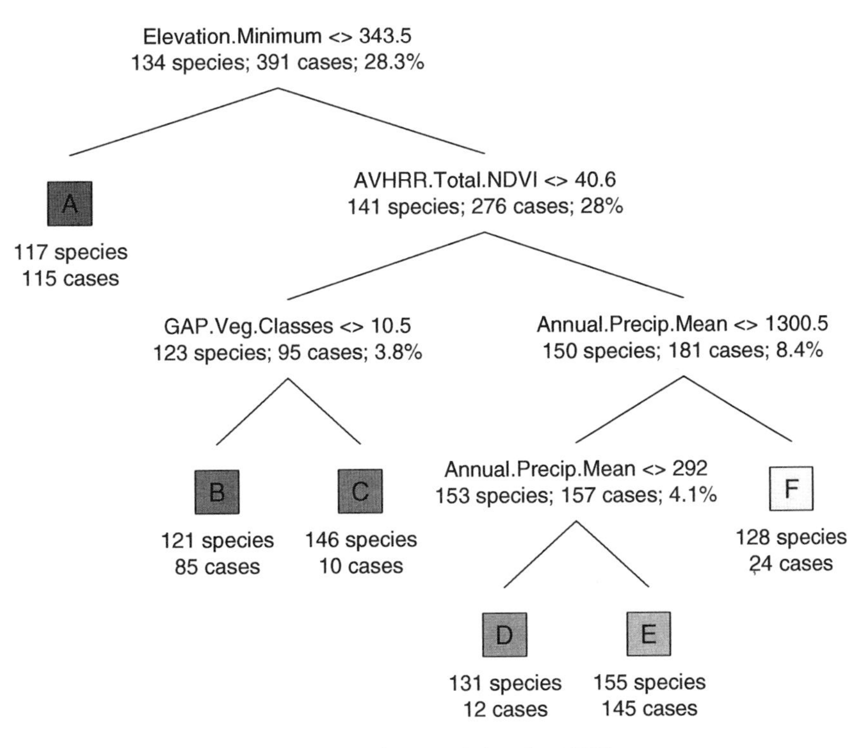

FIGURE 8.2. The regression tree for bird species richness has five splits using four different predictor variables resulting in six leaves. Each split indicates the splitting value of the splitting variable, the number of cases considered at the split, the mean value of the response variable for those cases, and the amount of variation explained by the split.

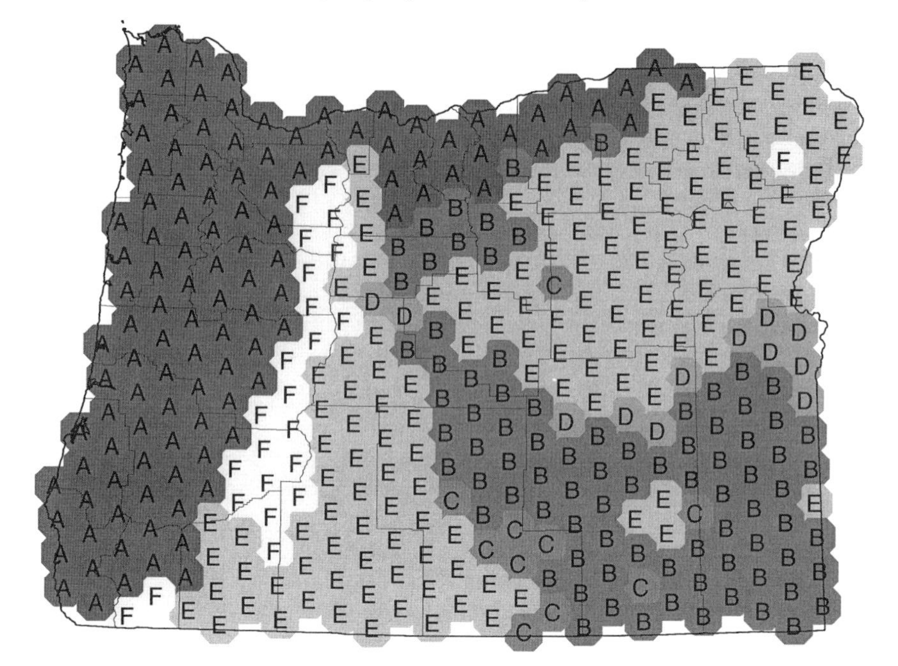

FIGURE 8.3. The map of the cases belonging to each leaf has a highly structured geography. The boundary between western and eastern Oregon is usually considered to be the crest of the Cascade Mountains. This boundary appears on the map, for the most part, as the boundary between "F" and "E" hexagon cells trending primarily north–south about one third of the distance from west to east across the state.

have been partially demonstrated by the split on Gap vegetation class richness.

So how is the prediction pattern to be explained? Leaves "A" and "E" covered much of the forested area in Oregon, but "E" hexagons had a mean response that was 38 species greater than "A" hexagons. One hypothesis is that, historically, conifer forest cover in western Oregon was so continuous and homogeneous in flora and structure that bird habitat was limited compared to the more open and varied habitat in eastern Oregon forests. The eastern Oregon areas represented by leaf "A"in the Columbia Plateau have an environment more like that of leaf "B" hexagons, and, in this study, had a similar richness level. Other splits in the tree suggest other mechanisms. The split on AVHRR Total NDVI is consistent with studies showing that species richness is positively correlated with higher available energy (e.g., Wright et al. 1993), indicated in this case by higher net primary productivity which is correlated with higher total NDVI (Reed et al. 1994). The first split on precipitation separated high precipitation, higher elevation hexagons much of whose precipitation is in the form of snow; their lower richness values compared to the other half of the split may correspond not to absolute precipitation but to cooler temperatures and reduced winter habitat due to snow cover. The second split on precipitation separated very dry hexagons with lower richness values from the large group of high richness value hexagons. This split was consistent with theory.

Three anthropogenic variables were included in this study, human population density, road density, and number of introduced species, but none entered into the model. Reasons for the lack of association with indicators of human disturbance may be that the scales of disturbance do not coincide with the scale of study (for example, disturbances at a smaller grain than 640 km^2), that Oregon is not affected by such disturbance, or that Oregon is not yet affected by such disturbance. In a related study (Rathert et al., in press), richness in introduced fish species was positively associated with richness in native fish species. This finding could correspond to the hypothesis that in the initial stages of intensive human occupation (as in the past 150 years or so in the western United States) humans are positively associated with biodiversity since humans are attracted to the same places as many other vertebrate spaces. Only after humans come to dominate a landscape for some decades or perhaps centuries in duration, does their disturbance eventually reduce native species presence by a significant amount. If, during the initial period of human occupancy, species diversity is increased by increased habitat heterogeneity, or by introduced species such as in the fish example, then it may be reasonable to think of a regional analogue of the intermediate disturbance hypothesis (Connell 1978). Effects like these on diversity may sometimes be captured by statistical modeling techniques (Wickham et al. 1997).

Complementarity of Species Distributions

Studies of complementarity used species lists in cells of the hexagon grid in Oregon and Pennsylvania. In Oregon the study included all native vertebrate species, divided into two groups: fish (67 native species) and all others (457 native amphibian, reptile, bird, and mammal species, hereafter called "terrestrial,"recognizing that a number of species could also be considered aquatic, e.g., frogs). In Pennsylvania 323 native terrestrial vertebrate species (same definitions as in Oregon) were studied.

In the analysis for Oregon, the study computed complementarity for eight places for both fish and terrestrial vertebrates. In 8 places, about 96% of the terrestrial vertebrates were included but only about 76% of the fish species. The locations of the places that comprised the sets of eight places were different for fish than for vertebrates and were positioned in different parts of the state (Figs. 8.4 and 8.5). In complementarity analyses, there is often more than one optimal or exact solution for a given number of places.

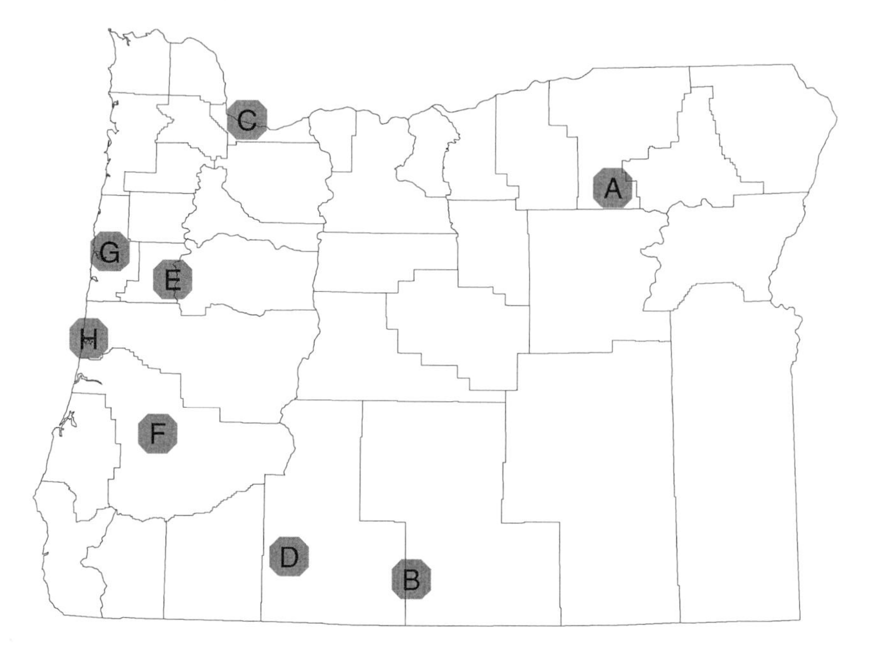

FIGURE 8.4. One combination of eight hexagons covers (i.e., contains in the joint species list) 76% of the native fish species in Oregon. There is no necessary priority to the eight hexagons; collectively their complement of species is the greatest for any combination of eight hexagons. Of course, some hexagons have more unique species than others.

Thus the five hexagons labeled "H" and the two labeled "G" in Figure 8.5 made similar, though not necessarily identical, contributions to the joint species richness. Each combination that has the highest joint richness included only one of the two "G" hexagons and one of the five "H" hexagons. When the solutions for the sequence of one place, two places, and following are examined, the pattern often resembles a recursive partitioning of the study area (see Kiester et al. 1996). These optimal coverage solutions had a quite different geography than the sets of the eight highest individually rich hexagons. For terrestrial vertebrates, for example, the eight richest hexagons were all in the south central and southwestern part of the state (Fig. 8.6). Also, the total coverage for the eight richest hexagons was substantially less than the optimal coverage: 72% for vertebrates and 31% for fish.

Because the optimal solutions for terrestrial vertebrates are located in different places than those for fish, it is fair to ask how well do the solutions for one set of species cover the other. For a single group, the percentage of species covered by the sequence of solutions from one place to the number of places required to totally contain all species increases steeply in the initial steps but levels off as the total number of species is approached. The first step accounts for the most species, and each additional step captures successively fewer. Now, in each step of the solution for one group of species, for example, terrestrial vertebrates, we can compute how many species of the other group are covered in the hexagons that comprise the solution, and vice-versa (Figs. 8.7 and 8.8). These "sweep" analyses (Kiester et al. 1996) tell different stories. Terrestrial vertebrates are completely covered in 20 hexagons, but a mean of only about 50% of the fish species are covered in the set of solutions for complete terrestrial coverage (Fig. 8.7). Conversely, in the set of solutions that completely cover all fish species, also requiring 20 hexagons, coincidentally, a mean of about 93% of the terrestrial vertebrate species are covered (Fig. 8.8). However, the number of species not covered is about the same in both cases: 34 terrestrial vertebrate species not covered by fish, and 33 fish species not covered by terrestrial vertebrates.

In Pennsylvania, the solution for six places for terrestrial vertebrates spreads the solution sites across the state much as in Oregon (Fig. 8.9). In this case, there are also multiple combinations, using one of the two "E" hexagons and one of the four "F" hexagons. (All combinations of the

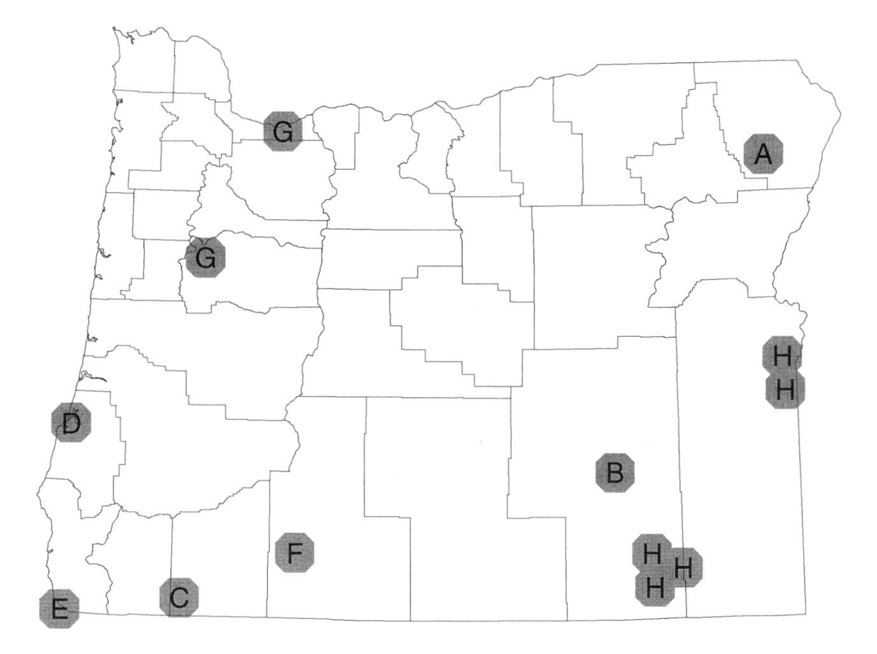

FIGURE 8.5. There are multiple combinations of eight hexagons that cover 96% of the native terrestrial vertebrate species in Oregon. Hexagons with the same letter comprise a group from which only one participates in any of the optimal combinations. Often, but not always, members of such groups are located in proximity.

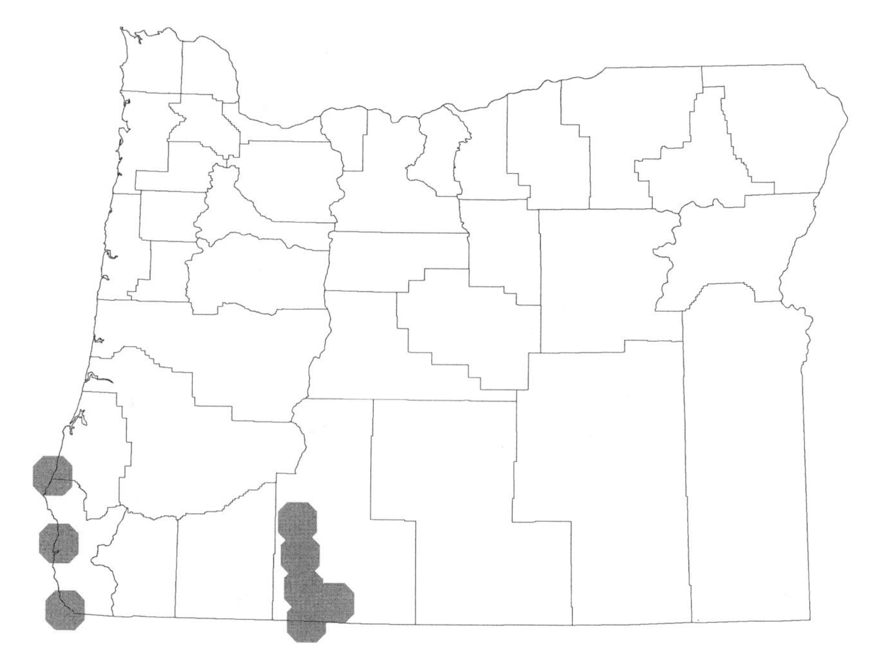

FIGURE 8.6. The eight hexagons with the greatest richness values for native terrestrial vertebrate species in Oregon are concentrated in the Klamath Falls area and along the south coast. The maximum value of richness is 275 species.

identified hexagons may not be maximal solutions; in the Pennsylvania example, there are eight possible combinations of six hexagons, taking one from each letter group, but only six of these contain the maximum number of species.)

Because the size of the accounting units in the complementarity studies are a compromise, some species will be better represented through this process than others. The solutions are not absolute spatial locations for conservation activities, but initial approximate representations of complementary biodiversity assemblages. Thus, it is important to ask what happens next after producing these solutions. We will focus on two studies at a finer scale whose locations in Oregon and Pennsylvania were partly identified through complementarity analyses at the larger extents. The target areas are suggested by the "E" hexagon in northeastern Pennsylvania (Fig. 8.9) and by the "E" hexagon of the fish solution in west central Oregon (Fig. 8.4).

Assessment of Alternative Future Landscapes

Monroe County is located in the Poconos region of Pennsylvania (Fig. 8.1). This region has been a vacation and second home destination since the nineteenth century and has recently begun to increase more rapidly in human population through suburban development. Projections to the year 2020 suggest that population in the county may double. These changes may threaten some of the natural values of the region, including its contribution to state level biodiversity as indicated in Figure 8.9. With respect to these projected changes, the impacts of possible future land development patterns on biodiversity were investigated (White et al. 1997).

Land cover data for this study included a remote sensing derived map of the current habitat in the county developed initially at Cornell University (Smith and Richmond 1994) and refined at Harvard University (Steinitz et al. 1994). The Harvard group also developed six maps of future habitat distributions resulting from different land development scenarios envisioned in consultation with stakeholders within the county. Biodiversity was represented as lists of 147 bird, 44 mammal, 20 reptile, and 20 amphibian species, all considered to be native to the study area, and the habitat associations for these species (obtained from Smith and Richmond 1994). In addition, White et al. (1997) estimated breeding area requirements for

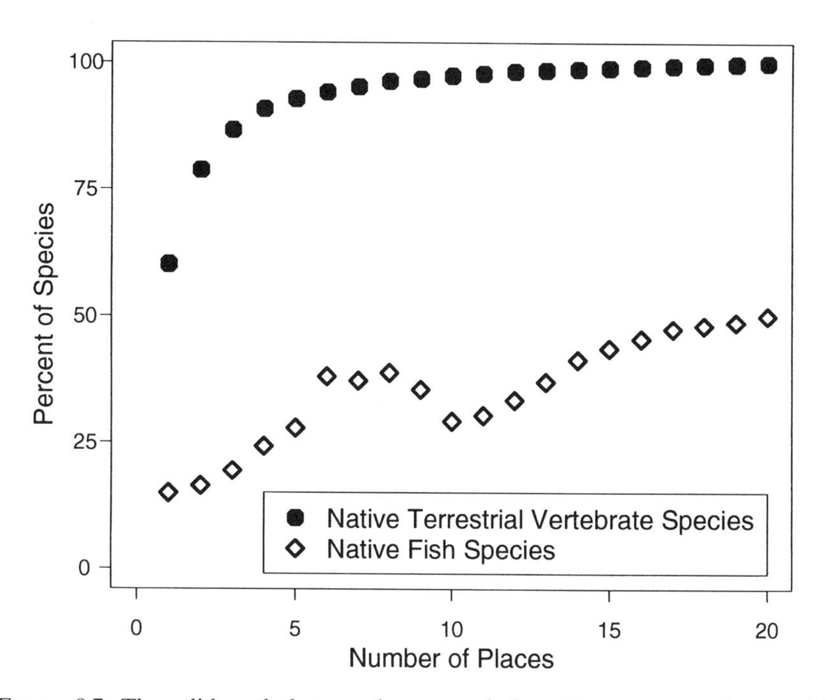

FIGURE 8.7. The solid symbols trace the accumulation of percentages of terrestrial vertebrate species covered in successive optimal solutions for joint species richness. With one hexagon approximately 60% of the species are covered; with two approximately 79%, and so on to 100%. In the hexagons that comprise these solutions, the corresponding percentages of fish species covered ("swept" along and symbolized in the open symbols) are 15%, 16%, and so on, up to a maximum of 50%.

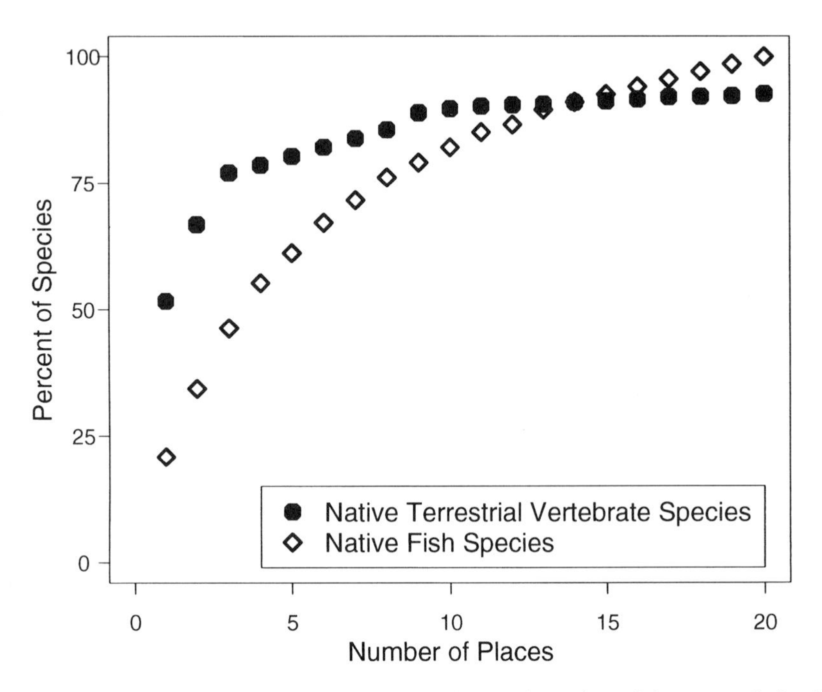

FIGURE 8.8. The counterpart to Figure 8.7 shows that when fish are optimized, coverage starts at about 21%, then 34%, and so on to 100%. The corresponding percentages of vertebrates swept along are about 52%, then 67%, and so on up to a maximum of 93%.

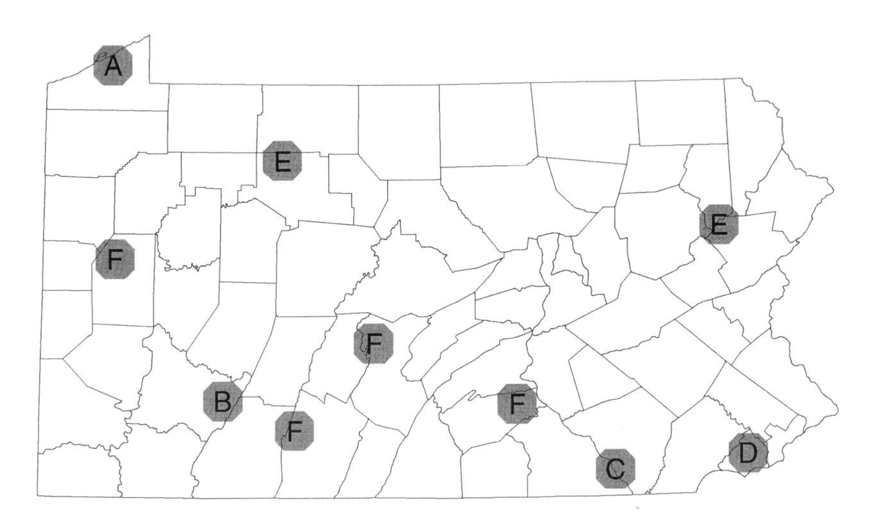

FIGURE 8.9. The optimal set of solutions for six hexagons in Pennsylvania consists of six solutions of six hexagons having the maximum number, about 95%, of the native terrestrial vertebrate species.

each species using home ranges, sampled population densities, or genetic area requirements that incorporated dispersal distances.

In this study, measures of biodiversity were species richness and indices of population abundance based on availability of suitable habitat. Indices of population abundance for each species were calculated in two ways. First, total habitat area was computed for each species in each landscape. Second, the number of breeding habitat units for each species in each landscape was calculated by dividing the size of each habitat patch in the landscape by the area requirement and summing over all patches. Species richness was based on presence of habitat so that species became locally extinct in the landscape only if they had no habitat area or no habitat units. For each species, ratios of abundance in each future scenario to abundance in the present were computed. The ratio of future to present species richness was also computed. Summary statistics were calculated across all species and subtracted from one to obtain a measure of risk.

Species richness changed little from present to future. However, there were distinctly greater risks to habitat area in scenarios that extrapolated from present trends or zoning patterns (Plan Trend and Build Out) as opposed to scenarios in which land development activities were designed to follow more constrained patterns (Township, Spine, Southern, Park). All taxonomic groups followed similar trajectories; amphibians had the greatest risk across all scenarios (Fig. 8.10). These results were similar for both indices of population abundance. Sensitivity analyses indicated that the results were robust to errors in the estimates of area requirements. Studies in progress refine the initial approach to include the use of habitat quality metrics in the species-habitat association matrix and a more restrictive definition of suitable habitat in relation to area sensitivity and interior/edge habitat preferences of some forest bird species.

The other study took place in the Muddy Creek watershed in western Oregon (Fig. 8.1). This 320 km^2 watershed includes commercial forest land in mid and high elevations, Christmas trees, vineyards, orchards, pasture, and mixed woodlands in midelevations, and primarily grass seed agriculture plus a wildlife refuge in low elevations. Current human settlement consists of about 1,200 households, located in one village in the lowlands and scattered rural residences in the low and mid-elevations. Anticipated growth to the year 2025 is between one third and two thirds of current population. (More complete descriptions of all aspects of this project are [in 1998] at the world wide web site http://ise.uoregon.edu).

For this study, University of Oregon researchers assembled a map of current conditions from several sources of remote sensing and from field work, as well as maps of other physical features, state land use zones, and land ownership. They then worked closely with stakeholders in the watershed to formulate a sequence of possible future landscape scenarios, arranged in a gradient from a high development oriented scenario to a high conservation oriented scenario. The midpoint in the gradient was

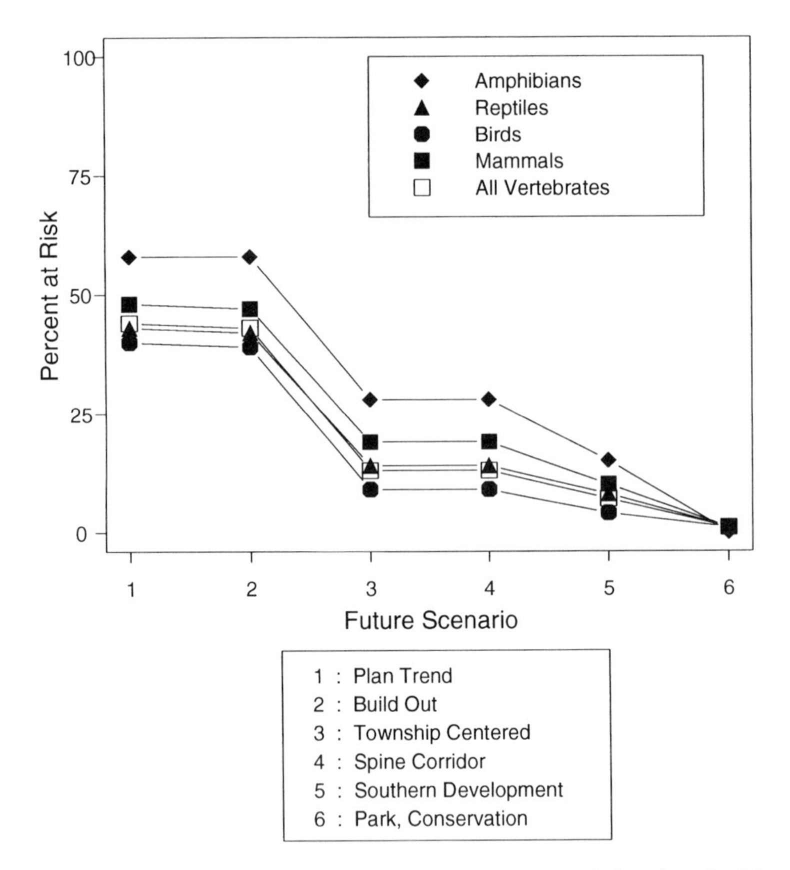

FIGURE 8.10. The percentage of habitat area at risk of being lost in Monroe County relative to the present conditions varies by taxonomic group and by future scenario.

considered the most likely scenario and labeled Plan Trend. The scenarios incorporated projected human population growth ranging from 10% to 100%. In addition to these future scenarios, the project acquired a map of presettlement vegetation for the watershed that was interpreted and interpolated by the Oregon Natural Heritage Program from General Land Office surveyors' notes (Christy et al. 1996–1997).

In consultation with local experts, Freemark et al. (1996) compiled lists of historical and current breeding species for the watershed (including 135 bird, 71 mammal, 16 reptile, and 14 amphibian species), and a species-habitat association matrix. Of the 236 species, 1 amphibian, 3 bird, and 4 mammal species had been permanently extirpated from the watershed since the time of European settlement; 8 bird and 2 mammal species native to the watershed were deemed rare (including currently extirpated); 1

amphibian, 1 reptile, 6 bird, and 6 mammal species were introduced. Using the methodology of the Monroe County study, risks for each species were calculated from habitat area in the future (or pre-settlement past) compared to the present, for various groups of species, subset by taxonomy, conservation status, and ecological function.

For all native species groups except reptiles, risk was greatest in the high development scenario and declined across the gradient of future scenarios (Fig. 8.11). Values for nonreptile species for the two conservation scenarios and for the presettlement scenario were negative, indicating improvement over the present. The trend across future scenarios was similar for all taxa except reptiles; amphibians had the greatest change, from risk in high

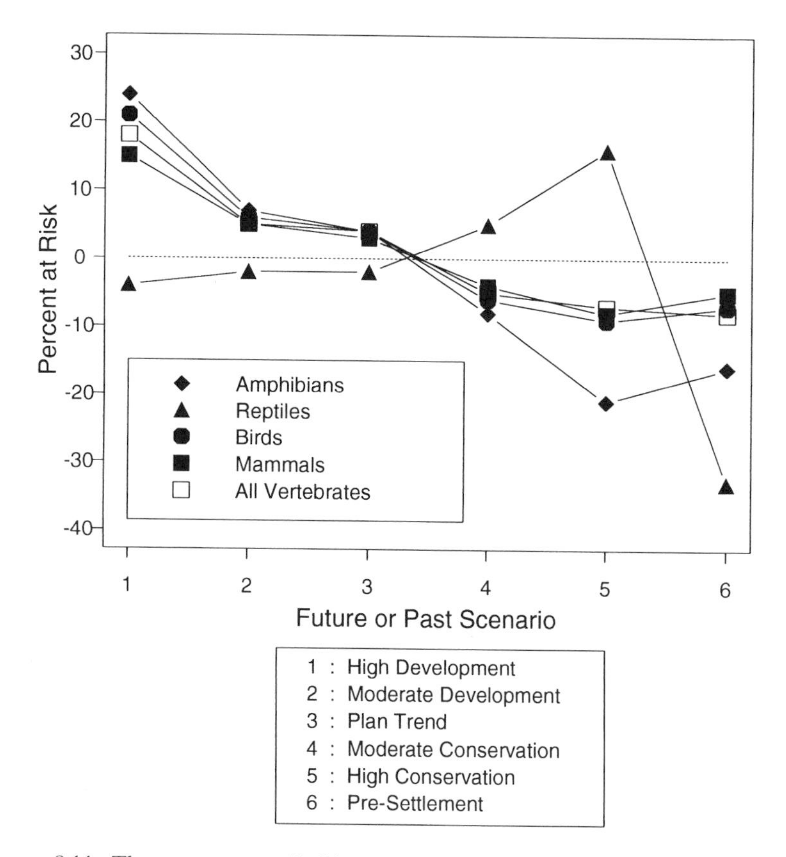

FIGURE 8.11. The percentage of habitat area at risk of being lost in the Muddy Creek watershed also varies by taxonomic group and by future scenario. All groups show a loss, indicated as a negative risk, between the reconstructed past landscape pattern and the present. Reptiles in Muddy Creek respond oppositely to the other taxonomic groups.

development to improvement in high conservation. Reptiles had the greatest loss from the presettlement to the present. The contrary response of reptiles is due primarily to their preferential assignment to nonforested, open habitat types. These habitat types were less abundant in the conservation scenarios than in other scenarios and much more abundant in the presettlement scenario.

These two studies demonstrate methods to discriminate the effects of potential changes in land development on biodiversity and thereby help inform the decision-making process. It is important to note that the modeling methodology begs the question of the viability of populations for any particular species. Abundant species with small ranges, or small area requirements for breeding, may be assessed much better than rare species or those with wide ranges or large area requirements. To look at all species adequately requires a hierarchical scope of study ranging from local to global. To look at the viability of any individual species in detail requires demographic modeling (Dunning et al. 1995).

Summary

The hierarchical framework presented suggests that understanding the distribution of species over large areas and then selecting important subareas for conservation actions can be usefully followed by looking at the consequences of possible landscape changes in those important subareas. Consequences at the landscape level can help to inform policy decisions over the larger area by providing additional information on risks for specific subareas. When a complementarity analysis over the larger area includes options of multiple subareas, understanding specific consequences can help to further prioritize where to initiate actions.

We believe the hierarchical framework makes several contributions to biodiversity conservation. First, it helps improve knowledge regarding the importance of different areas and environmental factors in contributing to the biodiversity of species, habitats, and ecosystems at different spatial scales. Many investigators have studied the distribution and possible causes of biodiversity. Our work has had a specific focus of understanding the hierarchical structure of prediction and the geography of explanatory relationships. Within Oregon, for example, the relationship between bird species distributions and environmental factors appears to depend more on regional history and mechanisms than on global patterns. An important research direction is the identification of the possible effects of human activities over a national or continental extent (O'Connor et al. 1996).

Second, the framework helps to identify species and regions that are poorly represented by current conservation activities and that may benefit from applying integrated planning for biodiversity conservation. This aspect of our work is very similar to goals and methods of the USGS Biological Resources Division's Gap Analysis Program (Scott et al. 1993; Csuti and Kiester 1996). Complementarity analyses explicitly show important areas of biodiversity. Combining this with knowledge of existing protection areas reveals gaps in coverage and targets for conservation planning.

Third, the focus of biodiversity conservation is extended beyond rare, threatened or endangered species or ecosystems to more comprehensive sets of species. This methodology has included comparing the effectiveness of determining conservation priorities by one taxonomic group versus another. For states for which there are vegetation maps, similar analyses can be conducted to determine how conservation priorities based on habitat coverage compare to animal or plant species priorities. In this way the approach contributes to integrating species-based and ecosystem-based assessments.

Fourth, the kinds of assessments are expanded to include the evaluation of alternative conservation strategies through collaboration with landscape planners (Rookwood 1995). These alternative future scenario projects provide a reasonable cost method for considering future impacts of human activities on biodiversity. The concerns of local stakeholders can be incorporated into the future visions and a range of future options can be considered, including restoration to earlier more pristine conditions or development to total urbanization, as appropriate. In addition to synthesized designs for alternative future scenarios, more formal allocation rule systems offer a more objective and controlled method for generating future possibilities (Bettinger et al. 1996).

Additional research projects associated with the work reported here seek to improve knowledge of the economic feasibility and sociopolitical acceptability of alternative conservation strategies (protection, mitigation, restoration). Another project is combining complementarity analyses of species distributions with economic constraints to address the question of where are the best places to conserve biodiversity under a limited budget (see Chapter 9).

Finally, these ideas provide a conceptual and spatial framework for decentralizing resource management decision making to more local levels, while maintaining the larger spatial perspectives necessary for sustainable resource use. This hierarchical perspective and framework for science, policy, and management responds, we believe, to the challenge of developing more comprehensive strategies for conservation of biodiversity.

Acknowledgments. We thank Blair Csuti, Larry Master, Raymond O'Connor, Tom Loveland, Frank Davis, David Stoms, Bruce Jones, Jim Wickham, Carolyn Hunsaker, Kevin Sahr, Manuela Huso, and Jean Sifneos, collaborators in the Biodiversity Research Consortium (BRC), for sharing ideas, methods, and results. The work in this chapter has represented several major themes of the research agenda of the BRC, a group of research scientists in U.S. federal agencies, universities, and nongovernmental organizations. A Memorandum of Understanding about this research was originally signed by the U.S. Environmental Protection Agency, the USDA Forest Service, the U.S. Geological Survey, the U.S. Fish and Wildlife Service (the part of which is now in the USGS Biological Resources Division), and the Nature Conservancy. Specific work reported in the chapter was funded, in part, through cooperative agreement CR821795 between US EPA and Environment Canada (KEF), cooperative agreement PNW 92-0283 between US Forest Service and OSU (DW), interagency agreement DW12935631 between US EPA and USFS (ARK, DW), and DOD SERDP Project #241-EPA (EMP, ARK, DW). This research has been officially reviewed by US EPA and approved for publication. The conclusions and opinions are solely those of the authors and are not necessarily the views of the agency. We also thank two anonymous reviewers for helpful comments.

References

Allen, T.F.H., and T.W. Hoekstra. 1992. Toward a Unified Ecology. Columbia University Press, New York.

Bettinger, P., K.N. Johnson, and J. Sessions. 1996. Forest planning in an Oregon case study: defining the problem and attempting to meet goals with a spatial-analysis technique. Environmental Management 20:565–577.

Breiman, L., J.H. Friedman, R.A. Olshen, and C.J. Stone. 1984. Classification and Regression Trees. Chapman & Hall, New York.

Brown, J.H. 1995. Macroecology. University of Chicago Press, Chicago.

Brown, J.H., and A.C. Gibson. 1983. Biogeography. C.V. Mosby, St. Louis.

Bush, G.L. 1993. A reaffirmation of Santa Rosalia, or why are there so many kinds of small animals? In: Evolutionary Patterns and Processes, pp. 229–249. D.R. Lees and D. Edwards (eds.). Academic Press, London.

Caldecott, J.O., M.D. Jenkins, T.H. Johnson, and B. Groombridge. 1996. Priorities for conserving global species richness and endemism. Biodiversity and Conservation 5:699–727.

Church, R.L., D.M. Stoms, and F.W. Davis. 1996. Reserve selection as a maximal covering location problem. Biological Conservation 76:105–112.

Christy, J.A., E.R. Alverson, M.P. Dougherty, and S.C. Kolar. 1996–1997. Presettlement vegetation of the Willamette Valley, Oregon. Version 1. Oregon Natural Heritage Program, The Nature Conservancy of Oregon.

Clark, L.A., and D. Pregibon. 1992. Tree-based models. In: Statistical Model, pp. 377–419. S.J.M. Chambers and T.J. Hastie (eds.). Wadsworth & Brooks, Pacific Grove, CA.

Connell, J.H. 1978. Diversity in tropical rain forests and coral reefs. Science 199:1302–1310.

Conroy, M.J., and B.R. Noon. 1996. Mapping of species richness for conservation of biological diversity: conceptual and methodological issues. Ecological Applications 6:763–773.

Csuti, B., and A.R. Kiester. 1996. Hierarchical Gap analysis for identifying priority areas for biodiversity. In: Gap Analysis: A Landscape Approach to Biodiversity Planning, pp. 25–37. J.M. Scott, T.H. Tear, and F.W. Davis (eds.). American Society for Photogrammetry and Remote Sensing, Bethesda, MD.

Csuti, B., S. Polasky, P.H. Williams, R.L. Pressey, J.D. Camm, M. Kershaw, A.R. Kiester, B. Downs, R. Hamilton, M. Huso, and K. Sahr. 1997. A comparison of reserve selection algorithms using data on terrestrial vertebrates in Oregon. Biological Conservation 80:83–97.

Daly, C., R.P. Neilson, and D.L. Phillips. 1994. A statistical–topographic model for mapping climatological precipitation over mountainous terrain. Journal of Applied Meteorology 33:140–158.

Davis, F.W., and D.M. Stoms. 1996. A spatial analytical hierarchy for Gap analysis. In: Gap Analysis: A Landscape Approach to Biodiversity Planning, pp. 15–24. J.M. Scott, T.H. Tear, and F.W. Davis (eds.). American Society for Photogrammetry and Remote Sensing, Bethesda, MD.

Dobson, A.P., J.P. Rodriguez, W.M. Roberts, and D.S. Wilcove. 1997. Geographic distribution of endangered species in the United States. Science 275:550–553.

Dunning, J.B., D.J. Stewart, B.J. Danielson, B.R. Noon, T.L. Root, R.H. Lamerson, and E.E. Stevens. 1995. Spatially explicit population models: current forms and future uses. Ecological Applications 5:3–11.

Faith, D.P., and P.A. Walker. 1996. How do indicator groups provide information about the relative biodiversity of different sets of areas?: on hotspots, complementarity and pattern-based approaches. Biodiversity Letters 3:18–25.

Freemark, K. 1995. Assessing effects of agriculture on terrestrial wildlife: developing a hierarchical approach for the US EPA. Landscape and Urban Planning 31:99–115.

Freemark, K.E., J.B. Dunning, S.F. Hejl, and J.R. Probst. 1995. A landscape ecology perspective for research, conservation and management. In: Ecology and Management of Neotropical Migratory Birds, pp. 381–427. T. Martin and D. Finch (eds.). Oxford University Press, New York.

Freemark, K.E., C. Hummon, D. White, and D. Hulse. 1996. Modeling risks to biodiversity in past, present, and future landscapes. Technical Report No. 268, Canadian Wildlife Service, Headquarters, Environment Canada, Ottawa K1A 0H3.

Harms, B., J.P. Knaapen, and J.G. Rademakers. 1993. Landscape planning for nature restoration: comparing regional scenarios. In: Landscape Ecology of a Stressed Environment, pp. 197–218. C.C. Vos and P. Opdam (eds.). Chapman & Hall, London.

Heywood, V.H., and R.T. Watson. 1995. Global Biodiversity Assessment. Cambridge University Press, New York.

Kagan, J., and S. Caicco. 1992. Oregon actual vegetation. Map compiled by B. Harmon, edited by B. Csuti. Idaho Cooperative Fish & Wildlife Research Unit, Moscow, ID.

Kiester, A.R., J.M. Scott, B. Csuti, R. Noss, B. Butterfield, K. Sahr, and D. White. 1996. Conservation prioritization using GAP data. Conservation Biology 10:1332–1342.

Loveland, T.R., J.W. Merchant, D.O. Ohlen, and J.F. Brown. 1991. Development of a land-cover characteristics database for the conterminous US. Photogrammetric Engineering and Remote Sensing 57:1453–1463.

Lubchenco, J. 1995. The role of science in formulating a biodiversity strategy. BioScience Supplement: 7–9.
Marks, D. 1990. The sensitivity of potential evapotranspiration to climate change over the continental United States. In: Biospheric Feedbacks to Climate Change: The Sensitivity of Regional Trace Gas Emissions, Evapotranspiration, and Energy Balance to Vegetation Redistribution, pp. IV-1 to IV-3. H. Gucinski, D. Marks, and D.P. Turner (eds.). EPA/600/3-90/078. U.S. Environmental Protection Agency.
Master, L. 1996. Predicting distributions for vertebrate species: some observations. In: Gap Analysis: A Landscape Approach to Biodiversity Planning, pp. 171–176. J.M. Scott, T.H. Tear, and F.W. Davis (eds.). American Society for Photogrammetry and Remote Sensing, Bethesda, MD.
McNeely, J.A., K.R. Miller, W.V. Reid, R.A. Mittermeier, and T.B. Werner. 1990. Conserving the world's biological diversity. International Union for the Conservation of Nature, Gland, Switzerland; the World Resources Institute, Conservation International, World Wide Fund for Nature, and the World Bank, Washington, DC.
O'Connor, R.J., M.T. Jones, D. White, C. Hunsaker, T. Loveland, B. Jones, and E. Preston. 1996. Spatial partitioning of environmental correlates of avian biodiversity in the conterminous United States. Biodiversity Letters 3:97–110.
Pickett, S.T.A., J. Kolasa, and C.G. Jones. 1994. Ecological Understanding. Academic Press, San Diego.
Pratt, J.R., and J. Cains, Jr. 1992. Ecological risks associated with the extinction of species. In: Predicting Ecosystem Risk, pp. 93–117. J. Cairns, Jr., B.R. Niederlehner, and D.R. Orvos (eds.). Princeton Scientific Publishing, Princeton, NJ.
Pressey, R.L., C.J. Humphries, C.R. Margules, R.I. Vane-Wright, and P.H. Williams. 1993. Beyond opportunism: key principles for systematic reserve selection. Trends in Ecology and Evolution 8:124–128.
Pulliam, H.R., and B. Babbitt. 1997. Science and the protection of endangered species. Science 275:499–500.
Rathert, D., D. White, J. Sifneos, and R.M. Hughes. Environmental associations of species richness in oregon freshwater fishes. Journal of Biogeography, in press.
Reed, B.C., J.F. Brown, D. VanderZee, T.R. Loveland, J.W. Merchant, and D.O. Ohlen. 1994. Measuring phenological variability from satellite. Journal of Vegetation Science 5:703–714.
Rojas, M. 1992. The species problem and conservation: what are we protecting? Conservation Biology 6:170–178.
Rookwood, P. 1995. Landscape planning for biodiversity. Landscape and Urban Planning 31:379–385.
Scheiner, S.M., and J.M. Rey-Benayas. 1994. Global patterns of plant diversity. Evolutionary Ecology 8:331–347.
Scott, J.M., F. Davis, B. Csuti, R. Noss, B. Butterfield, C. Groves, H. Anderson, S. Caicco, F. D'Erchia, T.C. Edwards, Jr., J. Ulliman, and R.G. Wright. 1993. Gap Analysis: a geographic approach to protection of biodiversity. Wildlife Monographs No. 123. Supplement, Journal of Wildlife Management 57.
Smith, C.R., and M.E. Richmond. 1994. Conservation of biodiversity at the county level: an application of Gap analysis methodologies in Monroe County, Pennsylvania. Report to the Environmental Services Division, Region 3, US EPA. New

York Cooperative Fish and Wildlife Research Unit. Department of Natural Resources, Cornell University, Ithaca, NY.

Steinitz, C., E. Bilde, J.S. Ellis, T. Johnson, Y.Y. Hung, E. Katz, P. Meijerink, A.W. Shearer, H.R. Smith, A. Sternberg, and D. Olson. 1994. Alternative futures for Monroe County, Pennsylvania. Unpublished report. Harvard University Graduate School of Design, Cambridge, MA.

Stoms, D. 1994. Scale dependence of species richness maps. Professional Geographer 46:346–358.

Turner, M.G. 1989. Landscape ecology: the effect of pattern on process. Annual Review of Ecology and Systematics 20:171–197.

White, D., A.J. Kimerling, and W.S. Overton. 1992. Cartographic and geometric components of a global sampling design for environmental monitoring. Cartography and Geographic Information Systems 19:5–22.

White, D., P.G. Minotti, M.J. Barczak, J.C. Sifneos, K.E. Freemark, M.V. Santelmann, C.F. Steinitz, A.R. Kiester, and E.M. Preston. 1997. Assessing risks to biodiversity from future landscape change. Conservation Biology 11.

White, D., and J.C. Sifneos. 1997. Mapping multivariate spatial relationships from regression trees by partitions of color visual variables. Proceedings, Auto-Carto 13, American Congress on Surveying and Mapping, pp. 86–95.

Wickham, J.D., J. Wu, and D.F. Bradford. 1997. A conceptual framework for selecting and analyzing stressor data to study species richness at large spatial scales. Environmental Management 21:247–257.

Wiens, J.A. 1989. Spatial scaling in ecology. Functional Ecology 3:385–397.

Williams, P.H., and C.H. Humphries. 1994. Biodiversity, taxonomic relatedness, and endemism in conservation. In: Systematics and Conservation Evaluation, pp. 269–287. P.L. Forey, C.J. Humphries, and R.I. Vane-Wright (eds.). Systematics Association Special Volume No. 50, Clarendon Press, Oxford.

Williams, P., D. Gibbons, C. Margules, A. Rebelo, C. Humphries, and R. Pressey. 1996. A comparison of richness hotspots, rarity hotspots, and complementary areas for conserving diversity of British Birds. Conservation Biology 10:155–174.

Wright, D.H., D.J. Currie, and B.A. Maurer. 1993. Energy supply and patterns of species richness on local and regional scales. In: Species Diversity in Ecological Communities, pp. 66–74. R.E. Ricklefs and D. Schluter (eds.). University of Chicago Press, Chicago.

9
Conserving Biological Diversity with Scarce Resources

Stephen Polasky and Andrew R. Solow

Over the past decade, the conservation of biological diversity has become a central environmental policy concern. Although considerable effort to conserve biological diversity has been undertaken, budgets of public and private conservation organizations fall far short of being able to fund all worthwhile conservation projects. Even though conserving biological diversity is important, it must compete with other important or popular social goals, such as improved education, health care, and material well-being. For the foreseeable future, resources devoted to conservation will remain limited. Therefore, it is necessary to set conservation priorities to allocate resources where they will do the most good. There is an active research program by conservation biologists and others to assess conservation priorities (e.g., Vane-Wright et al. 1991; Groombridge 1992; Reid et al. 1992; Scott et al. 1993).

This chapter describes a methodology for setting biodiversity conservation priorities and demonstrates how the methodology can be applied to conservation problems. The next section outlines a simple decision-making framework for assessing conservation strategies. The conservation problem can be broken down into three parts: (1) defining a measure of biological diversity, (2) assessing the probable biological effects of alternative conservation strategies, and (3) assessing the probable net cost of alternative conservation strategies. The third section discusses alternative diversity measures and gives an example of their use. The fourth section discusses assessing the probable biological and economic effects of alternative conservation strategies. This discussion focuses on the "reserve network selection" problem and provides an example of the problem using biogeographic data on terrestrial vertebrates distributions in Oregon. Both the third and fourth sections discuss what information is necessary to implement various approaches and what can be done when only limited information is available.

Most large-scale priority setting analyses, such as the one for Oregon, abstract from much of the detail that is often the focus of landscape level analyses. The final section discusses the importance and the difficulties of incorporating detailed site and species specific information into large-scale priority setting analysis. Some thoughts on how to bridge the gap between landscape level and larger regional or national level analyses are offered.

A Conservation Decision-Making Framework

This section outlines a simple conservation decision-making framework that was first described in Solow et al. (1993). Under this approach, the objective is to maximize expected biological diversity conserved given a limited budget. Formally, the optimization problem is:

$$\text{Max} \sum_{x \in X} D(x)Ps(x)$$
$$\text{s.t. } C(S) \leq B.$$

where $D(x)$ is a measure of biological diversity of outcome $x \in X$, $P_S(x)$ is the probability that outcome x will occur under conservation strategy S, $C(S)$ is the cost of implementing conservation strategy S, and B is the size of the conservation budget. There are three important components of this framework:

1. The biodiversity measure of various outcomes, $D(x)$.
2. The biological consequences of implementing a certain conservation strategy, $P_S(x)$.
3. The cost of implementing a conservation strategy, $C(S)$.

Finding reasonable definitions for these functions and gathering information to estimate the value of these functions raises difficult issues. In the third and fourth sections the issues involved and the types of information required to implement this approach are discussed.

Virtually any conservation strategy can be analyzed in this framework. Presented in this general way, though, it may not be easy to see how to apply the framework to actual conservation decision making. To help make this clear, suppose the conservation strategy under consideration is one of setting up a system of reserves to conserve species. A conservation strategy, S, is a list of properties to be included in the reserve network, $C(S)$ is the cost of purchasing and maintaining this land. An outcome, x, is a set of species conserved in the reserve network and $D(x)$ is the biodiversity measure of that set of species. Finally, $P_S(x)$ is the probability that a given set of species will be conserved if the set of properties in the reserve network is S. The optimal reserve network is the one that yields the greatest expected biodiversity from the set of affordable reserve networks. An example of this particular problem is given in the section 4.

Many policies that may impact biodiversity conservation can be thought of in a similar fashion. Policies including forestry policy, agricultural policy, infrastructure development, and zoning, affect the pattern of land use, which has both biological and economic consequences. For example, building roads to a particular region or increasing agricultural subsidies may increase conversion of land from natural habitat to human use. One could consider the problem of how to maintain the maximum amount of biodiversity given a certain level of economic development activity. In other words, where or how should economic development activity be channeled so that it has the least negative impact on biodiversity conservation?

An important point to note about the framework is that it is specified with a fixed budget constraint. This allows the conservation decision-making problem to be cast in terms of a cost-effectiveness analysis rather than a cost-benefit analysis. Cost-effectiveness analysis requires only that a decision maker can compare whether an outcome provides more or less biodiversity than some alternative. In other words, the measure of relative value of outcomes can be done in biological terms, such as the number or variety of species conserved, rather than in monetary terms.

Measures of Species Diversity Based on Presence/Absence

Biological diversity can refer to diversity at different scales from genetic, species, to ecosystem level. In the second section, the discussion was kept general so that diversity at any level could be incorporated into the framework. This section focuses on measures of species diversity, and, in particular, on measures based on the presence or absence of species. The simplest such measure is species richness. Recent work on such measures attempts to account for the dissimilarity among species as well as the number (e.g., Vane-Wright et al. 1991; Faith 1992; Weitzman 1992; Faith and Walker 1993; Polasky et al. 1993; Solow et al. 1993; Solow and Polasky 1994). Faith (1994) contains a review of many of these diversity measures. It is important to note that these measures do not depend on the relative abundance of species. Measures of diversity based on relative abundance are also common in the literature (e.g., Magurran 1988).

Defining an appropriate species diversity measure (or a biodiversity measure) depends in large part on the goals of conservation, that is, on what gives species (biodiversity) conservation value? A number of different goals for conservation at the species level have been articulated including: (1) conserving the greatest number of species, (2) conserving the greatest number of different higher taxonomic groups, (3) conserving the greatest dissimilarity among species, (4) conserving evolutionary potential, and (5) conserving species that generate the greatest wealth. What follows ties various presence/absence measures of species diversity to these various goals.

Other goals commonly mentioned in the conservation literature, such as rarity, vulnerability, and endemism will not be discussed directly (Margules and Usher 1981). In our framework, these goals are more closely tied to statements about the probability of survival under alternative conservation strategies, $P_S(x)$, than to statements about the measure of diversity, $D(x)$. A rare or endemic species may make it important to conserve a particular location that contains that species, yet the species itself is presumably not more valuable than a common species simply because it is rare or endemic. Likewise, the probability of survival of vulnerable species may be low without aid of a targeted conservation effort, but vulnerability per se does not make a particular species valuable.

If the goal is to conserve the greatest number of species, the measure of diversity that should be used is species richness. Implicit in this goal is the assumption that all species are equally valuable to conserve, which may not be true. For example, a set consisting of four beetle species is in some sense less diverse than a set consisting of one beetle species, one aphibian species and one bird species, even though the latter set has only three species. Intuitively, dissimilarity among species as well as the number of species is of concern. In practice, species richness is often used as a species diversity measure. The information necessary to quantify the dissimilarity among species may not be readily available, or as with methods that quantify genetic dissimilarity, may be quite costly to obtain.

One way to capture some of the importance of dissimilarity among species is to base the measure of diversity on the number of higher taxonomic groups conserved (e.g., number of families) rather than on the number of species conserved. Alternatively, as in the node counting method of Vane-Wright et al. (1991), it is possible to weight differences between species based on whether or not species are in the same genus, family, order, and so on. These approaches incorporate dissimilarity among species in a relatively crude manner. Williams and Gaston (1994) argue for the higher taxon approach on different grounds. They claim that it is easier to count (or estimate) the number of higher order taxa present in an area than it is to count (or estimate) the number of species present. The problem of incomplete biogeographic information will be discussed further later in this chapter.

Much of the recent work on species diversity measures attempts to incorporate the degree of dissimilarity among species directly. A reason for caring about dissimilarity among species is that dissimilar sets of species conserve a greater collection of potentially valuable characteristics, such as genetic traits. If the underlying characteristic are ultimately of value, conserving the most dissimilar set of species is the appropriate conservation goal. In some cases, which species contain which characteristics may be known. Portions of the genetic code for some species have been studied. Also, the characteristics of interest may have morphological differences that can be readily observed. In such cases, one can directly observe the number of characteristics represented in a set of species and choose the set that represents the greatest collection of characteristics (Faith 1992).

In general, though, information on which species contain which characteristics will not be available. In some instances, a summary measure of the dissimilarity, or distance, between pairs of species may be available, that may be based on genetic dissimilarity. Sibley and Alquist (1990) have compiled such measures for a wide variety of bird species. It is also possible to construct such measures on the basis of morphological differences (the example using glucosinolate-producing plants below takes this approach). In other cases, it simply may not be feasible to get measures of pairwise dissimilarity. If so, relying on simpler approaches will be necessary. The following discussion assumes that such pairwise distance measures are available.

Obtaining a diversity measure of a set of species beginning with pairwise distances between species is the problem analyzed by Weitzman (1992). Weitzman begins by assuming that the diversity of a set consisting of a single species is 0. He defines the distance between a single species and a set of species as the distance between the species and its nearest neighbor in the set. It seems natural to require that including an additional species to a set of species increases the measure of diversity by the nearest neighbor distance. Doing so would also provide a convenient algorithm for computing diversity of any set of species. Starting with any species, the measure of diversity could be calculated by adding species one at a time and incrementing the measure of diversity by the nearest-neighbor distance. The results of this calculation, however, depends on the order in which the spceies are added. To remove this ambiguity, Weitzman defines his measure of diversity, referred to as $W(x)$, as the maximum over all possible ways to construct the diversity measure for the set of species. Weitzman shows that $W(x)$ has a number of desirable and intuitive properties.

Solow and Polasky (1994) derive a diversity measure based on an "option value" argument. Species contain various characteristics, some of which may turn out to be valuable. One value of conserving species is to conserve the option of being able to use a particular set of characteristics in the future. Without imposing a specific probabilistic model, the probability that a set of species will contain a specific characteristic cannot be determined only from the knowledge of the probability that any given species has the characteristic and the pairwise distances among species. Solow and Polasky (1994), however, show that a lower bound on the probability that a set of species contains a specific characteristic is proportional to the summed elements of the inverse of a matrix, whose i-jth element is defined by a declining function of the pairwise distance between species i and j. We refer to this measure of the lower bound on the probability as $V(x)$. As a diversity measure, $V(x)$ has some appealing properties. If the distance between species is essentially infinite (i.e., all species are uncorrelated in the probability of having a given characteristic), then $V(x)$ is equal to the number of species in the set. On the other hand, if all species are very similar so that they approach being perfectly correlated in the characteristics they contain,

$V(x)$ approaches 1. In a sense, $V(x)$ can be interpreted as the "effective number of species" in the set.

Solow and Polasky (1994) assume that having more than one species with the characteristic is no better than having a single species with it. Polasky and Solow (1995) discuss a model in which finding an additional species with the characteristic can be valuable. Finding additional species containing the characteristic may yield a compound of superior quality or one that costs less to collect or process.

Solow and Polasky (1994) present an application of the diversity measures $W(x)$ and $V(x)$ using data from Rodman (1991) on pairwise distances for a set of 26 glucosinolate-producing plants. The 26 species are listed in Table 9.1 and a graph of their locations (in two dimensions) is presented in Figure 9.1. In Table 9.2, the values of $W(x)$ and $V(x)$ are given for three sets of species, x_1 = {all 26 species}, x_2 = {FLA, RES, GER}, and x_3 = {CAP, RES, TOV, BRA, KOE, GYR, BAT, SAL, MOR}. In calculated $V(x)$, we used an exponential function, $e^{-\theta d}$, where d is the pairwise distance between species and θ is a parameter, to define the elements of the matrix. We let θ take on values 0.1 and 0.5. Larger values of θ tend to give more weight to the number of species and less weight to their dissimilarity. In this sense, $W(x)$ is similar to $V(x)$ when θ is small. Note that $W(x)$ and $V(x)$ give the same ranking of x_2 and x_3 for $\theta = 0.1$. In contrast, the ranking is different for $\theta = 0.5$, because $V(x)$ is more strongly influenced by the number of species. In fact, for $\theta = 0.5$, the species in x_2 are effectively independent, so that their effective number is equal to their actual number.

In special cases, a direct relationship can be made between a diversity measure defined on pairwise distances, the exact number of characteristics represented, and a taxonomic tree for a set of species. Weitzman (1992) and Faith (1992) consider variants of a model in which all species start from a

Table 9.1. Taxa of glucosinolate-producing plants and putative relatives.

Taxon	Code	Taxon	Code
Akaniaceae	AKA	Tovariaceae	TOV
Bataceae	BAT	Tropaeolaceae	TRO
Brassicaceae	BRA	Balsaminacea	BAL
Bretschneideraceae	BRE	Celastraceae	CEL
Capparaceae	CAP	Centrospermae	CEN
Cariacaceae	CAR	Dilleniaceae	DIL
Drypetes	DRY	Euphorbiaceae	EUP
Gyrostemonaceae	GYR	Flacouticeae	FLA
Limnanthaceae	LIM	Geraniaceae	GER
Moringaceae	MOR	Koeberliniaceae	KOE
Pentadiplandraceae	PEN	Oxalidaceae	OXA
Resedaceae	RES	Passifloraceae	PAS
Salvadoraceae	SAL	Sapindaceae	SAP

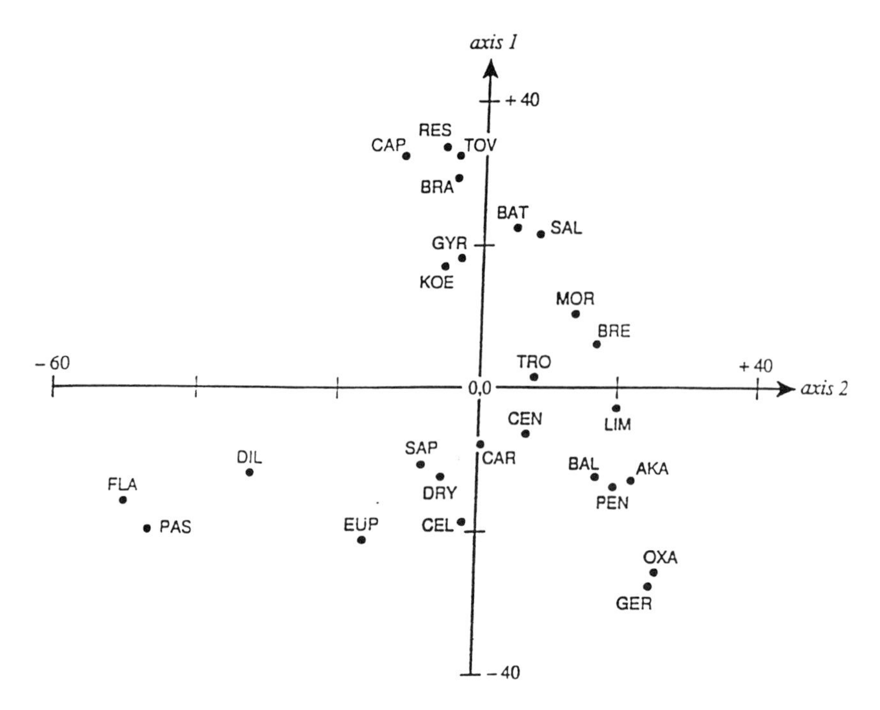

FIGURE 9.1. Locations of 26 species of glucosinate-producing plants along the first two principle axes.

common ancestor and add traits over time. Using cladistic analysis, Faith (1992) constructs a taxonomic tree most consistent with the pairwise data. Faith proposes a measure of "phylogenetic diversity" that corresponds to the branch length of the constructed taxonomic tree. Weitzman (1992) shows that when distances are "ultrametric" (the middle pairwise distance equals the maximum pairwise distance for any set of three species), his diversity measure corresponds to the branch length of a tree where the branch length is proportional to the time to the split from a common ancestor. Longer branch lengths correspond to greater evolutionary time and genetically more unique species.

TABLE 9.2. Values of $W(x)$ and $V(x)$ for: x_1 = all 26 species; x_2 = (FLA, RES, GER); x_3 = (CAP, RES, TOV, BRA, KOE, GYR, BAT, SAL, MOR).

x	$W(x)$	$V(x)$ $\theta = 0.1$	$V(x)$ $\theta = 0.5$
x_1	106.73	2.62	10.55
x_2	38.82	2.32	3.00
x_3	25.81	1.50	3.66

On the other hand, some have argued that it is important to conserve species in dynamic evolutionary lineages (Erwin 1991). Because these lineages evolve species relatively quickly, species within these lineages will have many close relatives and will have low diversity value according to the measures described above. One measure that can take account of aspects of both the desire to conserve evolutionarily dynamic lineages as well as maximum dissimilarity is one proposed in Solow et al. (1993). Starting from the set of currently existing species, the loss of diversity is defined by summing the nearest neighbor distance from each species that goes extinct to the set of surviving species. As long as one member of an evolutionarily dynamic lineage remains, the loss of other nearby species will not lead to large diversity loss. However, the loss of the entire lineage will lead to a large diversity loss.

The above discussion focused on diversity measures in which all species where identical except in their relationship to other species. This approach is reasonable in cases such as genetic prospecting where it is not known a priori which species or characteristics will turn out to be of value. In other cases, however, knowledge about which species or characteristics are of value to society, either in agriculture, pharmaceuticals, tourism, or other use, already exists. It is possible to combine a diversity measure, with information about the worth of specific species to approximate the total value of a set of species. In order to adequately estimate the total value, some decision must be made on the relative importance of conserving characteristics that may be of value in the future versus the value of conserving characteristics of known current value. Several attempts have been made to quantify the "option value" of genetic prospecting for pharmaceutical products (e.g., Mendelsohn and Balick 1995; Simpson, Sedjo and Reid 1996). In addition, there have been attempts to quantify the value of "ecotourism" (e.g., Maille and Mendelsohn 1993; Menkhaus and Lober 1996) and the value of "ecosystem services." The estimates of option value can then be weighed against estimates of current use value. In addition to these values, one may also wish to weigh other values, such as "existence value," in deciding which elements of biodiversity are the most important to conserve. Norton (1987) contains a lengthy discussion of the various sorts of value that may accrue from conserving biodiversity.

Although the discussion in this section has been cast almost entirely in terms of measures of species diversity, much of the analysis applies to other levels of diversity as well. One reason for caring about dissimilarity among species was motivated by the goal of conserving genetic characteristics. In a similar fashion, one could apply a measure of diversity at the ecosystem level with the goal of conserving underlying species diversity. For many regions of the world, biogeographic information is so limited that it does not make sense to try to use species lists as the basis of conservation policy. An alternative approach is to construct measures of dissimilarity among ecosystems using measures of environmental variables, such as temperature, precipitation, elevation, topography, and soil type, which are readily available for many areas. One can then apply a diversity measure and choose a set of reserves based on maximizing ecosystem diversity. Finally, one could attempt to aggregate the value of conservation at the various different levels, although as in the problem of aggregating option and current use values, a decision on the relative weight to assign to each type of value would need to be addressed.

Assessing Biological and Economic Effects of Alternative Conservation Strategies

Perhaps the greatest challenge to defining reasonable conservation policy is understanding the link between particular conservation strategies and the likely biological and economic effects. There are numerous potential strategies to conserve biodiversity, such as preventing the spread of exotics, restricting harvest, and reducing pollution. Perhaps the greatest current threat to biodiversity, especially in countries with rapid population growth and economic expansion, is habitat loss and fragmentation. For this reason, the discussion in this section focuses on strategies to conserve habitat. In particular, it begins with an analysis of the reserve network selection problem. It is important to keep in mind, however, that the framework outlined in Section 2 is quite general. Given adequate information, a wide variety of conservation strategies beyond protected areas strategies can be analyzed.

Reserve Network Selection as an Example of Conservation Strategy

Suppose there is a finite budget that can be allocated to conserving biodiversity through the instrument of habitat conservation. In the simplest approach to this problem, the following assumptions are made:

1. The biodiversity measure is species richness.
2. A complete biogeographic database exists for the species and the area of interest.
3. Species represented in at least one site selected as a reserve will survive.
4. Species not represented in any site selected as a reserve will become extinct.
5. All potential reserve sites are equally costly to conserve.

Given these five assumptions, the conservation problem is to select an affordable set of reserves that represents the greatest number of species at least once. This problem is easy to formulate and has been well studied in operations research where it is known as the maximal coverage problem (Church and ReVelle 1974).

Even though it is simple to state, no simple method guarantees that an optimal solution to the problem can be found in a reasonable amount of time. There exist linear programming based branch and bound algorithms that can often find an optimal solution fairly quickly even when the problem is of reasonably large size. Use of specialized algorithms and intelligent preprocessing of the data can often help in finding an optimal solution quickly (Camm et al. 1996). Recently, several analyses of the reserve selection problem have used linear programming based branch and bound methods to solve for the optimal collection of reserve sites (Cocks and Baird 1989; Saetersdal et al. 1993; Church et al. 1996; Pressey et al. 1996; Csuti et al. 1997). In the worst possible case, it may be necessary to check all possible combinations of sites in order to guarantee finding an optimal solution, but this is infeasible except for conservation problems with a small number of potential or actual reserve sites. For example, if 10 sites out of 100 sites can be included in a reserve network, there are 17.3 trillion possible combinations to check.

Because methods from operations research are not widely known in the conservation field and because the methods may not work if the problem is too large or complex, a number of different heuristic algorithms have been used. A heuristic algorithm, which is usually simple to understand and to use, need not generate an optimal solution. One popular method that does not work well is choosing hotspots, defined as locations containing a large number of species. To see this, consider the example given in Table 9.3. Suppose there are four potential sites, labeled A through D, of which two can be included in a reserve network. Species, which are labeled 1 through 6, that inhabit each site are listed in the column for that site. In Table 9.3, sites A and B are hotspots because they have four species each, while the other sites only have three. Choosing sites A and B represents species 1 through 4 in the reserve network. The optimal reserve network for two sites, however, is to choose sites C and D. In this case, all six species are represented in the reserve network. Because choosing hotspots fails to incorporate the pattern of overlap among species between sites (complementarity), it may fail to protect sites that are not species rich but have many endemic species.

A better approach is to incorporate the complementarity of species between sites through the use of iterative algorithms that choose sites that

TABLE 9.3. Example showing failure of the hotspots approach.

Potential reserve sites	A	B	C	D
Species list	1	1	1	3
	2	2	2	4
	3	3	5	6
	4	4		

add the greatest diversity to what is already included in other conserved sites (Pressey et al. 1993). One such iterative algorithm is the "greedy algorithm." The "greedy algorithm" begins by selecting the site with the greatest species richness. Then, it selects sites sequentially that add the most additional species to those already represented. A variant of this approach weights species by their rarity and sequentially adds sites that give the greatest increase in the rarity weight (Margules et al. 1988). Richness based "greedy algorithms" and rarity based algorithms usually work fairly well but do not necessarily yield a reserve network that represents the maximum number of species possible.

As part of the Oregon Gap Analysis Program, in conjunction with the Biodiversity Research Consortium, a data set was assembled on species distributions for all terrestrial vertebrates that breed in the state of Oregon. In Csuti et al. 1997, a linear programming based branch and bound algorithm as well as several heuristic algorithms were used to solve for a reserve network that represents the maximum the number of species for a fixed number of sites. The state was overlaid with a hexagon grid (640 km^2), which created a set of 441 potential sites. Within each of these 441 sites, the likelihood of finding a species at a site was put in one of four categories:

1. Confident: there has been a verified siting of the species at the site;
2. Probable: the site contains suitable habitat and there have been verified sitings in nearby sites;
3. Possible: habitat of questionable suitability;
4. Not present: habitat is unsuitable for the species.

Species were assumed to be present at the site if they fell into the confident or probable categories at the site Csuti et al. (1997). Species were assumed not to be present if they fell into the other two categories. In this way, a list of species thought to be present at each of the 441 sites was compiled.

In Table 9.4, the number of species represented in the solution of various algorithms is given. The number of sites included in the reserve network was varied from one site up to the number it took to fully represent all species. There are several things to note in Table 9.4. First, there is a rapid accumulation of species in the first few sites. Fully 90% of species are represented in the first five sites, and 95% are represented in ten sites. Terrestrial vertebrates in Oregon have fairly wide geographic distribution. One would not expect such rapid accumulation to occur for other taxa, many plants for example are more narrowly distributed, nor for other regions, such as the tropics. Second, many of the heuristic algorithms, while not optimal, perform fairly well. In particular, the "greedy algorithm" does well, except when only a few sites are included and when complete coverage is desired. The "greedy algorithm" requires 27 sites for complete coverage as compared to 23 sites in the optimal solution. In contrast, the "rarity based" algorithms do well in solving for complete coverage (24 sites) but do not necessarily do well when only a small number of sites are included.

TABLE 9.4. Oregon terrestrial vertebrate species accumulation.

Sites (#) Algorithm	Greedy algorithm	Rarity-based algorithm	Linear programming branch and bound
1	254	254	254
2	306	303	318
3	347	343	356
4	365	360	374
5	379	377	384
6	388	387	390
7	394	392	395
8	398	396	400
9	401	400	403
10	404	403	406
11	406	405	408
12	408	407	410
13	410	409	412
14	412	411	414
15	414	413	416
16	415	415	418
17	416	417	419
18	417	419	420
19	418	421	422
20	419	422	423
21	420	423	424
22	421	424	425
23	422	425	426
24	423	426	—
25	424	—	—
26	425	—	—
27	426	—	—

Information can also be gained by looking at the geographic pattern of the sites chosen. Figure 9.2 depicts combinations of sites to represent all 426 species in the state using only 23 sites. There are 144 possible combinations of sites that do the trick. The fact that there are multiple combinations that yield an optimal solution shows that there is at least some flexibility over which sites to include. The importance of flexibility, because of constraints on actual conservation decision-making not included in the analysis, has been stressed in Pressey et al. (1993). Surprisingly, any combination that includes one of two sites, labeled with 50%, one of three sites (33%), one of four sites (25%), and one of six sites (17%) works. The mix and match outcome of this problem does not hold in general and describing all optimal combinations is usually more complex. Figure 9.3 shows the contrast in the pattern of sites between the optimal solution with ten sites and the ten hotspots in Oregon (ten sites with the greatest species richness). The hotspots are geographically clustered in one area. Because the optimal reserve network incorporates the principle of

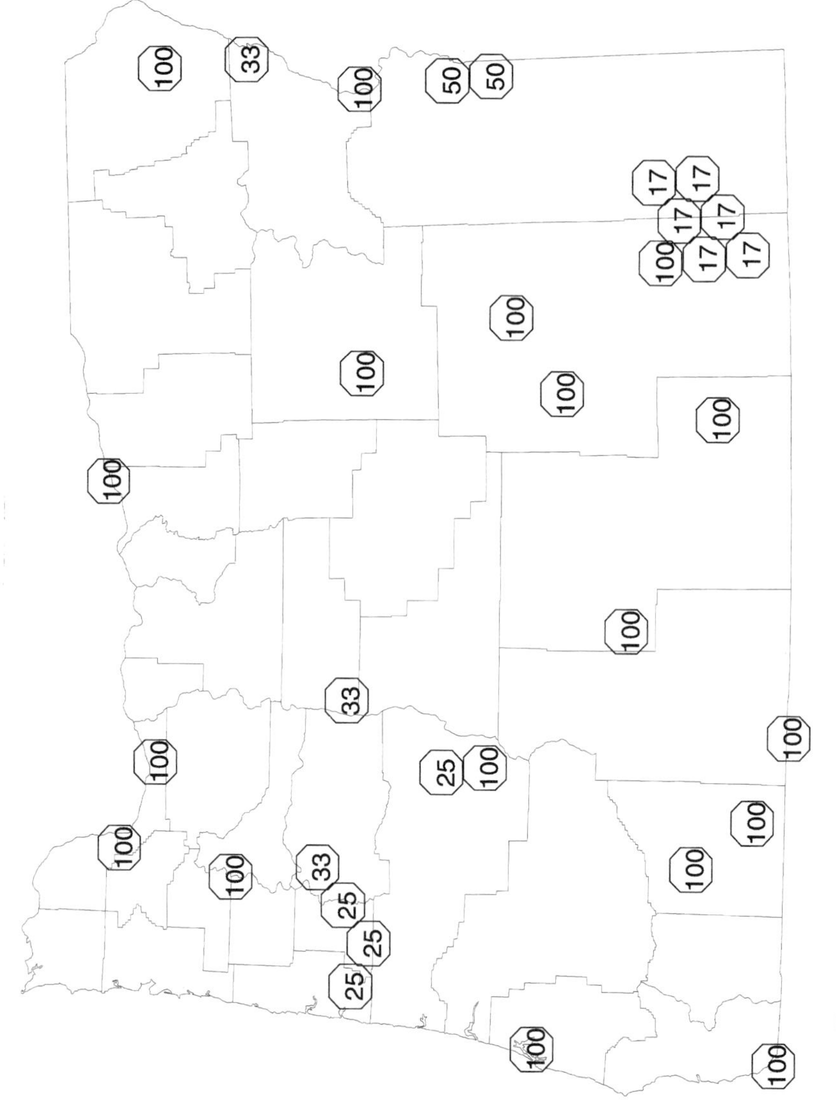

FIGURE 9.2. Complete representation of terrestrial vertebrate species in Oregon in 23 sites.

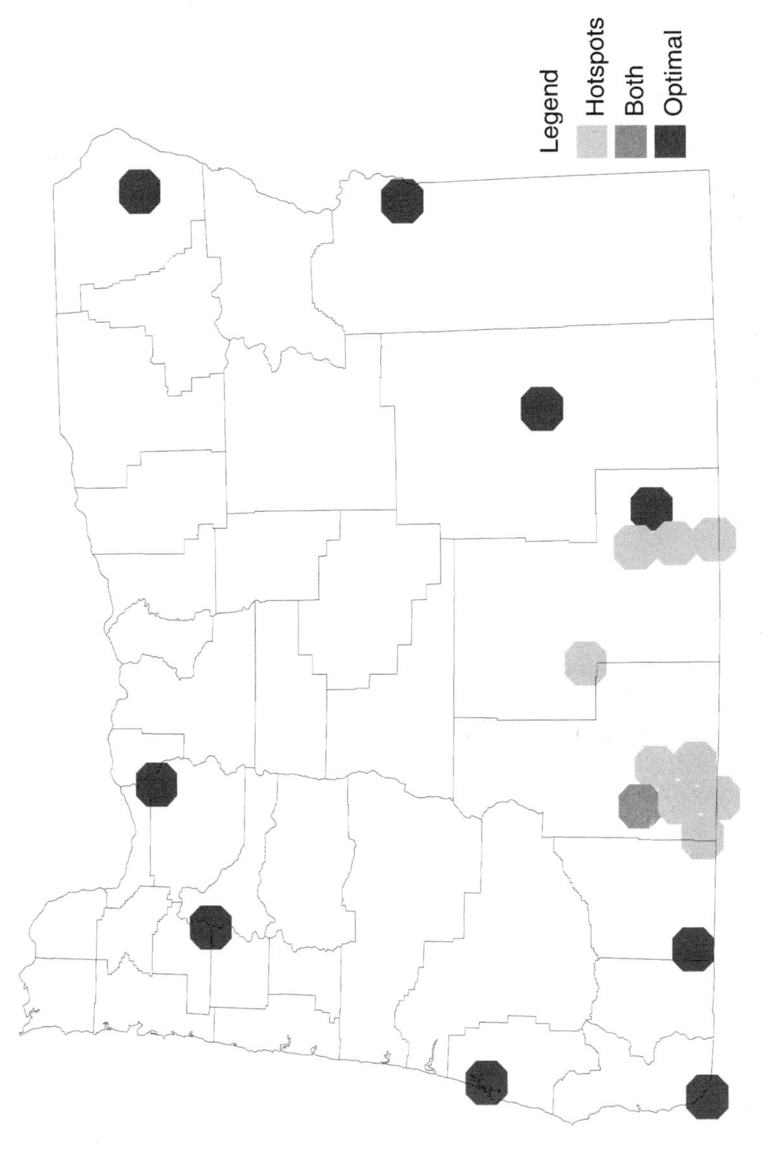

FIGURE 9.3. Comparison of hotspots and optimal reserve network for 10 sites.

complementarity, however, its sites are spread across all major ecosystem types in the state.

Although this analysis is informative of which areas in the state may be of high priority for conservation, it abstracts from a number of important issues. First, the analysis assumes that information about species locations are known with certainty. Second, the analysis is actually an exercise in representation but not necessarily in conservation. There is no analysis of the survival requirements of various species or of habitat quality. Third, the analysis does not include an analysis of management options or of the costs involved. The next subsection discusses some important economic considerations that should factor into conservation decision making, while a later subsection discusses some of the complications that arise from considering the first two points.

Opportunity Cost of Alternative Conservation Strategies

Conservation strategies, such as selecting reserve networks, have economic effects as well as biological effects. Incorporating cost estimates of alternative strategies is necessary for understanding what strategies are feasible given a limited budget. Further, incorporating cost estimates of particular actions may influence the design of optimal conservation policy. For example, in the reserve network selection problem, incorporating land cost may shift conservation efforts away from expensive areas toward cheaper areas where the conflict between conservation and economic activity is not so great. In some respects, the problem for economists is easier than that for biologists. Many of the issues raised in counting the costs, or benefits, of conservation strategy do not differ markedly from cost accounting for other projects or policy analyses. The economic analyst can rely on an existing tool kit rather than having to improvise or invent methods, as must be done to answer some of the biological questions.

If the conservation strategy involves setting aside land as nature reserves, much of the cost of the strategy is the opportunity cost of not allowing economic activity to occur. Estimates of the opportunity cost may be developed in several ways. In most states, land is subject to property tax. For tax purposes, assessed land values are publicly available. In theory, land values represent the capitalized value of the stream of net benefits that accrue to ownership of the land. If land is set aside for a conservation reserve, much of the economic returns that may have come from use of the land will cease. If all economic returns cease with reserve designation, the current land value is the appropriate economic opportunity cost for reserve designation of the site. In some cases, there will be only a partial decrease in the economic return from the land from a conservation strategy. For example, conservation easements restrict only some but not all potential uses of the property. The appropriate opportunity cost of the conservation easement is the difference between the value of the land with and without the restrictions. Obtaining estimates of this difference may come from analysis of similar prior conservation easements or from estimating the economic returns available from the land with the restrictions in place. In places without land value records, or on public lands, estimates of opportunity cost must be derived from estimates of the present value of the highest value economic use of the land. For example, if the highest economic returns come when the land is utilized for timber, estimates of timber productivity, timber prices, and costs of harvest can be used to generate an estimate of the present value of timber operations on the land. In addition to the opportunity cost calculation, there are additional maintenance and operations costs of implementing a conservation strategy that have to be tabulated.

There may be important recreational or ecological service flows that increase with habitat conservation. Including these benefits will decrease the net cost of conservation and may be sufficient to make conservation have no net opportunity cost. Some have claimed that there is great economic value in maintaining the resilience of ecosystems through conservation (Perrings et al. 1995). In addition, there may be economic value from ecotourism, maintaining aesthetic values, as well as value from genetic prospecting, as discussed earlier.

Although there is great hope that conservation will be economically rewarding for landowners, there is little evidence to this point that this is actually so. A particularly difficult problem in making conservation financially rewarding for landowners is that the rents the landowner can capture from conservation may be only a fraction of the total social value. If so, a landowner may not have an incentive to conserve even when that may be the socially preferred course of action. (Defenders of Wildlife 1993; Keystone Center 1995; Polasky et al. 1997). In theory, it is important to include all of the potential costs and benefits accruing from conservation strategy. In practice, it will be difficult to accurately account for nonmarket and speculative (potential) costs and benefits as well as account for the distribution of those costs and benefits across various members of society.

Accounting for Data Uncertainties and Probabilities of Survival

In setting out the reserve network selection problem, it was assumed that complete biogeographic data was known, that species would definitely survive within reserves, and that species would definitely go extinct outside of reserves. Of course, all of these assumptions are incorrect. Below, is described how to conduct priority setting analysis when each of these assumptions is relaxed. The discussion begins with the problem of incomplete biogeographic data.

For most taxa and for most regions of the world, particularly the developing world, knowledge of species distributions are quite limited. Even with the Oregon data there was a fair amount of guess work involved in completing the species lists for all potential sites, with much of the data falling into the "probable" or "possible" categories. Suppose that instead of acting as if the species were present and absent were known, a probability based on the best available information of a species being present at a site was assigned. Then, the probability that a species is represented somewhere in the reserve network is one minus the probability that it is not in any reserve.

With probabilistic data, the objective of the reserve network problem is modified to maximize the expected number of species represented within a budget constraint. This problem is analyzed in Polasky et al. (1996). Using probabilistic data versus converting the data into presence/absence data can make a big difference in the results. Consider an example from Polasky et al. (1996) with 20 species, labeled A through T, and ten sites, which is shown in Table 9.5. At the first five sites, there are potentially many species present, but each species has only a 0.4 probability of being present. At sites 6 through 10, there are fewer species that may be present, but each of these species has a higher probability of actually being present. Assuming that the species is present if the probability is greater that 0.5, and is absent otherwise, sites 1 through 5 have no apparent conservation value as there are no species coded as being present. Sites 6 through 10 do have species coded as being present. When converting the data to presence/absence data, site 6 is the highest priority site followed by sites 7, 8, 9, and 10. Under the probabilistic approach, however, site 1 is the most important, followed by site 2, 3, and so on. Of course, this example was contrived to make a point that the two approaches need not give similar answers. An important question is how far apart the two approaches are likely to be in real biogeographic data. Returning to the Oregon terrestrial vertebrate data, Polasky et al. (1996) show that different reserve networks are often chosen when probabilistic rather than presence/absence data are used.

The probabilistic approach to conservation issues has additional advantages besides being able to represent more accurately the existing state of knowledge. First, by varying probability assessments, the sensitivity of conservation strategies to particular assumptions can be gauged. In particular, this may be useful in assessing what information is of greatest value to collect. Limited biological survey resources can then be directed to collect information about particular taxa or particular areas where the information gain would be most valuable.

Second, a large concern of many conservation analysts is the degree of threat (vulnerability) and how that threat is reduced with the conservation strategy. For example, in the context of the reserve network selection problem, species will not necessarily survive even if they are represented in the reserve network, nor will they necessarily go extinct if they are not represented. Some areas face little threat from development. Some species thrive in developed areas. Species conservation efforts should go towards

TABLE 9.5. Synthetic data.

Site	Species																			
	A	B	C	D	E	F	G	H	I	J	K	L	M	N	O	P	Q	R	S	T
1	0.4	0.4	0.4	0.4	0.4	0.4	0.4	0.4	0.4	0.4	0.4	0.4	0.4	0.4	0.4	0.4	0.4	0.4	0.4	0.4
2	0.4	0.4	0.4	0.4	0.4	0.4	0.4	0.4	0.4	0.4	0.4	0.4	0.4	0.4	0.4	0.4	0.4	0.4	0.4	0
3	0.4	0.4	0.4	0.4	0.4	0.4	0.4	0.4	0.4	0.4	0.4	0.4	0.4	0.4	0.4	0.4	0.4	0.4	0	0
4	0.4	0.4	0.4	0.4	0.4	0.4	0.4	0.4	0.4	0.4	0.4	0.4	0.4	0.4	0.4	0.4	0.4	0	0	0
5	0.4	0.4	0.4	0.4	0.4	0.4	0.4	0.4	0.4	0.4	0.4	0.4	0.4	0.4	0.4	0.4	0	0	0	0
6	0.6	0.6	0.6	0.6	0.6	0	0	0	0	0	0	0	0	0	0	0	0	0	0	0
7	0	0	0	0	0	0.6	0.6	0.6	0.6	0	0	0	0	0	0	0	0	0	0	0
8	0	0	0	0	0	0	0	0	0	0.6	0.6	0.6	0	0	0	0	0	0	0	0
9	0	0	0	0	0	0	0	0	0	0	0	0	0.6	0.6	0	0	0	0	0	0
10	0	0	0	0	0	0	0	0	0	0	0	0	0	0	0.6	0	0	0	0	0

those species facing the greatest risks, which is not necessarily the same thing as trying to represent species in a reserve network.

Third, probabilities can be interpreted as the probability of survival, which allows one to combine ideas from the reserve design and reserve selection literatures in conservation biology. The size and spatial pattern of the reserves is clearly an important factor influencing the viability of species within reserves. In theory, incorporation of reserve design questions into the framework discussed here is straight-forward, though in practice the data requirements, especially on the biological side, may be large.

Integrating Landscape and Regional Level Analyses

The goal of conservation policy is to insure the survival of various elements of biodiversity not merely to represent them. To insure survival of species, some understanding of population dynamics must be added to the knowledge of the current location of species. For example, conserving habitat that serves as a sink rather than a source area for a species will not be an effective strategy to conserve the species over the long term. Studies of population dynamics, however, are often complex and time consuming. Not many species have been studied in depth. Even for those species that have been extensively studied (e.g., the northern spotted owl), many unanswered questions remain. There is a pressing need to incorporate some aspects of population dynamics into large-scale regional priority setting exercises. What is needed are methods that can deliver good, although not necessarily perfect, estimates of population survival probabilities relatively quickly. In many ways the needs here are analogous to the needs for gathering biogeographic data in many tropical countries. The response in that case was to try to develop rapid assessment methods that could give a reasonable estimates to important biogeographic data.

There are several ways that efforts to incorporate some estimate of survival probabilities may be pursued. One example of an approach to this problem is given in White et al. (1997). In this paper, area requirements for a variety of species were used to see whether these species would survive under alternative landscape patterns for Monroe County, Pennsylvania. Another approach is to utilize a spatially explicit model incorporating fecundity, mortality, dispersion, and habitat suitability that can be adapted for use on a variety of species.

Without incorporating analysis of survival probabilities, large scale priority setting analyses will remain exercises in representation rather than conservation. Without the bigger picture provided by regional or national priority setting analysis, landscape level analysis cannot be effective in setting conservation strategy or for allocating conservation efforts to high priority projects. It is important to work out ways to incorporate elements of both scales of analysis to inform conservation decision making.

References

Camm, J.D., S. Polasky, A. Solow, and B. Csuti. 1996. A note on optimization models for reserve site selection. Biological Conservation 78:353–355.

Church, R., and C. ReVelle. 1974. The maximal coverage location problem. Papers of the Regional Science Association 32:101–118.

Church, R.L., D.M. Stoms, and F.W. Davis. 1996. Reserve selection as a maximal coverage problem. Biological Conservation 76(2):105–112.

Cocks, K.D., and I.A. Baird. 1989. Using mathematical programming to address the multiple reserve selection problem: an example for the Eyre Peninsula, South Australia. Biological Conservation 49:113–130.

Csuti, B., S. Polasky, P.H. Williams, R.L. Pressey, J.D. Camm, M. Kershaw, A.R. Kiester, B. Downs, R. Hamilton, M. Huso, and K. Sahr. 1997. A comparison of reserve selection algorithms using data on terrestrial vertebrates in Oregon. Biological Conservation 80:83–97. Some of the art in this chapter is from this article, with kind permission from Elsevier Science Ltd, The Boulevard, Langford Lane, Kidlington 0X5 1GB, UK.

Defenders of Wildlife. 1993. Building Economic Incentives into the Endangered Species Act, 2nd ed. Defenders of Wildlife, Washington, DC.

Erwin, T.L. 1991. An evolutionary basis for conservation strategies. Science 253, 16 August, 750–752.

Faith, D.P. 1992. Conservation evaluation and phylogenetic diversity. Biological Conservation 61:1–10.

Faith, D.P. 1994. Phylogenetic pattern and the quantification of organismal biodiversity. Philosophical Transactions of the Royal Society of London Series B: Biological Sciences 345:45–58.

Faith, D.P., and P.A. Walker. 1993. DIVERSITY: Reference and Users Guide. CSIRO Division of Wildlife and Ecology, Canberra, Australia.

Groombridge, B. (ed.). 1992. Global Biodiversity: Status of the Earth's Living Resources. Chapman & Hall, London.

Keystone Center. 1995. The Keystone Dialogue on Incentives for Private Land owners to Protect Endangered Species, Final Report. Keystone Center, Keystone, CO.

Magurran, A.E. 1988. Ecological Diversity and Its Measurement. Princeton University Press, Princeton, NJ.

Maille, P., and R. Mendelsohn. 1993. Valuing ecotourism in Madagascar. Journal of Environmental Management 39:213–218.

Margules, C.R., A.O. Nicholls, and R.L. Pressey. 1988. Selecting networks of reserves to maximize biological diversity. Biological Conservation 43:63–76.

Margules, C.R., and M.B. Usher. 1981. Criteria used in assessing wildlife conservation potential: a review. Biological Conservation 21:79–109.

Mendelsohn, R., and M.J. Balick. 1995. The value of undiscovered pharmaceuticals in tropical forests. Economic Botany 49(2):223–228.

Menkhaus, S., and D.J. Lober. 1996. International ecotourism and the valuation of tropical rainforests in Costa Rica. Journal of Environmental Management 47: 1–10.

Norton, B.G. 1987. Why Preserve Natural Variety? Princeton University Press, Princeton, NJ.

Perrings, C., K.-G. Mäler, C. Folke, C.S. Holling, and B.-O. Jansson. 1995. Biodiversity Loss: Economic and Ecological Issues. Cambridge University Press, Cambridge.

Polasky, S., R. Ding, A.R. Solow, J.D. Camm, and B. Csuti. 1996. Choosing reserve networks with incomplete species information. Biological Conservation.

Polasky, S., H. Doremus, and B. Rettig. 1997. Endangered species conservation on private land. Contemporary Economic Policy 15(4):66–76.

Polasky, S., A. Solow, and J. Broadus. 1993. Searching for uncertain benefits and the conservation of biological diversity. Environmental and Resource Economics 3:171–181.

Polasky, S., and A.R. Solow. 1995. On the value of a collection of species. Journal of Environmental Economics and Management 29(3):298–303.

Pressey, R.L., C.J. Humphries, C.R. Margules, R.I. Vane-Wright, and P.H. Williams. 1993. Beyond opportunism: key principles for systematic reserve selection. Trends in Ecology and Evolution 8(4):124–128.

Pressey, R.L., H.P. Possingham, and C.R. Margules. 1996. Optimality in reserve selection algorithms: when does it matter and how much? Biological Conservation 76:259–267.

Reid, W., C. Barber, and K. Miller. 1992. Global Biodiversity Strategy: Guidelines for Action to Save, Study, and Use Earth's Biotic Wealth Sustainably and Equitably. WRI, IUCN, and UNEP, Washington, DC.

Rodman, J.E. 1991. A taxonomic Analysis of glucosinolate-producing plants, Part 1: phenetics. Systematic Botany 16:598–618.

Saetersdal, M., J.M. Line, and H.J.B. Birks. 1993. How to maximize biological diversity in nature reserve selections: vascular plants and breeding birds in deciduous woodlands, Western Norway. Biological Conservation 66:131–138.

Scott, J.M., F. Davis, B. Csuti, R. Noss, B. Butterfield, C. Groves, H. Anderson, S. Ciacco, F. D'Erchia, T.C. Edwards, Jr., J. Ulliman, and R.G. Wright. 1993. Gap Analysis: A Geographic Approach to Protection of Biological Diversity. Wildlife Monograph No. 123. The Wildlife Society. Washington, DC.

Sibley, C.G., and J.E. Ahlquist. 1990. Phylogeny and Classification of Birds: A Study of Molecular Evolution. Yale University Press, New Haven.

Simpson, R.D., R.A. Sedjo, and J. Reid. 1996. Valuing biodiversity for use in pharmaceutical research. Journal of Political Economy 104(1): 163–185.

Solow, A.R., and S. Polasky. 1994. Measuring biological diversity. Environmental and Ecological Statistics 1:95–107. Some of the art in this chapter is from this article, with permission from Chapman & Hall.

Solow, A., S. Polasky, and J. Broadus. 1993. On the measurement of biological diversity. Journal of Environmental Economics and Management 24:60–68.

Vane-Wright, R.I., C.J. Humphries, and P.H. Williams. 1991. What to protect? —Systematics and the agony of choice. Biological Conservation 55:235–254.

Weitzman, M.L. 1992. On diversity. Quarterly Journal of Economics 107:363–405.

White, D., P.G. Minotti, M.J. Barczak, J.C. Sifneos, K.E. Freemark, M.V. Santelmann, C.F. Steinitz, A.R. Kiester, and E.M. Preston. 1997. Assessing risks to biodiversity from future landscape change. Conservation Biology 11(2):349–360.

Williams, P.H., and K.J. Gaston. 1994. Measuring more of biodiversity: can higher-taxon richness predict wholesale species richness? Biological Conservation 67:211–217.

10 Spatial Concepts, Planning Strategies, and Future Scenarios: A Framework Method for Integrating Landscape Ecology and Landscape Planning

Jack Ahern

Landscape planning can be defined as the practice of planning for the sustainable use of physical, biological, and cultural resources. It seeks the protection of unique, scarce, and rare resources, avoidance of hazards, protection of limited resources for controlled use, and accommodating development in appropriate locations (Fabos 1985). Sustainable landscape planning has been strongly supported through major international policy agreements, and can be generally defined as "a condition of stability in physical and social systems achieved by accommodating the needs of the present without compromising the ability of future generations to meet their needs" (IUCN 1980; WCED 1987). Increased international interest in sustainable landscape planning has stimulated much discussion at professional conferences and symposia and in recent publications (Lyle 1994; Forman 1995). More significantly, and in the context of this chapter, this challenge for sustainable landscape planning has also inspired a dialogue between ecologists and landscape planners within the discipline of landscape ecology (Forman 1990a; Golley and Bellot 1991; Vos and Opdam 1993; Hersperger 1994; Langevelde van 1994).

There are multiple dimensions to sustainability including, economic, social, ethical, and spatial. Landscape planning is most fundamentally linked with the latter, the spatial dimension, and predominantly at the scale of the landscape. Landscape plans are actually hypotheses of how a proposed plan (i.e., landscape structure) will influence landscape processes. The landscape plan offers specific recommendations regarding land-use allocation, designation of levels of protection and management, and setting a strategy to "undo" negative changes in the landscape from the past. If the planning recommendations are implemented, the plan, as a landscape ecological hypothesis, becomes a field experiment from which landscape ecologists may gain new knowledge (Golley and Bellot 1991). This model of landscape planning and landscape ecological cooperative interaction follows the concept of adaptive management (Holling 1978). All of these applied landscape ecological activities engage the pattern:process dynamic that is at the core of landscape ecology (Turner 1989). Thus the activity of landscape planning can be seen as a primary basis for collaboration and knowledge exchange between landscape planners and landscape ecologists.

The landscape scale is appropriate for sustainable planning because it is sufficiently large to contain a heterogeneous matrix of landscape elements that provide a context for mosaic stability (Forman 1990a, 1995). The definition of landscape by Forman and Godron (1986, p. 594) is referenced in this context. "A heterogeneous area composed of a cluster of interacting ecosystems that are repeated in similar form throughout. Landscapes vary in size, down to a few kilometers in diameter." A landscape has at least a theoretical potential to support disturbance regimes, landscape succession, and changes in land use while maintaining some level of "mosaic stability." The ecosystem scale is by definition vulnerable to irrecoverable disturbance or "permanent" change because of building projects and, therefore, is not an appropriate scale for sustainable landscape planning. The ecosystem is a useful spatial unit to understand vertical or topological relationships, but is spatially too limited to understand the "horizontal" or chorological patterns and processes (Zonneveld 1995). At the other end of the scale continuum, the biosphere is perhaps the ultimate ecological scale, the scale in which all ecological processes are involved. Humans are only beginning to understand the global dimensions of ecology, let alone attempting to conceive and implement plans at the biosphere scale. The landscape scale is consistent with the scale of human perception, decision making, and physical management; the biosphere scale is not. At least in conceptual terms, the landscape is probably the optimal scale for sustainable landscape planning.

Landscape ecology has provided a terminology and taxonomy for describing landscapes and their associated patterns and processes. Hierarchy theory has established a conceptual means for understanding the interdependence of patterns and processes within a system of nested scales. First principles are emerging to inform and guide planning. Landscape ecology has thus established a theoretical foundation for clear communication of research results and for application to decision making.

Landscape ecologists and planners are united by a common interest in the pursuit of sustainable landscapes. Beyond this rather obvious and intuitive common pursuit, and the common interest in the fundamental interaction of landscape pattern:process, just how can the scientific and the applications "sides" communicate, or even better collaborate?

This chapter attempts to answer this recurring question. It will do this first by reviewing some basic theory and methods from landscape planning, and proposing essential attributes of a landscape ecological-based planning framework method. It will then introduce the idea of spatial concepts, which acknowledges the centrality of the spatial dimension of sustainable landscape planning (Forman 1990a; Zonneveld 1991). The idea of spatial concepts moves the sustainability discussion away from abstract theory toward specific solutions by integrating landscape ecological principles and knowledge with creative solutions appropriate to a specific spatial context. A thesis of this paper is that landscape ecology can assist in the conception and evaluation of spatial concepts, and that the implementation of spatial concepts in landscape plans represents a basis for field experiments which can, in turn, generate new knowledge.

A well-conceived spatial concept for landscape planning requires a strategic approach to develop and implement the actual plan. Scenarios can be employed in strategic planning to achieve surprising, yet plausible plans and unexpected results (Hirschorn 1980). The use of scenario studies has proven effective, especially in the Netherlands, to communicate the spatial consequences to the landscape of specific policy decisions (Steinitz et al. 1994; Schooenboom 1995; Veenenklaas 1995).

This chapter contains numerous references to Dutch landscape planning and landscape ecology. This is, in part, the result of the author's continued involvement with the Wageningen Agricultural University, but more so because the Netherlands is a landscape under severe ecological stress. By a number of measures of ecological integrity, the Netherlands is in a nonsustainable condition[i]. In the face of this dire situation, and following a tradition of ambitious and innovative responses to profound challenges, the Dutch have embarked on a national plan for a sustainable environment that is globally unprecedented. For this reason the Netherlands has been labeled an "experimental garden" and was the subject of a major edited book on landscape ecology (Vos and Opdam 1993). There is a great deal to be learned from the Dutch landscape experience, particularly with respect to the manner in which landscape planning and landscape ecology are integrated in theory and application.

The chapter presents an application of the proposed framework method to a landscape plan for the town of Orange in Massachusetts. The plan emphasizes open space planning, which was broadly interpreted to include the abiotic, biotic, and cultural resources and issues. The planning process employed spatial concepts, defined strategies, and offered scenarios as a basis for community decision making. The application illustrates the proposed framework method for landscape ecological planning.

Landscape Planning Theories and Methods

Landscape planning is an activity that promotes the wise and sustainable use of resources, hazard avoidance, and management of the process(es) of landscape change. It determines the capacity and limits of natural resources and the effects of changes. Landscape planning has been described as "the process of choice based on knowledge about people and land" (Steiner 1991, p. 520). McHarg defines ecological planning as "that process whereby a region is understood as a biophysical and social process comprehensible through the operation of laws and time. This can be reinterpreted as having explicit opportunities and constraints for any particular human use. A survey will reveal the most fit locations and processes" (McHarg 1997, p. 321). Landscape planning "cuts across" numerous planning sectors, and is performed at multiple scales and governmental levels (Kiemstedt 1994). As a professional activity, it has roots in landscape architecture and physical planning (Fabos 1985). As these definitions and statements suggest, landscape planing is an inherently interdisciplinary field with biological, physical, and social science components, as well as strong connections with the creative traditions of landscape architecture.

In the latter part of the twentieth century, as humankind gained an unprecedented perspective on the global environment through remote sensing, landscape planning gained visibility and a sense of urgency. Many have called for, or offered, perspectives on the theory which could guide the development, teaching and practice of landscape planning (Lynch 1985; Berger 1987; Young and Zube 1988; Steinitz 1990; Ndubisi 1997). Many have also acknowledged the significance of the landscape ecological perspective on landscape planning, and a resulting need for reconsideration of planning theory and methods (Berger 1987; Steiner and Osterman 1988; Golley and Berlot 1991; Hersperger 1994; van Langevelde 1994; Ahern 1995; Forman 1995). Through landscape ecology, the limitations of some previous landscape planning theories and methods were revealed, and new methods have been proposed to apply the knowledge generated from landscape ecology to landscape planning.

Ndubisi (1997) defines two fundamental theories in landscape planning: substantive and procedural. Substantive theories originate in the natural and social sciences and provide descriptive and predictive information. Procedural theories concern the methodology of planning. The interaction of the two theoretical types produces a tension that both challenges and rewards interdisciplinary research. Hersperger (1984) suggests that in true landscape ecological planning, the distinction between substantive and procedural theories might blur.

Steinitz (1990) argues the need for a more robust theory among all those professions involved with altering landscapes. His six-step framework is based on discrete models for representation, process, evaluation, change, impact, and decision. This framework can be divided into two major parts the descriptive/evaluative and the prescriptive/planning components. The descriptive/evaluative part has strong parallels with landscape ecology in that it deals with articulating the fundamental landscape pattern:process dynamic. In Steinitz' framework, this is included in the framework's representation, process, and evaluation models. Turner (1989) describes how spatial pattern influences many ecological processes, and how landscape planning and management, in turn, influence landscape pattern. Turner argues that as landscape pattern and process are dynamically interrelated, the landscape pattern:process dynamic forms a basic tenet of landscape ecology. In the prescriptive/planning part of Steinitz' framework, change, impact, and decision models are included with a landscape ecological perspective. These models largely cover the domains of landscape planning and design, where alternative actions are conceived, evaluated, and implemented. Steinitz' framework provides a basis for continued advancement of theories and methods, and identifies the major questions and knowledge gaps to which interdisciplinary collaboration between landscape ecologists and planners may be directed.

Recently, several landscape planning methods have been advanced that explicitly include a landscape ecological perspective (Berger 1987; Steiner 1991; Van Buuren and Kerkstra 1993). A number of methodological changes can be identified in these methods, as contrasted with earlier methods. They are characterized by an interdisciplinary approach, they address landscape pattern:process at multiple scales, and they include a human ecological component.

One distinction between landscape ecology and earlier approaches to landscape planning is the integration of topological and chorological perspectives. Topological analysis is a parametric approach which describes and analyzes the "vertical" relationships between many factors that occur at a given location, be it a patch of wetland, a forest edge, or a residential neighborhood. The topological approach popularized by McHarg (1969) builds a "layer cake" of factors which collectively describe a place or places. Factors considered often start at the "bottom" with bedrock geology and are followed by surficial geology, soils, subsurface and surficial hydrology, vegetation, wildlife, and climate. In many methods, these factors are overlaid with aggregate values derived from the combinations of spatially concurrent factors.

In landscape ecological planning, the topological approach is complemented, not replaced by, a chorological approach which describes and anaylzes horizontal relationships and flows (Zonneveld 1995). It can describe dynamic spatial processes and particularly horizontal relationships such as hydrological dynamics between land uses, nutrient flows, metapopulation dynamics in fragmented landscapes, and human transportation.The chorological perspective is still in the process of integration with landscape planning.

Berger (1987) proposes a landscape planning framework that links the traditional physical, biological, legal, and economic tools of planning with a humanistic view of how people use, perceive, and shape a landscape. He suggests that a "land use ecology" could link the earth and life sciences with resource management. The framework is inherently interdisciplinary and presumes complementarity of the knowledge bases of the collectively disciplines that are involved with landscape planning. Berger's framework includes contributions from environmental history to analyze changes in the regional social and natural context; cultural ecology to derive a view of patterns of use and traditions as a basis for siting new uses; and cultural history to evaluate the ability of the environment to provide basic human needs. He argues for a humanistic view,to understand how ordinary landscapes fit into everyday life, as well as a knowledge of local attitudes towards the planning process.

In the "landscape planning working method," Steiner and Osterman (1988) and Steiner (1991) argue for an interdisciplinary and integrated approach, coordinated across scales, with significant public participation and a human ecological perspective. Their method explicitly links landscape pattern and process with planning goals and continues through to implementation and monitoring of the plan. This method advances the established practice of ecologically based planning, although its explicit consideration of three nested scales simultaneously and through its focus on the human ecological component of planning.

Lyle (1994) argues for a systemic planning process based on an ecosystem model. He attributes the failure of much past planning to its fundamental starting premise. Since the passage of the U.S. National Environmental Policy Act of 1969, much planning has been focused on attempting to reduce adversity, rather than in seeking systemic solutions. "We must learn to deal with environmental problems at the systemic level; if we heal the trunk and branches, the benefits to the leaves will follow naturally" (Karl-Henrik Robert, in Lyle 1994, p. 10).

Lyle's planning approach is based on three modes of ecosystem order: structure, function, and location. Integral with his plans are models of energy and nutrient flows and transformations. In this sense Lyle, like Steinitz, engages the fundamental landscape pattern:process dynamic. He also presents a somewhat radical yet optimistic perspective in which the creative intellectual capability of humans is understood to be a powerful ecological force itself, capable of reconceiving human environments according to an ecosystem model. Human intervention is not viewed as an activity whose impacts need to be controlled, but rather for its potential to create human ecosystems that are physically and biologically sustainable, which may add cultural meaning and may express the concept of sustainability in physical form in the landscape. This idea is closely related to spatial concepts discussed later in this chapter.

From this brief review of recent landscape planning theory and methods, the essential attributes of a landscape ecological planning method can be identified. The theory reviewed indicates an increasing awareness of landscape ecology as an essential basis for landscape planning, but also calls for more attention to the cultural aspects of landscape planning, and emphasizes its creative dimension.

Essential Attributes of a Landscape Ecological Planning Method

Landscape planning methods vary widely as a function of physical scale, planning objectives, time frame, goals, political support, availability of data and knowledge, level of participation in the planning process, and driving issues. The preceding review of planning theory identifies a clear interest in integrating landscape ecology with landscape planning. The following attributes are proposed as essential to a landscape ecological planning method, independent of the variables mentioned above. These attributes are integral to the framework method proposed in this chapter (Fig. 10.1).

1. The planning process is inherently interdisciplinary and integrates public and expert participation and advice.

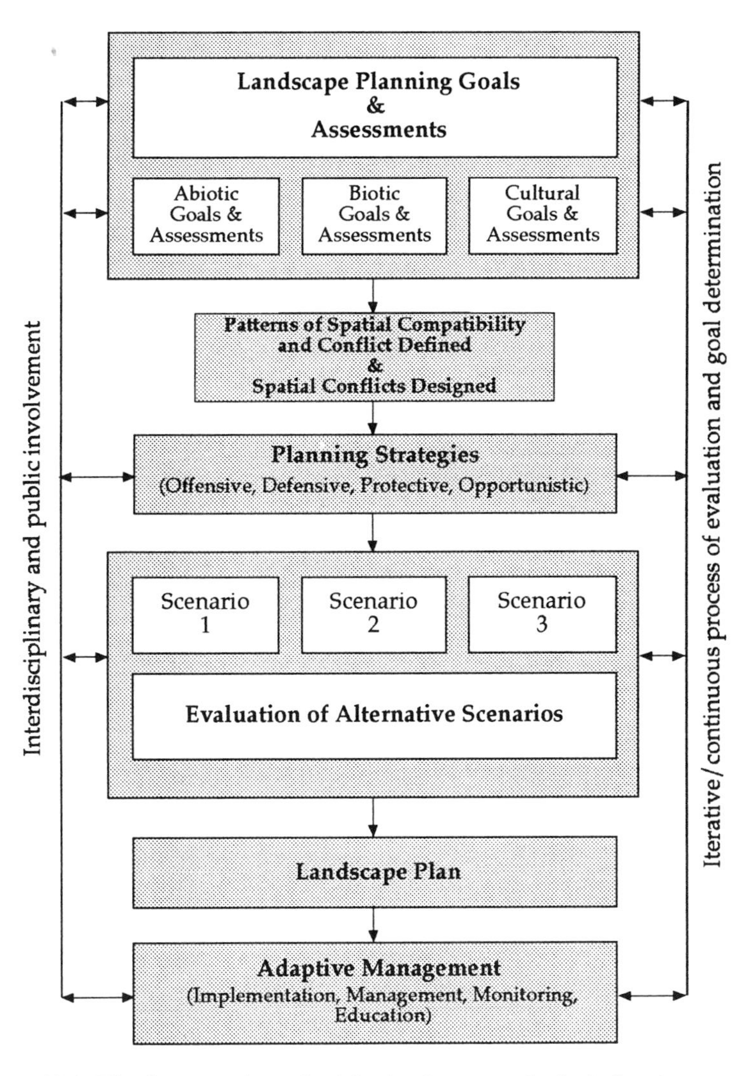

FIGURE 10.1. The framework method for landscape ecological planning a continuous, participatory, and interdisciplinary process.

The challenge of sustainable development depends on a constant infusion of knowledge from scientific research and monitoring. This expert knowledge is central to understanding the fundamental landscape pattern:process dynamic and must be integral to a landscape ecological planning process. Scientists representing the abiotic, biotic, and cultural disciplines should be fully integrated with the planning method.

A landscape plan is different from a research project. It will promote recommendations that may be implemented and will influence residents and stakeholders. Therefore, a participatory process involving nonexpert public officials, local inhabitants, and special interest representatives is essential. This type of planning process promotes "mutual learning" (Friedmann 1973) through which experts and participants are jointly involved in process leading to goal determination, integration of local landscape knowledge, perceptions, and values; evaluation of alternatives; and ultimately, implementation, monitoring, and management.

2. The dynamic relationship between landscape pattern and process is fundamental to the planning process.

If the ecosystem is understood as the basic unit of landscapes, then ecosystem structure and function are essential to understand in landscape planning (Turner 1989). The dynamic relationship of landscape pattern to process is therefore fundamental to both landscape ecology and to landscape planning. In current landscape ecological knowledge, the landscape pattern:process dynamic is perhaps best understood in its abiotic and biotic dimensions (Quinby 1988; Turner 1989; Forman 1990b; Schreiber 1990). Often landscape ecological plans focus only on goals articulated in terms of biodiversity, water quality, or soil protection. Although these goals are undeniably essential components of a sustainable landscape, the cultural component is also important, but has been given less attention in research and publications (Berger 1987; Schreiber 1990; Golley and Bellot 1991). Ndubisi et al. (1994) call for an integration of abiotic, biotic, and cultural understanding of landscapes as a basis for landscape planning at multiple scales. The cultural component should also consider the appearance of the landscape because people express understanding and preferences about ecological quality from the look of the land, and this, in turn is vital in landscape planning (Nassauer 1992).

When a landscape plan is based on an understanding of pattern: process dynamics, the plan can be explicitly linked with the consequences, and the plan itself, when implemented, may be considered a field experiment.

3. The planning process is explicit and replicable.

For the process to be equitable, rigorous, and defensible, the methods and processes must be explicit and replicable. This is particularly relevant in the era of GIS-based spatial analysis, through which methods and procedures can be subjected to rigorous tests of accuracy and replication. When the planning process is explicit and transparent, and the assumptions, variables, and goals clearly presented, several benefits can be realized. Interdisciplinary collaboration is facilitated, nonexpert participation is enabled, and alternatives can be generated that demonstrate the spatial and ecological consequences of modification of planning assumptions. This often leads to the generation of alternative planning scenarios, which will be discussed later.

4. The process should integrate knowledge, goals, and spatial concepts in a strategic manner.

Landscape planning is inherently a strategic process in that it attempts to understand and manage the elements and forces that are the causes of landscape changes rather than employing tactics to respond to the changes themselves (Sijmons 1990; Ahern 1995). Planning is, by definition, proactive, but not all planning is strategic. Effective strategic planning requires integration of interdisciplinary knowledge to define strategic goals consistent with political expectations, economic factors, and the reality of the existing landscape condition. Strategic landscape planning requires a balance of knowledge, vision, and political skills.

Spatial concepts and strategies are also ways to build systemic solutions to complex problems, to achieve symbiotic synthesis of rational and intuitive thought that collectively can identify spatial concepts that may get ahead of problems that landscape plans address.

5. Landscape planning is an iterative process integrating adaptive management.

As ecological knowledge has become more routinely integrated in planning, a common dilemma regarding the accuracy and certainty of the knowledge recurs. The planners ask legitimate questions, such as "How wide does the corridor need to be?" and the ecologist replies, "It is impossible to generalize this type of information—detailed, site-specific research is the only path to the answer." The concept of adaptive management (Holling 1978; Gunderson et al. 1995). addresses this dilemma by reconceptualizing the "problem" of making planning decisions with imperfect knowledge as an opportunity. In addition to contributing the best current knowledge available to a planning decision, the ecologist provides guidelines for implementation and monitoring, through which the planning decision may become a field experiment from which new knowledge may be generated.

Proposed Framework Method

The following process is presented as a working method to integrate landscape ecology and landscape planning. It is a framework in that it identifies the major components (steps) in the process. In actual use the framework will be complemented with additional steps to complete the planning process, including goal definition, resource assessments, alternative plan generation, and implementation.

The framework method is graphically presented as a linear process (Fig. 10.1). In use, however, the process is intended to be nonlinear, cyclical, and iterative, and it may be initiated at any stage. For example, the planning process may start with an evaluation of a preexisting plan, followed by a revision of earlier goals, and resource assessments.

In Figure 10.1 the process begins with a determination of sustainable landscape planning goals, defined strategically to match the public will, economic climate, and existing landscape condition. In a participatory process including interdisciplinary experts and stakeholders, specific goals are proposed for abiotic, biotic, and cultural resources. A synthesis of these assessments defines areas of potential spatial conflict and compatibility and is used to design spatial concepts. Next, appropriate planning strategies are selected from protective, defensive, offensive, or opportunistic approaches (Ahern 1995). The spatial concepts are used to design a number of scenarios to illustrate possible futures, including the means to their realization. With expert and stakeholder participation, the scenarios are evaluated and ultimately revised or modified into a dynamic landscape plan. When the plan is implemented, a policy of adaptive management is followed, based on monitoring to yield new knowledge for continuing the planning process.

The framework method includes several key ideas/steps that will be discussed in greater detail: planning strategies, spatial concepts, and future landscape scenarios. An Open Space Plan for Orange Massachusetts illustrates the application of the framework method.

Spatial Concepts

> Make no little plans for they have no magic to stir men's blood, and probably themselves will not be realized. Make big plans: aim high in hope and work, remembering that a noble, logical diagram, once recorded, will never die.
>
> Daniel Burnham

A spatial concept expresses through words and images an understanding of a planning issue and the actions considered necessary to address the issue. Spatial concepts are related to the proactive, or anticipatory, nature of landscape planning, in that they express solutions to bridge the gap between the present and the desired future situation. Spatial concepts are often carefully selected metaphors, for example "Green Heart" or "Stepping Stones" which communicate the essence of the concept clearly, to build consensus, and as a basis for more concrete planning decisions (Steiner 1991; Zonneveld 1991; van Langevelde 1994). Spatial concepts structure the planning process. Five functions of spatial concepts can be defined: (1) the cognitive in which interdisciplinary knowledge is synthesized; (2) the intentional to manifest the creative insights of planners and designers; (3) the institutional to influence landscape regulation; (4) to improve communication between experts, stakeholders, and special interests; and (5) action to influence the achievement of planning objectives (Zonneveld 1991). Translation of knowledge of landscape pattern and process is a key value of landscape ecology to spatial concept development. Spatial concepts often manifest basic assumptions upon which more specific decisions can be based.

Although scientific input from landscape ecology is essential to conceive spatial concepts, its potential is limited. Many scientists are reluctant to make the "leaps of faith" that are essential to conceive spatial concepts. There is an essential element of creativity in the design of spatial concepts. They represent an interface of empirical and intuitive knowledge. If human intellectual and spiritual activities are accepted as valid ecological elements, then, clearly spatial concepts are a legitimate part of the planning process (Lyle 1994; Zonneveld 1995). Through spatial concepts, rational knowledge is complemented with creative insights. Spatial concepts in landscape planning can be thought of as design concepts—essential ideas that transcend basic knowledge and result in successful solutions. In site-scale landscape architecture, design concepts are the basis for giving physical form in response to; goals, resource assessments, and the designer's creative insight. In landscape planning, spatial concepts are the basis for giving form to landscapes in like manner, in either a generic or a spatially specific manner. Figure 10.2 presents a series of spatial concepts that have been used in landscape planning. Some are intentional, others result from the long-term interaction of physical, biological, and cultural forces. All can be linked with metaphors and synonyms that aid in their imageability by scientists, planners, and those involved with, or affected by, a plan.

The Netherlands has a rich tradition of landscape planning and has long employed spatial concepts in the planning process. The "Green Heart" is a good example of a spatial concept in Dutch landscape planning. It is a spatial strategy to maintain a "green core" of agriculture, forests, and recreation within the densely populated western Netherlands. The core is surrounded by the Randstad (Ring City), which is a reciprocal strategic spatial concept. The "green heart" concept has significantly guided Dutch planning and development strategies since the 1950s, during a major period of population growth and land-use change.

The framework concept is a more contemporary spatial concept for landscape planning in the Netherlands. It is based on the paradox of time and uncertainty. Change and uncertainty are both fundamental in natural and cultural systems. The landscape is no different—change is fundamental—and uncertainty is a "given." This is the paradox of time in landscape planning. Some key ecological processes, like groundwater recharge, require a certain level of stability to function within acceptable limits. These

FIGURE 10.2. The landscape of Orange, Massachusetts, displays spatial heterogeneity within a forested matrix.

are the "low dynamic" functions. In the case of groundwater recharge, a degree of stability is necessary in terms of vegetative cover, soil stability, and nutrient inputs to maintain a renewable supply of clean groundwater. Other processes in the landscape, driven by social and economic forces, are more uncertain and "high dynamic." (such as land-use change) and require flexibility. The framework provides nature a long-term stability and allows more flexibility for land use change in the other areas. This is the framework's quid pro quo (Ahern and Kerkstra 1994). In the Netherlands the framework concept is known as "casco" in reference to an architectural practice in which buildings are designed with only a main structural framework, allowing occupant modification. In the context of this discussion, casco is both a spatial and conceptual framework for landscape planning (Hamhuis et al. 1992). The framework concept promotes a spatially integrated network of lands, managed for "low dynamic" functions and uses, based primarily on abiotic factors. It is spatially defined by the hydrologic landscape structure, in which discrete geohydrological units can be identified (Kerkstra and Vrijlandt 1990; Buuren van and Kerkstra 1993). Within this network structure, which is reserved for "low dynamic" functions, are opportunities for "high dynamic" functions and uses (Fig. 10.2).

Forman's spatial solution (1995) is a generic spatial concept based on the assertion that certain indispensable landscape patterns exist in all landscapes and that a spatial solution can be defined to support these patterns and their associated ecological functions. The indispensable patterns are large patches, riparian corridors, bits of nature, and connecting corridors. Forman adds a dynamic dimension to his spatial solution, the "jaws" model, through which the presence of the indispensable patterns is maintained for the maximum time through a process of landscape transformation. The jaw's model succeeds in translating fundamental landscape ecological knowledge into the spatial language of planning.

Greenways have become a popular spatial concept in North America in the last decade. It is a spatial concept based on the particular advantages of linked linear systems. It has captured popular attention in a manner unprecedented for a landscape planning issue (Little 1990). The President's Commission on Americans Outdoors (1987) provided the metaphor of a "the giant circulating system" capturing the public attention and presenting just the type of powerful, clear logical diagram that Burnham describes. Greenways are multipurpose and often favor recreation over ecological goals. The ecological potential of greenways has been increasingly recognized and has been the subject of significant recent research (Smith and Hellmund 1993; Fabos and Ahern 1995).

Landscape ecology can assist in the conception and evaluation of spatial concepts. It can identify indispensable patterns that support a ecological functions, it can demonstrate the strategic benefits of connectivity, particularly the movement of organisms (Soulé 1991; Vos and Opdam 1993). It also provides a basis for understanding the frequency and distribution of disturbances, defining a minimum dynamic area, and linking other forms of pattern:process knowledge with the landscape planning process.

Planning Strategies

Landscape planning is an inherently strategic activity. It strives to craft policies and actions that systematically address the trends and forces that shape and change landscapes. Strategic planning is driven by goals that are focused, linked with implementation, and presumed to be achievable. When strategic planning is informed by a landscape ecologically informed understanding of pattern:process dynamics, and is guided by appropriate spatial concepts, it may form a sound basis for plan development and implementation. There are a number of fundamental strategies that can be employed, including protective, defensive, offensive, and opportunistic (Ahern 1995).

When the existing landscape supports the abiotic, biotic, and cultural resource goals, a protective planning strategy may be employed. Essentially this strategy articulates the spatial pattern that is desirable and protects it from change. Conversely, it defines the areas in the landscape where change can be accommodated. The protective strategy is useful in relatively undisturbed landscapes and can often be applied at low cost. Ironically, it is difficult to promote politically because, by definition, it is used when the landscape is already functioning well. While landscape planners attempt to be forward thinking and anticipatory, human nature is often reactive. In this case, education and public awareness are useful to promote understanding of the issues and strategic options available.

When the existing landscape is already in a spatial configuration that is negatively impacting abiotic, biotic, or cultural resources, a defensive strategy is needed. This strategy seeks to control and arrest the negative processes of landscape change (i.e., fragmentation, dissection, perforation, or attrition) (Forman 1995). As a last resort, the defensive strategy is often appropriate, but it can also be described as reactionary and ineffective. By definition, a defensive strategy attempts to "catch up with" or "put on the brakes" against the inevitable process of landscape change. When the root causes of negative landscape change remain active, the defensive strategy will never be completely effective and best delays the inevitable change in defense of an ever-decreasing nature (Sijmons 1990).

In marked contrast with the defensive, the offensive strategy is inherently proactive in nature. It is appropriate when the landscape is already deficient with respect to supporting biotic, abiotic, or cultural resources. It promotes a "possible" future landscape that can be realized only through restoration. Since, by definition, it cannot be guided by an assessment of existing resources, it must be based on a spatial concept crafted by a combination of rational and creative processes. The offensive strategy relies on knowledge from; landscape ecology, planning, and ecological restoration. It is costly and uncertain. This strategy is often practiced in Europe, where centuries of use have produced a cultural landscape with limited opportunities for protection or defense of desirable landscape patterns and associated processes.

Often landscapes contain unique elements or configurations of elements that allow for opportunistic landscape planning. These unique elements may or may not be optimally located, but represent positive opportunities, nonetheless. This strategy is dependent on the presence of certain unique landscape elements, which are often in the configuration of a corridor (e.g., abandoned railroad lines, transmission line corridors) or as a remnant environmental resource patch (Forman and Godron 1986). This strategy involves recognition of such special opportunities and integrating them with other planning strategies, often with the opportunistic strategy.

These four strategies collectively constitute a typology (Ahern 1995). The typology can promote more accurate communication between landscape ecologists, planners, and stakeholders in the planning process. The strategies are not mutually exclusive—they are more often used in an integrated manner. The strategies are a key link between abiotic, biotic, and cultural resource assessments, spatial concepts and scenarios with a resulting landscape plan.

Future Landscape Scenarios

Scenarios are important tools for landscape planning and are integral with the framework landscape ecological planning method. They provide a perspective that is not constrained by the present situation. Scenarios have been used in corporate and governmental decision making since the 1970s because of their inherent advantages over expert judgments and other planning approaches. In landscape planning, scenarios are well suited to linking goals and assumptions with potential future spatial changes. A complete scenario should include a description of the current situation, a potential future state, and a means of implementation. Without all three of these elements, scenarios can be faulted as utopian. Scenarios are different from forecasts that attempt to predict the expected future. In contrast, scenarios pose, and answer a series of "if then" questions. Scenarios may be based on mathematical models. Other models can be used in scenarios or they may be more normative. The scenario approach is more appropriate when there is a great deal of uncertainty concerning the future or when there is a general dissatisfaction with the present. Trend breaks are one reason that scenarios may be more useful than forecasts. Changes in technology or global economics can cause a paradigm shift that can alter the most fundamental assumptions in a planning activity (Schooenboom 1995; Veenenklaas 1995).

Two fundamental types of scenarios can be defined—state and process. A state scenario simply describes a future situation without articulating the steps or events needed to get there. A process scenario provides a "road map" of assumptions, events and steps linking the present with the future (Hirschorn 1980). Process scenarios are the most appropriate for landscape planning. Two fundamental types of process scenarios can be identified. A "forecasted," or "beginning state driven," scenario projects current trends and control practices to produce a trajectory on which a possible future may be conceived. A common forecasted scenario in landscape planning is the "build-out", in which current land use controls are used to determine a theoretical or maximum level of development. Build outs do not represent predictions, but establish a theoretical maximum level of development as base line for comparison of other alternatives. A "backcasted," or "end state driven," scenario, in contrast, is based on an idealized spatial concept, or vision, of what the future could be. Backcasted scenarios are often designed to articulate and visualize the spatial consequences of planning goals or assumptions and the steps necessary to realize them (Schoonenboom 1995). Hirschorn (1980) proposes that "developmental scenarios" are most useful for planning. They are process based, beginning state scenarios containing "chains of cause and effect," with explicit decision rules to link the present with a possible future that is both plausible and surprising.

In many instances, alternative scenarios are intentionally generated with the explicit purpose of demonstrating a range of alternatives. This has been described as identifying the four corners of an abstract frame within which a more balanced or compromised alternative may be selected (Harms et al. 1993). Or as the four points of a tetrahedron, indicating a more dynamic third-dimensional aspect to the alternatives (Forman 1995). Scenarios are not predictions; they are vignettes of possible futures, of what could be given specific assumptions and actions, as opposed to what will be.

Scenarios may also be classified according to such motivating factors as biodiversity protection (Harms et al. 1993); development control (Steinitz et al. 1994); visual impact management, resource allocation, or integratation of several goals. Scenarios should be methodologically explicit and replicable to facilitate rigorous tests of accuracy and replication. They are common in European and particularly in Dutch planning and were the theme of a recent international symposium (Schoute et al. 1995)

Application of the Method: Open Space Plan for Orange, Massachusetts

The Open Space Plan for the town of Orange was conducted in a graduate landscape architecture studio directed by the author at the University of Massachusetts in 1996 (DLARP 1996). A participatory planning process, based on the framework method proposed in this chapter, was conducted with significant involvement of town officials, representatives of special interests, stakeholders, and the general public.

The goal of the plan was to provide the town with a basis for long-term, land-use decision making, with an emphasis on open space and recreation. The open space goal was broadly interpreted to comprehensively address the abiotic, biotic, and cultural issues that affect landscape pattern and process. Thus development, biodiversity, and economic planning were incorporated into the open space plan. The recreational component of the plan was also conceived at the landscape scale, focusing on those "extensive" recreational activities that can benefit from, and successfully coexist with, more conservation-based landscape planning and management.

Abiotic Resource Goals and Assessments

The town is located the northeast of the United States in North Central Massachusetts, on the Millers River, a major tributary of the Connecticut River. The town is bordered to the east and north by significant upland schist ridges with less resistant gneiss valleys and alluvial riparian corridors and wetlands. The soils formed on this rugged geologic base of Orange have limited agricultural potential.

An assessment of abiotic resources was conducted and identified the following as key resources to be protected in the open space plan: the upland ridges that define the character of the town, provide important wildlife habitat, and recreational opportunity; the Millers River and floodplain that provide flood protection, a wildlife habitat, and recreational and amenity benefits; a system of forested and nonforested wetlands that provide multiple benefits and functions; two lakes important for wildlife, recreation, and cultural amenity; and a large glacial outwash plain that provides a major portion of the town's drinking water.

Biotic Resource Goals and Assessments

The land cover of Orange is predominately forested (85%) with significant spatial heterogeneity (Fig. 10.3). Forest ages vary as a function of timing of agricultural abandonment, natural disturbance, and hurricanes and periodicity of forest harvesting activities. Forest composition includes large patches (>1000 hectares) of hardwoods (*Acer*, *Quercus*, *Fraxinus*) softwoods (*Pinus*, *Tsuga*), and mixed hard and softwoods. The extensive forest in Orange is part of a large regional forest extending across north central Massachusetts into southern New Hampshire. Maintaining linkages with the regional forest was identified as a priority in the open space plan.

The biodiversity component of the plan relied largely on a target species approach. A target species is one determined to be an appropriate goal or "target" in biodiversity planning by virtue of its habitat requirements, position in the food chain, and compatibility with human occupation and disturbance. A target species is not the same as an indicator species, but is an acceptable representative of a particular habitat type, and associated species can be reliably assumed to also be present. The selection of target species for this plan was made in consultation with wildlife biologists from the University of Massachusetts and the U.S. Fish and Wildlife Service (DeGraff, personal communication).

Two target species were selected for this study, one for each of the town's nondeveloped extensive landscape types—forests and wetlands/riparian zones. The pileated woodpecker (*Dryocopus pileatus*) was selected as a target species for the town's forested landscapes, and the mink (*Mustela vison*) for the wetlands and riparian areas, including lakes rivers, and forested and nonforested wetlands. Both species are indigenous to the area and are present in viable populations.

In assessing the habitat quality for the pileated woodpecker, the following criteria were applied: 20 hectare minimum forest patch size, with a preference for wetland adjacency (Bent 1992). Lower suitability assessments resulted from smaller forested patch sizes and when the adjacent land use was cleared, agriculture or development. The assessment indicated that abundant areas of high quality habitat were present, especially on the town's upland forested ridges and lower riparian zones.

Spatial Concept	Examples & References	Metaphors & Synonyms	Diagram
Containment	• cloister • fortification • greenbelt • refuge	• border • barrier • wall • harness • levee	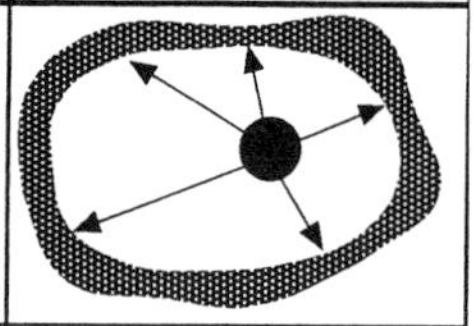
Grid	• U.S. 1785 Land Ordinance Survey • International School	• network • rational • authority • eglatarian • anthropocentric	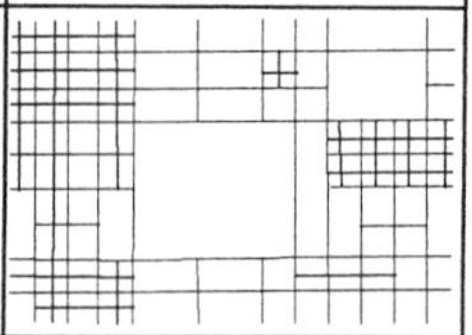
Interdigitation	• New Exploration (MacKaye 1962) • Pattern Language (Alexander et al 1977) • Forman 1990a	• symbiosis • harmony • biocentric • interdependent	
Segregation	• Compartment Model (Odum 1969) • Euclidian Zoning • MAB Biosphere Reserves	• controlled • strategic • compromise • *quid pro quo*	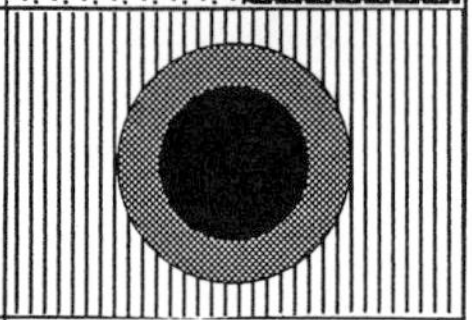
Network	• National Ecological Network (Netherlands) • U.S. Interstate Highway System	• integrated • linked • nodes & corridors • stepping stones	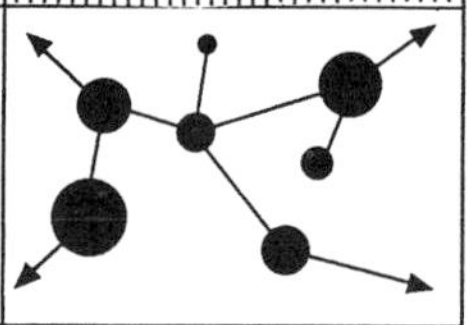
Framework	• CASCO, Plan Stork (de Bruin et al 1987) • hydrological framework (van Buuren & Kerkstra 1993)	• integrated network • topological & chorological • low dynamic	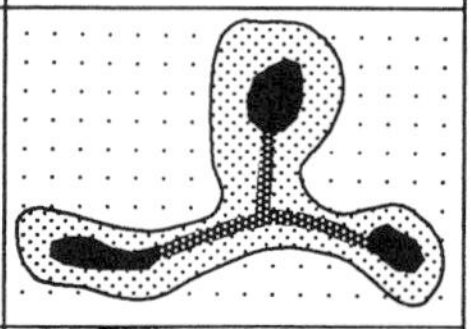
***Laissez faire* (defacto no strategy)**	• suburban sprawl • Megalopolis (Gottman 1961) • Edge City (Garreau 1991)	• mosaic • individualistic • dynamic • free-market • competitive	

FIGURE 10.3. Spatial concepts for landscape planning.

The mink is a predatory semiaquatic mammal that prefers riparian margins, lake shores, and marshes for habitat. Mink are moderately adaptable and will change their habitats for food and cover in response to human disturbance. The following criteria were applied in assessing mink habitat:

most suitable habitats include palustrine wetlands over 400 hectares with permanent water and >100 meter woody buffer; palustrine, lacustrine, or riverine wetlands with <100 meter woody buffer were secondary; palustrine wetlands with <9 months of permanent water were considered marginally suitable (Allen 1986). In applying these criteria, significant amounts of habitat were found, and a basis for planning for protection was established.

Cultural Resource Goals and Assessments

The town of Orange is seeking a vision for its future that will protect its historical resources, and rural character, while providing a healthy economic future. To identify the town's cultural resources, a two part interview process was conducted. The goal of the interviews was to identify the town's demographic composition, to define special interests, and to spatially define areas of interest of individuals and special interest groups. The interviews were conducted with key informants—those known to represent special interests or those with special knowledge of the town's historical and cultural resources. Spontaneous interviews were also conducted to obtain a representative understanding of the resident's attitudes and values towards recreation, open space planning, and development. The interview results were compiled and cross-tabulated to identify and map trends and areas of particular concern. The findings emphasized the residents' value of the town's center, historical resources, and remote natural areas, chiefly along the upland ridges. To determine the distribution and extent of land available for future development, an eliminative process was applied. Development planning is often absent from open space planning, resulting in a conservation bias that causes problems with public acceptance of the plan. The Net Usable Land Area Process NULA (Fig. 10.4) first eliminates lands that are already developed, protected, or regulated against development. Next critical resources are eliminated. In Orange, these included aquifers and their recharge areas, prime agricultural soils, sand and gravel deposits, and endangered species habitats. The rationale for eliminating critical resources was based on a participatory process through which consensus was built on resource protection. Thus the definition of critical resources reflects the community's values. The last stage in the eliminative process involves the removal of hazardous areas including, 1% probability floodplain, slopes over 25%, soils with seasonally high water table, and soils with very poor drainage. These areas are eliminated because development would result in hazards to individuals or to society. The NULA process is a simple method for focusing the planning activity on those areas where land use competition or conflict may exist. All of the lands that pass through the NULA process are potentially available for development.

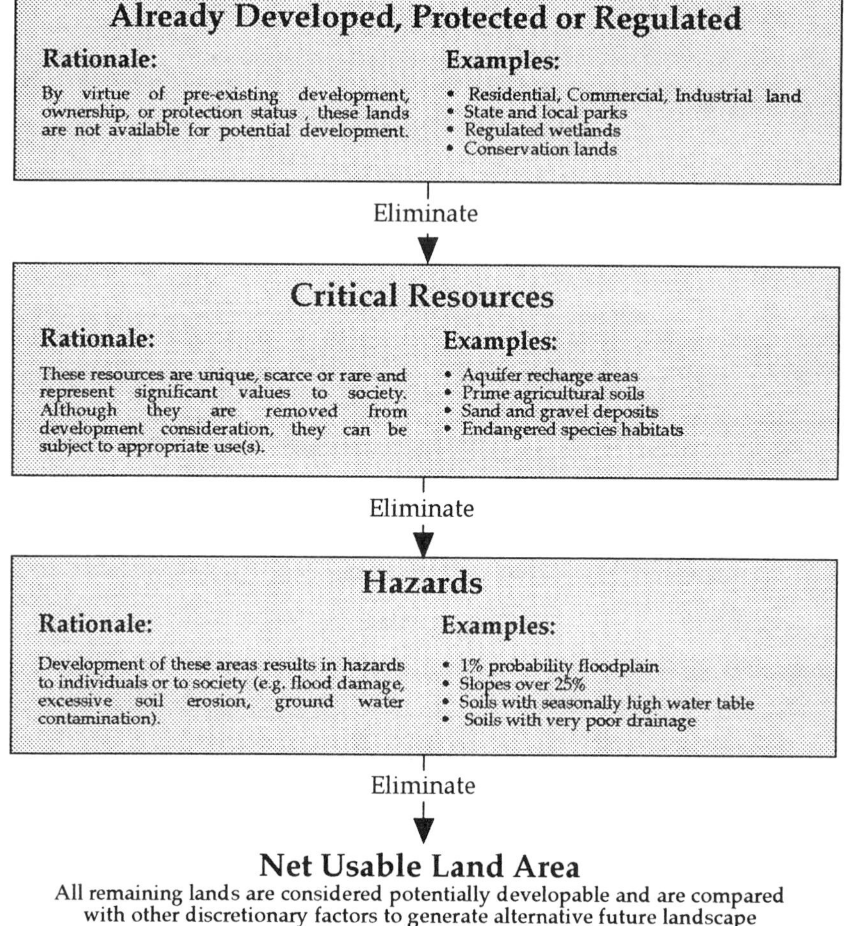

Planning Scenarios

The comparison of mapped assessments of biotic, abiotic, and cultural resources identifies patterns of compatibility or conflict and provides a basis for the design of spatial concepts. Following the framework method, next alternative three scenarios with associated spatial concepts were generated (see Fig. 10.5). These scenarios illustrate the consequences of varied resource assessments and spatial concepts.

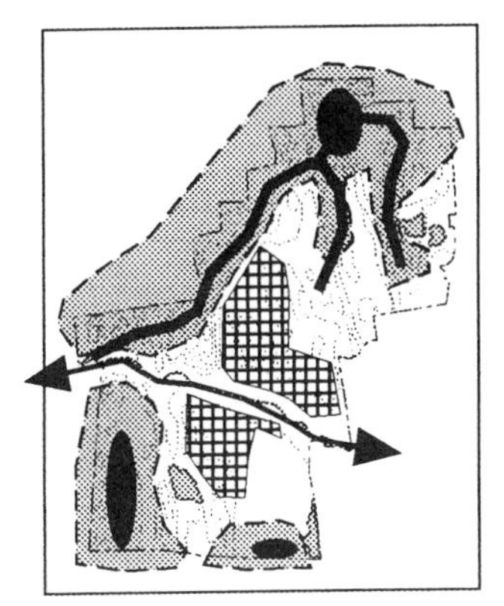

Biodiversity Scenario

Goal: To protect the town's unique biological resources, and to link with regionally significant wildlife habitats and movement corridors

Spatial Concept: Segregation of protection and development, buffering and linkage of large forest and wetland patches.

Implementation Strategies: Acquire easements for large forest patchesand buffers, prepare forest harvest management plans, identify wetland and water resource management "gaps", integrate with educational and recreational programs.

Recreation Scenario

Goal: To support a variety of recreational activitities that benefit from the town's natural and cultural resources, to provide linkages with regional recreational resources, and to promote the economic potential of recreation in the town.

Spatial Concept: Biotic, Abiotic and Cultural linkages.

Implementation Strategies: Easements for trail and hunting access, public ownership of shorelines, protect historic districts, adopt zoning to encourage infill development and to protect scenic road corridors.

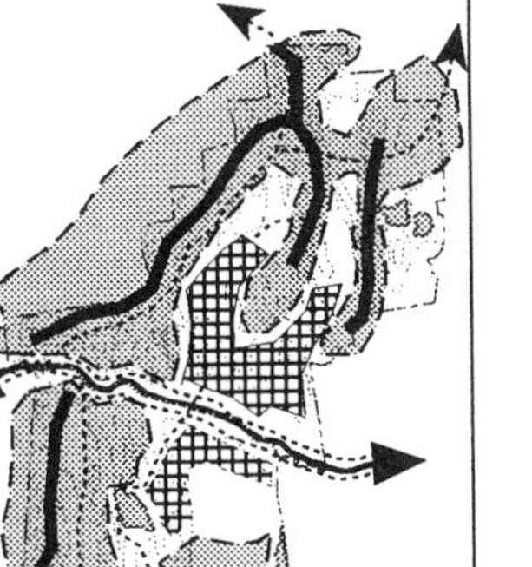

Citizen's Scenario

Goal: To integrate the varied interests of the town, balancing development and protection, to preserve the town's rural character, to encourage regional linkages and to be a catalyst for economic development.

Spatial Concept: linkage and integration

Implementation Strategies: Increased public access and linkage to recreational resources, implement the town's greenway and heritage trail plans, encourage infill development and open space zoning.

FIGURE 10.5. Landscape planning scenarios for Orange, Massachusetts.

The biodiversity scenario is based on a spatial concept of segregation of protection and development. Large patches of forest and wetland habitat for the indicator species are identified for protection. Regional linkages are planned. Recreational access is carefully controlled, and development is concentrated adjacent to the existing town center, enabled by additional urban infrastructure. In outlying areas of the town, development is restricted in wetlands and along lake shores to protect habitat. The scenario can be implemented through a major program of land acquisition and easements. Forest harvesting is to be carefully managed to preserve large habitat patches. An educational program is planned to assist understanding of the value of biodiversity protection.

The recreation scenario emphasizes recreational opportunity, both within the town and the adjacent protected areas. The spatial concept is one of linkage of abiotic, biotic, and cultural resources. A planned greenway on the Millers River is promoted and is linked with other trails throughout the town. Development is encouraged in a more dispersed manner than in the biodiversity scenario, emphasizing the recreational value of the protected areas to the town's residents. Priorities identified through interviews have been integrated with the plan. The overall pattern of protection mirrors the town's distinct pattern of ridges, riparian corridors, and wetlands. The scenario can be implemented largely through easements to improve access to protected lands, with new public acquisitions proposed for areas of high recreational value, such as lakefronts. The potential economic benefits of increased recreational activity have been recognized in this scenario.

The citizen's scenario is a based on the spatial concept of linkage and integration. It reflects more directly the ideas and values obtained from the citizen's during the interview process. The scenario includes an ecological framework to protect the habitats of the indicator species, but unlike the biodiversity scenario, integrates recreational uses (trails, hunting) with the habitat areas. The "fingers" of protected land that extend into the town are key routes to provide recreational access, while the larger patches to the north can remain less disturbed. The recreational component of this scenario emphasis on increased public access and linkages. An innovative "heritage trail" is proposed to link the town's historical, cultural, and physical resources into a series of trails, for auto, bicycle, and foot travel. This trail benefits from the context of protected lands, and provides the greatest potential for recreationally related economic benefits for the town. Open space development is proposed as a means of financing the protection of large patches of forest and wetland and to manage the rural character of the town.

The scenarios were evaluated and a preliminary open space plan resulted. Specific recommendations were made for implementation and monitoring (i.e., to learn if the habitat suitability assessment is appropriate for the target species selected). Additional measures were identified to monitor the abiotic resources, particularly soil resources and surface and groundwater quality and quantity. A final step in this exercise was the initiation of an education program to raise awareness of landscape planning issues and to identify areas of mutual interest and opportunity.

Summary

This application illustrates how the proposed framework method can be applied to a routine landscape planning exercise. Through the abiotic, biotic, and cultural resource assessments, patterns of compatibility, and conflict were identified and served as a basis for the development of spatial concepts and scenarios. Each scenario includes strategies for implementation and collectively aided in the ultimate determination of the open space plan. Through specific recommendations for monitoring, adaptive management is promoted and new knowledge regarding the effectiveness of particular planning decisions is gained. This application illustrates but one application of the framework method, which is conceived so as to be adaptable across a range of scales, landscape contexts, and planning issues.

Conclusion

Several key principles have been presented concerning the evolving theory of landscape ecological planning. There is an assumed complementarity of rational and intuitive knowledge. Both types are essential to formulate and evaluate scenarios and plans for sustainable landscape. Spatial concepts provide a means of communicating the essence and intuitive intent of planners in a manner that informs and guides the rational component. Landscape ecological planning involves abiotic, biotic, and cultural resources. For each resource type, procedures for resource assessment and goal determination have been defined. This informs an awareness and understanding of a landscape's present and future pattern:process dynamic in a rigorous and informed manner.

Landscape ecological planning is a strategic process informed by landscape pattern:process knowledge and guided by goals determined by those with a stake in the outcome. Scenarios are useful in this strategic process to escape from thinking that is rooted in the status quo, and to offer alternative futures that may be both visionary and ecologically accountable.

Landscape ecological planning methods such as this proposed framework method can facilitate interdisciplinary cooperation between physical and social scientists and planners. This may result in more informed plans, but perhaps even more importantly, may advance the knowledge base as a new product of planning. An important next step in the evolution of this planning framework, and related methods, is to articulate the essential criteria for evaluating plans and scenarios against abiotic, biotic, and cultural resource management goals. Such criteria will, in turn, structure the evaluative models that are essential to realize sustainable landscape scenarios and plans.

Acknowledgments. This material is based upon work supported in part by the Cooperative State Research, Extension, Education Service, U.S. Department of Agriculture, Massachusetts Experiment Station, Project No. 761. The work presented in this chapter has benefited from numerous collaborations with colleagues at the University of Massachusetts and the Wageningen Agricultural University in the Netherlands. I would like to particularly acknowledge assistance from the following individuals: George Tukel, René nij Bijvank, Vera Kolias, and the students in the Orange landscape Planning Studio: Amy Ansell, Erica Broussard, Eric Freer, Mark Fisher, Megan Gardner, Timothy Gerrish, Ingrid van Herel, Ellen Jouret-Epstein, Michael Lally, Regina Leonard, Tim Silva, Kyle St. Germaine, Jay Vinskey, and Selene Weber.

References

Ahern, J. 1995. Greenways as a planning strategy. In: Greenways: The Beginning of an International Movement, pp. 131–155. J.G. Fabos and J. Ahern (eds.). Elsevier, Amsterdam.

Ahern, J., and K. Kerkstra. 1994. Time space ecology and design: landscape aesthetics in an ecological framework. In: Proceedings of the ASLA Symposium on Ecology, Aesthetics and Design, October 1994, San Antonio, TX.

Alexander, C., S. Ishikawa, and M. Silverstein. 1977. A Pattern Language. Oxford University Press, New York.

Allen, A.W. 1986. Habitat Suitability Index Models: Mink. Habitat Procedures Group. National Ecology Center, U.S. Fish and Wildlife Service. Biological Report 82, November. Washington DC.

Bent, A.C. 1992. Life Histories of Northern Woodpeckers. Indiana University Press, Bloomington and Indianapolis, IN.

Berger, J. 1987. Guidelines for landscape synthesis: some directions—old and new. Landscape and Urban Planning 14:295–311.

Bruin, D. de, D. Hamhuis, L. van Nieuwenhuijze, W. Overmars, D. Sijmons, and F. Vera. 1987. Ooievaar: De toekomst van het rivierengebied (Stork: The future of the rivers area) Stichting. Gelderse Milieufederatie, Arnhem, Netherlands.

Buuren, Michael van, and K. Kerkstra. 1993. The framework concept and the hydrological landscape structure: a new perspective in the design of multifunctional landscapes. In: Landscape Ecology of A Stressed Environment, pp. 219–243. C.C. Vos and P. Opdam (eds.). Chapman & Hall, London.

DeGraff, R. 1996. Personal communication. U.S. Fish and Wildlife Service, Amherst, MA. April.

Department of Landscape Architecture and Regional Planning. 1996. A Foundation for an Open Space and Recreation Plan for Orange, Massachusetts, unpublished report. University of Massachusetts, Amherst.

Fabos, J.G. 1985. Land-Use Planning: From Global to Local Challenge. Chapman & Hall, New York.
Fabos, J.G., and J. Ahern. 1995. Greenways: The Beginning of an International Movement, Elsevier, Amsterdam.
Forman, R.T.T. 1990a. Ecologically sustainable landscapes: the role of spatial configuration. In: Changing Landscapes: An Ecological Perspective, pp. 261–278. I.S. Zonneveld and R.T.T. Forman (eds.). Springer-Verlag, New York.
Forman, R.T.T. 1990b. The Beginnings of landscape ecology in America. In: Changing Landscapes: An Ecological Perspective, pp. 35–44. I.S. Zonneveld and R.T.T. Forman (eds.). Springer-Verlag, New York.
Forman, R.T.T. 1995. Land Mosaics: The Ecology of Landscapes and Regions. Cambridge University Press, Cambridge.
Forman, R.T.T., and M. Godron. 1986. Landscape Ecology. John Wiley & Sons, New York.
Friedmann, J. 1973. Retracking America: A Theory of Transactive Planning. Anchor Press, Garden City, NY.
Garreau, J. 1991. Edge City: Life on the New Urban Frontier. Doubleday, New York.
Golley, F.B., and J. Bellot. 1991. Interactions of landscape ecology, planning and design. Landscape and Urban Planning 21:3–11.
Gottman, J. 1961. Megalopolis: The Urbanized Northeastern Seabord of the United States. MIT Press, Cambridge, MA.
Gunderson, L.H., C.S. Holling, and S.S. Light (eds.). 1995. Barriers and Bridges to the Renewal of Ecosystems and Institutions. Columbia university Press, New York.
Hamhuis, D. van Nieuwenhuijze, and D. Sijmons. 1992. Het casco-concept. Ministerie van Landbouw, Natuurbeheer en Visserij. Utrecht, Netherlands.
Harms, B.H., J.P. Knaapen, and J.G. Rademarkers. 1993. Landscape planning for nature restoration: comparing regional scenarios. In: Landscape Ecology of a Stressed Environment, pp. 197–218. C.C. Vos and P. Opdam (eds.). Chapman & Hall, London.
Hersperger, A.M. 1994. Landscape ecology and its potential application to planning. Journal of Planning Literature, 9(1):14–29.
Hirschorn, L. 1980. Scenario writing: a developmental approach. APA Journal April: 172–183.
Holling, C.S. 1978. Adaptive Environmental Assessment and Management. John Wiley & Sons, New York.
IUCN. 1980. The World Conservation Strategy, IUCN, UNEP, WWF. Gland.
Kerkstra, K., and P. Vrijlandt. 1990. Landscape planning for industrial agriculture: a proposed framework for rural areas. Landscape and Urban Planning, 18:3–4, 275–287.
Kiemstedt, H. 1994. Landscape Planning: Contents and Procedures. The Federal Minister for Environment, Nature Protection, and Nuclear Safety, Bonn, Germany.
Langevelde, F. van. 1994. Conceptual integration of landscape planning and landscape ecology, with a focus on The Netherlands. In: Landscape Planning for Ecological Networks, pp. 27–69. E.A. Cook and H.N. van Lier (eds.). Elsevier, Amsterdam.
Lewis, P. 1996. Tomorrow by Design: A Regional Design Process for Sustainability. John Wiley & Sons, New York.

Little, C. 1990. Greenways for America. Johns Hopkins University Press, Baltimore.
Lyle, J.T. 1994. Regenerative Design for Sustainable Development. John Wiley & Sons, New York.
Lynch, K. 1985. Good City Form. MIT Press, Cambridge, MA.
MacKaye, B. 1962 (reprinted). The New Exploration: A Philosophy of Regional Planning 1928. University of Illinois Press, Urbana, IL.
McHarg, I.L. 1969. Design with Nature. Natural History Press, Garden City, New York.
McHarg, I.L. 1997. Ecology and Design. In: Ecological Design and Planning, pp. 321–332. George F. Thompson and Frederick R. Steiner (eds.). John Wiley & Sons, New York.
Nassauer, J.I. 1992. The appearance of ecological systems as a matter of policy. Landscape Ecology 6:4, 239–250.
Ndubisi, F. 1997. Landscape Ecological Planning. In: Ecological Design and Planning, pp. 9–44. George F. Thompson and Frederick R. Steiner (eds.). John Wiley & Sons, New York.
Ndubisi, F., T. deMeo, and N.D. Ditto. 1995. Environmentally sensitive areas: a template for developing greenway corridors. In: Greenways: The Beginning of an International Movement, pp. 159–177. J.G. Fabos and H. Ahern (eds.). Elsevier, Amsterdam.
Netherlands Ministry of Agriculture, Nature Management and Fisheries. 1990. Nature Policy Plan of the Netherlands, The Hague.
Odum, E.P. 1969. The Strategy of Ecosytem Development. Science 164:262–270.
President's Commission on Americans Outdoors. 1987. The Report of the President's Commission on Americans Outdoors: The Legacy, the Challenge. Island Press, Washington, DC.
Quinby, P.A. 1988. The Development of Landscape Ecology: A Brief History. Landscape Research 13:3, 9–11.
Schooenboom, I.J. 1995. Overview and State of the Art of Scenario Studies for the Rural Environment. In: *Scenario Studies for the Rural Environment*, pp. 15–24. J.F.Th. Schoute, P.A. Finke, F.R. Veeneklaas, and H.P. Wolfert (eds.). 1995. Proceedings of the Symposium: Scenario Studies for the Rural Environment, Wageningen, The Netherlands, 12–15 September, 1994. Kluwer Academic, Dordrecht.
Schoute, J.F., P.A. Finke, F.R. Veenenklaas, and H.P. Wolfert. 1995. Scenario Studies for the Rural Environment. Kluwer Academic, Dordrecht, Netherlands.
Schreiber, K.F. 1990. The History of Landscape Ecology in Europe. In: Changing Landscapes: An Ecological Perspective, pp. 21–33. I.S. Zonneveld and R.T.T. Forman (eds.). Springer-Verlag, New York.
Sijmons, D. 1990. Regional Planning as a Strategy. Landscape and Urban Planning. 18(3–4):265–273.
Smith, D.S., and P.C. Hellmund. 1993. The Ecology of Greenways. University of Minnesota Press, Minneapolis, MN.
Soulé, M.E. 1991. Land Use Planning and Wildlife Maintenance: Guidelines for conserving wildlife in an urban landscape. Journal of the American Planning Association 3:313–323.
Steiner, F. 1991. Landscape Planning: A Method Applied to a Growth Management Example. Environmental Management 15(4):519–529.

Steiner, F., and D. A. Osterman. 1988. Landscape Planning: a working method applied to a case study of soil conservation. Landscape Ecology 1(4):213–226.

Steiner, F., G. Young, and E. Zube. 1988. Ecological Planning: Retrospect and Prospect. Landscape Journal 7(1):31–39.

Steinitz, C. 1990. A Framework for Theory Applicable to the Education of Landscape Architects (and Other Environmental Design Professionals). Landscape Journal 9(2):136–143.

Steinitz, C., with students of the Graduate School of Design, Harvard University 1994. Alternative Futures for Monroe County, Pennsylvania. Harvard University Graduate School of Design, Cambridge, MA.

Turner, M.G. 1989. Landscape Ecology: The Effect of Pattern on Process. Annual Review Ecological Systematics 20:171–197.

UNESCO. 1970. Plan for a Long-Term Intergovernmental and Interdisciplinary Programme on Man and the Biosphere. Document 16 C/78, UNESCO, Paris.

Veenenklaas, F.R., and L.M. van den Berg. 1995. Scenario Building: art, craft, or just a fashionable whim? In: Scenario Studies for the Rural Environment, pp. 11–13. J.F.Th. Schoute, P.A. Finke, F.R. Veeneklaas, and H.P. Wolfert (eds.). 1995. Proceedings of the Symposium: Scenario Studies for the Rural Environment, Wageningen, The Netherlands, 12–15 September, 1994. Kluwer Academic, Dordrecht.

Vos, C.C., and P. Opdam. 1993. Landscape Ecology of a Stressed Environment. Chapman & Hall, London.

Vos, C.C., and J.L.S. Zonneveld. 1993. Patterns and processes in a landscape under stress: the study area. In: Landscape Ecology of A Stressed Environment, pp. 1–27. C.C. Vos and P. Opdam (eds.). Chapman & Hall, London.

World Commission on Environment and Development. 1987. Our Common Future. Oxford University Press, Oxford.

Zonneveld, I.S. 1995. Land Ecology. SPB Academic Publishing, Amsterdam.

Zonneveld, W. 1991. Conceptvorming in de Ruimtelijke Planning. Patronen en Processes (Vol. 1) and Encyclopedie van Planconcepten (Vol. 2). Planologische Studues 9, Universiteit van Amsterdam. Amsterdam, Netherlands.

Part IV
Concepts, Methodological Implications, and Numerical Techniques

11 Concepts of Scale in Landscape Ecology

Mark Andrew Withers and Vernon Meentemeyer

The number of distinct scales of length of natural patterns is for all practical purposes infinite.
Mandelbrot 1982, p. 1

Kuhn's concept of a paradigm shift is a useful way to interpret the annual meeting [1988] of the ESA. . . . Every symposium or session I attended featured, included, or was structured by the concepts of scale and spatial patterns. I left feeling I had observed one of those rare creatures of the intellectual bestiary, a paradigm shift. Why should we suddenly be interested in spatial patterns, and problems of shifting spatial scale?
F. Golley 1989, Vol. 3, pp. 65–66

"It is impossible to know yet whether the sudden expansion of interest in 'scale' in population and community ecology is a passing fashion, or the beginning of an enduring change in the way that ecological research is pursued."
Schneider 1994, p. 1

This chapter seeks to evaluate the degree to which issues of spatial scale and scaling problems have been incorporated into different research problems in landscape ecology and to identify research methods applicable to problems of spatial scale. Then, the study identifies research foci in which scale has been relatively little explored, in part to determine if some questions have been largely divorced from questions of scale or whether appropriate conceptual or methodological approaches are lacking. The results may shed some light on to what degree, and in what subdisciplines within landscape ecology, the "paradigm shift" noted by Golley (1989) has been embraced. Finally, a standard vocabulary is proposed for issues of scale in the field of Landscape Ecology.

Emergence of the Spatial Dimension in Ecology and Landscape Ecology

Scale has long been a vexing problem in ecology, cartography, geography, and many other disciplines. Much of the difficulty in treatment of "scale" is the great variability in the interpretation and meaning of "scale." Thus scale may refer to measurement scale (ratio, interval, etc., Stevens 1946); ecological scale (Calder 1983; Clark 1990); body size scale (Brown and Nicoletto 1991; Peters 1983); grain size and extent (e.g., Wiens 1989; Palmer and White 1994); map scales (Turner and Gardner 1991b); and relative versus absolute scale (Meentemeyer 1989). Further complications are created by the terminology: to ecologists, a "large-scale" study is one of a large area, with limited detail (as used, for example, by Stiling et al. 1991; Frelich and Reich 1995). In geography and cartography, "large-scale" means a large fractional scale of display (e.g., 1/24,000 versus 1/250,000), and thus a small area of greater detail.

Scale issues have a long history in ecological research and geography, although the great "bloom" has occurred since 1985 (see review by Schneider 1994). In geography, because of its ancient traditions in mapping and exploration, scale is an inherent part of the discipline. Geographers do not, however, state their scales explicitly with any more regularity than do researchers in other disciplines. Consequently there have been many appeals (e.g., Harvey 1968; Stone 1968; Nir 1987) for improved depiction of scale. Early attempts to incorporate scale in environmental biology include the work of Hutchinson (1953, 1971), and developments in allometry (Gould 1966) and sample design (Grieg-Smith 1952). In ecology the size of predator versus prey has always been important. From the early 1970s (e.g., Hutchinson 1971) to the 1980s, many calls were made by ecologists to address issues of spatial scale. By 1992, Levin argued that the "problem of pattern and scale is the central problem in ecology, unifying population biology and ecosystem science, and marrying basic and applied ecology."

Impetus for studies of and/or involving scale has come from many new sources, including remote sensing, global-scale modeling and monitoring initiatives, climate change impacts, conservation planning, and indeed from within the discipline of landscape ecology itself. By the mid-1980s, many ecologists had begun to incorporate scale effects into research designs. An important contribution was made by Allen and Starr (1982) who applied hierarchy theory to problems of ecological scaling. Later, Wiens (1989) developed a framework for "ecological scaling," and Gosz and Sharpe (1989) developed a scheme of "ecotone hierarchies." Ecologists continued to expand the spatial dimensions of their study areas throughout the 1980s, creating more need to address spatial scale. By the end of the decade, Golley (1989) had identified "one of those rare creatures of the intellectual bestiary, a paradigm shift," in which ecological research focused increasing attention on the problem of spatial scale.

Methods of Survey and Review

This study is based on a search of literature at two levels. First, the authors reviewed all research articles published in the journal *Landscape Ecology* from 1990 to 1995 (Volumes 4–10, editorials and commentary excluded). For each article, the following were identified:

1. The research focus/topic addressed (Table 11.1);
2. The conceptual approach employed to issues of scale (Table 11.2); and
3. The method and tools, if any, employed to deal with issues of spatial scale and scaling (Table 11.3).

The results were tallied by subdiscipline to assess (1) to what degree and (2) by what conceptual approach spatial scale has been incorporated into the various research foci identified in Table 11.1.

The research foci presented in Table 11.1 are by necessity subjective and arbitrary to some degree. However, they represent a reasonable cross-section of literature in the journal *Landscape Ecology*. Moreover, their sequence in Table 11.1 can be considered a continuum of sorts, such that adjacent entries are less clearly distinct than entries more distantly separated.

TABLE 11.1. Common research topics in articles found in the journal *Landscape Ecology*.

- Landscape structure: patch size, mosaic pattern, fragmentation
- Land-use/land cover classification and mapping
- Plant ecology and vegetation analysis: physiognomy, composition, mapping
- Landscape dynamics: succession, disturbance, stability, temporal sequences, geomorphology
- Habitat analysis: Patch sizes, fragments, utilization, movement rules
- Animal ecology: population biology, habitat selection, competition, coexistence, predation, food webs
- Biodiversity
- Biogeochemical cycles: energy/material balance, nutrient cycles, hydrology
- Global change: Ecosystems models, impacts, global feedbacks
- Landscape planning, design, and management
- Methodology
- Scale per se

TABLE 11.2. Continuum of conceptual approaches to scale in landscape ecology.

- Scale ignored
- Scale mentioned as a factor, or used in qualitative analysis or discussion
- Single-scale (specified) studies
- Replication at multiple scales and hierarchies (micro-, meso, and macro-scale cross studies)
- Scale used as a variable (discrete or specified)
- Scale used as a variable (continuous)
- Cross-scale extrapolation/interpolation
- Multiple phenomena and use of multiple scales
- Theories/hypotheses on scale as property/process/control

Articles that could fall into multiple categories were assigned uniquely to one category based on the main thrust of the article. For example, Lidicker et al.'s (1992) "Utilization of a Habitat Mosaic by Cotton Rats during a Population Decline," was classified as an animal ecology study, because the main thrust was the population biology of a species, rather than the structure of the habitat. Benninger-Traux et al.'s (1992) study of "Trail Corridors as Habitat and Conduit for Movement of Plant Species" was classified as a study of habitat structure because the main impetus was to gauge the degree to which trails serve as corridors for exotic species dispersal.

A computerized search of wider literature in more broadly defined subdisciplines of ecology identified other studies that address issues of scale. The keywords *scale* and *ecology* were used in a global search of the Current Contents Plus™ data base (for years 1990–1995). Finally, the results from the review of *Landscape Ecology* and the wider literature were reviewed.

TABLE 11.3. Selected tools used in the analysis of scale.

- Simple scaling/extrapolation
- Scaling using multiple variables
- Remote sensing
- Spatially explict models (SEMS)
- Geographic information systems (GIS)
- Fractal geometry/fractal dimension
- Spatial statistical methods (variogram, kriging, etc.)
- Simulation models
  - Neutral models
  - Simple simulation models
  - Dynamic simulation models
- Hierarchy theory

Scale in Research Designs: A Matter of Degree

Conceptual Approaches to Scale in the Literature

A great variety exists in conceptual approaches to issues of scale in landscape ecology research. The conceptual complexity in addressing spatial scale can be viewed as a continuum (Table 11.2). At the simplest level, scale is simply ignored as a factor or issue in research design and analysis. In such cases, it is for the reader to determine if spatial scale is a confounding or significant factor. Often it is not. Nearly a quarter of all articles published in *Landscape Ecology* did not address, even in a qualitative way, spatial (or temporal) scaling issues (*scale ignored*). At the other extreme, several authors have developed an elegant framework toward developing more formal concepts of scale, suggesting general findings, or presenting hypotheses for testing and analysis (*theories/hypotheses*). Between lies a continuum of increasingly sophisticated approaches to dealing with the problem of scale. Beyond ignoring scale, scale is mentioned and/or specified, often by making clear the size of the study area, while recognizing explicitly that the results may apply only at that scale (*single scale*). Further up the continuum, many authors include *replicate sampling scales*, or use some scale-related measure, such as area, grain, or study extent, as either a *discrete* or a *continuous variable*. Studies of cross-scale extrapolations and interpolations appear to be quite scale conscious and some studies involve the complexity of multiple phenomenon examined at multiple scales [*scaling* using multiple variables]. Finally some studies are directed to scale itself and how scale affects a particular landscape or ecological phenomenon.

Progress in the Treatment of Scale

To form an impression of the most common conceptual approaches and tools involving scale used in a number of the subdisciplines (parts) of landscape, we created a matrix. The 159 articles that appeared in *Landscape Ecology* during the years 1990 to 1995 were reviewed. We identified twelve research topics (the rows in Table 11.4) and nine conceptual approaches to scale in a continuum from "scale ignored" to "scale theory/hypothesis" (the columns in Table 11.4).

The research topics to be identified in *Landscape Ecology* were selected a priori via a broader review based on a search of Current Contents Plus™ (1990–1995). The most common study topic found was "Landscape Dynamics" with $N = 27$ (17.0%). Among these, scale was ignored in 22.3%. For 13.2% a single scale was selected but specified and 18.2% used cross-scale extrapolation (Table 11.4). In our sample, no study used replicate scales.

TABLE 11.4. Conceptual approach to spatial scale in subdisciplines of landscape ecology.

Research topic	N	% of all studies	Percentage of studies with specified approach to scale								
			I	Q	SS	RS	DV	CV	CSE	MVS	T/H
Landscape structure	22	13.8%	13.6	13.6	4.5	13.6	4.5	27.3	4.5	0.0	18.2
Land-use/land cover	18	11.3%	5.6	5.6	27.8	11.1	5.6	5.6	33.3	0.0	5.6
Plant ecology/vegetation analysis	6	3.8%	33.3	0.0	0.0	16.7	16.7	16.7	16.7	0.0	0.0
Landscape dynamics	27	17.0%	25.9	3.7	18.5	0.0	3.7	11.1	22.2	3.7	11.2
Habitat structure/corridors	24	15.1%	16.7	12.5	20.8	16.7	12.5	8.3	8.3	0.0	4.2
Animal ecology	13	8.2%	38.5	0.0	15.4	7.7	0.0	23.1	7.7	0.0	7.7
Biodiversity	6	3.8%	33.3	0.0	0.0	16.7	16.7	16.7	16.7	0.0	0.0
Biogeochemical cycles	10	6.3%	10.0	10.0	10.0	0.0	0.0	10.0	60.0	0.0	0.0
Global change impacts	13	8.2%	53.8	0.0	0.0	15.4	0.0	0.0	30.8	0.0	0.0
Land planning/mgmt.	9	5.7%	44.4	11.1	22.2	0.0	0.0	22.2	0.0	0.0	0.0
Methodology	4	2.5%	25.0	0.0	0.0	25.0	0.0	0.0	0.0	0.0	50.0
Scale per se	7	4.4%	0.0	0.0	0.0	0.0	0.0	28.6	14.3	0.0	57.1
Total	159	—	37	10	21	15	8	22	29	1	16
As percentage	—	100%	23.3	6.3	13.2	9.4	5.0	13.8	18.2	0.6	10.1

I, igonored; Q, qualitative; SS, single scale; RS, replicate scales; DV, discrete variable; CV, continuous variable; CSE, cross-scale extrapolation; MVS, multi-variable scaling; T/H, theory or hypothesis.

"Habitat structure/corridors" was the next most numerous topic with an $N = 24$. Here the most frequent conceptual approach to scale involved treatment of scale as a single scale, followed by replicate scales and scale ignored. For the conceptual approaches, the theory/hypothesis approach was, as should be expected, most commonly found in studies of "methodology" (50%) and "scale per se" (57.1%). Studies that use multiple variables and multiple scales (MVS) are still very rare (0.6%).

In conclusion, scale is ignored in nearly one fourth of all articles in landscape ecology, and some conceptual approaches to scale are not found in many research foci. This may not be merely an oversight: We cannot say with certainty that certain conceptual approaches are more/less appropriate to a particular research topic, but it does perhaps reveal openings for new mixes of research topics and scale concepts. Overall the progress in treatment of scale, identified by Levin (1992) as a central problem in ecology, has been substantial. Further development could possibly be enhanced if a common terminology of scale could be developed.

Some Tools Used to Analyze Scale

The explosion of interest in scale has involved several concurrent developments in digital spatial technologies, as well as the paradigm shift to spatial analyses and the emergence of landscape ecology itself. The tools used to explore scale are a curious mix of older methods, including statistical approaches and simulations, and newer approaches, including, for example, geographic information systems, remote sensing, simulation models, and neutral models (Table 11.3). Central to these new methods are prodigious steps in technology that have made possible collection of imagery, and the storage and quantitative analysis of data volumes that were unthinkable a decade ago.

A review of the abstracts of papers in *Landscape Ecology* and papers presented at meetings of the North American Chapter of the International Association of Landscape Ecology, indicate that much of the broader research agenda is driven by many of the newer tools. Here are some of the tools used in studies involving scale, together with examples of specific applications; the discussion is not inclusive. Many of the authors cited use several of these methods in combination, and could easily be cited under two or more categories.

"Simple" Scaling/Extrapolation

Extrapolation among different spatial scales is a central problem in landscape ecology and earth sciences (e.g., Avissar and Chen 1993; Pierce and Running 1995). This problem is particularly inherent in modeling and simulation experiments, that must bring together multiple phenomena, often measured at different resolution, into a common coordinate grid with a single raster size. Such discussions often revolve around the problems of the "bottom-up" versus "top-down" approaches, or the problem of "scaling up" versus "scaling down" (Vitousek 1993). Naturally there is great concern for how well predictions based on knowledge from one level, say plant physiological ecology, can be extrapolated to a higher level, such as an ecosystem. Likewise, loss of data and error propagation are major problems with changes in grain size and study extent (Turner et al. 1989a). Rules for the conservation of energy and matter often act as checks on the predictions among levels.

The explosive growth in digital data collected from satellites, and widespread distribution of global and regional data sets on the World Wide Web, will create great demand for development of methods for cross-scale extrapolation and error analysis. A particularly useful volume is *Scaling Ecophysiological Processes: From Leaf to Globe*, edited by Ehleringer and Field (1993).

Scaling Using Multiple Variables

In this conceptual approach, multiple variables/parameters are examined simultaneously, each at multiple scales. Occasionally temporal elements are also included as in the work of Holling (1992) who examined the processes that "structure ecosystems across scales in time and space."

The main thrust of such approaches is to identify phenomena that vary together at similar temporal and spatial scales (Levin 1992). To understand the natural world, and especially to explain or predict phenomena based on observations of related or dependent phenomena, researchers must look at those phenomena that vary on similar spatial and temporal scales (Meentemeyer and Box 1987). This conceptual approach, though relatively little developed within landscape ecology, offers considerable promise to aid in understanding diverse phenomena at multiple levels of resolution. For example, Devasconcelos et al. (1993) developed an ecosystem model that simulates landscape dynamics by incorporating hierarchical structures that operate simultaneously at multiple spatial and temporal resolutions. Other methodological papers of the "multidimensional" type include Rastetter et al. (1992), which explores methods that involve multiple parameters at multiple scales.

Remote Sensing

Remotely sensed data offer the potential to develop detailed spatial data bases and maps for virtually any spectral phenomenon, or for any phenomenon that can be linked reliably to spectral data. Extensive effort has been directed toward relating landscape spectral characteristics to such phenomena as landscape structure (Simmons et al. 1992), biological richness (Stoms and Estes 1993), photosynthesis and net primary productivity (NPP) (Begue et al. 1994), land cover classes (Duncan et al. 1994; Malingreau and Belward 1992), disturbances in forests (Boose et al. 1994), and so on. But using remotely sensed imagery to estimate ecologically meaningful data remains inadequate and often inappropriate for many phenomena.

Remote sensing data are highly sensitive to scale issues. Data are generalized to the grid cell size and may thus lead to statistical fallacies (Fisher 1991), inaccuracies (Elston and Buckland 1993), and underrepresentation of rare phenomena (Johnson 1990). Development of methods to provide meaningful output and interpretation of remotely sensed images (e.g., Duncan et al. 1994), analysis of their effectiveness (e.g., Davis and Goetz 1990), issues of resolution and grain size (e.g., Malingreau and Belword 1992; Benson and McKenzie 1995), and extrapolation/aggregation across scale (e.g., Iverson et al. 1994) are therefore among the leading research topics in landscape ecology.

To date, few methods of analysis and interpretation are available that can be standardized from one location, or one phenomenon, to another. Rather, each application requires independent calibration, verification, and error analysis. Thus, it appears that fully automated processing and interpretation of remotely sensed data remains an elusive goal. However, given the vast accumulation of comprehensive data of global coverage, repeated at predicted intervals, combined with rapid technological advances in GIS, data storage capabilities, computational power, and automated optical processing capabilities, remote sensing appears to offer staggering potential for landscape ecology over a wide range of spatial scales.

Spatially Explict Models (SEMs)

Spatially explicit models (SEMs) are simply maps of actual or simulated phenomena, usually population dynamics of a single species or community superimposed on a landscape. With such a map, "the locations of habitat patches, individuals, and other items of interest are explicitly incorporated into the model, and the effect of changing landscape features on population dynamics can be studied" (Dunning et al. 1995). The map may be a remotely sensed image, various map layers entered into a GIS (e.g., Kesner and Meentemeyer 1989), or even a topographic sheet. Consequently spatial scale issues abound.

Often the key objective of these models is to predict spatial patterns of species, populations and vegetation types, to identify critical habitat for endangered species, or to simulate the effects of landscape change scenarios on individual species. Most typically, SEMs are raster-based models constructed at a single spatial scale. However, Cherrill et al. (1995) used a hierarchical approach to multiple scales to predict plant species distributions across a region in Northern England and Holt et al. (1995) have linked different SEMs to demonstrate scale sensitive problems.

Useful reviews of SEMs include Turner and Gardner's (1991a) *Quantitative Methods in Landscape Ecology*, and the special issue of *Ecological Applications* on SEMs (*Ecological Applications*, Vol. 5, No. 1, 1995). SEMs often employ cartographic modeling, a method common in geography: that is, they attempt to predict spatially a phenomenon for which there is poor information using spatial data for which there is much information (e.g., elevation). Certainly these disciplines share the ancient goal of predicting spatial patterns and are all therefore geographic models sensitive to scale.

Geographic Information Systems (GIS)

One of the most extensively used tools in modern landscape ecology is the GIS. GIS is a set of tools for storage, analysis, and display of spatial data. Often the GIS is used simply to store or map data, but it can also be used to measure area, examine spatial intersection or overlap, identify temporal change, for proximity analyses, to calculate fractal dimensions and produce output from SEMs (Johnson 1990). Like SEMs, the GIS application must resolve scale issues, including the effects of grain size and extent, number of layers on machine time (Kesner and Meentemeyer 1989), and cross-scale compatibility of multiple data layers.

The technique of cartographic modeling allows map layers to be overlaid in various arithmetic or functional relationships to portray complex statistical or mechanistic simulations in geographic space (Tomlin 1991). Such uses have been applied repeatedly to map species habitats (e.g., Davis and Goetz 1990); to assess landscape dynamics and stability (Friedman and Zube 1992); identify boundaries of ecosystems, ecozones, or priority conservation targets (Kolejka 1992; Fernandes 1993); and for biodiversity gap analysis (Bojorqueztapia et al. 1995). Such applications require input data in compatible raster units, and thus rescaling is inevitably required (Dumanski et al. 1993).

A GIS provides an ideal framework to develop landscape simulation models and spatially explicit population models. GIS has been used as a tool to display results of statistical analysis or dynamic or stochastic models, such as landscape change, disturbance regime, and succession (Johnson 1990). GIS databases have also been used to produce input for model calculations (Johnson 1990; Johnston 1992). Several authors have used GIS simulations to assess the effects of scale change and grain size/resolution on the interpretation of landscape structure (Turner 1990; O'Neill et al. 1992), disturbance regimes (Boose et al. 1994), and other phenomena. Scale-sensitive analyses are relatively straightforward in most GIS packages, which provide built-in routines for resampling. Band and Moore (1995) provide a useful discussion of scaling landscape-scale hydrological phenomena to broader spatial extent in a GIS. Dumanski et al. (1993) outline a procedure for rescaling multiple data layers influencing crop growth through a nested, hierarchical GIS database.

Fractal Geometry/Fractal Dimension

No treatment of spatial scale in landscape ecology would be complete without a discussion of fractals. Fractal geometry was developed to quantify the infinite (Mandelbrot 1982), and to address irregular geometric shapes that classical Euclidean geometry does not. One of the founding problems of fractal geometry is the scale dependence of coastline and boundary lengths: the measured length of coastline increases, indefinitely, as resolution increases (Mandelbrot 1982). Mandelbrot developed measures of fractal length and fractal dimension to characterize the scale dependence of such irregular and seemingly patternless shapes. As such, fractal geometry has been employed in landscape ecology to assess scale dependence of many phenomena (Palmer 1988; Milne 1992; With 1994a, 1994b).

The fractal dimension, D, is usually calculated as:

$$L(\in) \sim F\in^{1-D}$$

where $L(\in)$ is the length or perimeter of a patch, F is an empirical constant, and D is the fractal dimension (Mandelbrot 1982). The fractal dimension D is a noninteger typically varying from 1 to 3, and is a measure of landscape patch complexity (Milne 1992). Typically D is independent of measurement scale, and the landscape is said to be "self-similar" over the range of scales that D remains constant (after Mandelbrot 1982; Sugihara and May 1990; Xu et al. 1993; Leduc et al. 1994; Vedyushkin 1994). There are, however, a number of alternative fractal measures and various methods to estimate fractal dimension (see e.g., Mandelbrot 1982; Clark 1986; Milne 1988, 1991; Klinkenberg and Goodchild 1992; Gallant 1993; Riitters et al. 1995).

Thus, fractal dimension has appeal as a means to describe a landscape and as a scale-invariant property of landscape boundaries. An early paper on landscape fractal dimensions was produced by Burrough (1981); Milne (1988, 1991) has contributed much on how to calculate and interpret landscape fractals. Other useful reviews of fractals include those from the geographic point of view (Goodchild and Mark 1987), ecology (Sugihara and May 1990), geophysics (Lovejoy and Schertzer 1991), and geomorphology (Snow and Mayer 1992; Xu et al. 1993).

Spatial Statistical Methods

Ecologists have employed classical statistics for many years. More recently, advances in spatial statistics and geostatistics have provided further opportunity to explore spatial and scale-related issues in landscape ecology. Among these statistical and numerical techniques are measures of landscape structure (e.g., fractal dimension, contagion), resampling or "blocking" techniques, measures of spatial dependence and autocorrelation, moving windows techniques, regression tree analysis, and geostatistical modeling (kriging) (see Turner and Gardner 1991a, Cullinan and Thomas 1992; Riitters et al. 1995; for reviews and applications; see Cressie 1993 for a treatise on spatial statistical methods).

Several measures of landscape pattern are used to reveal the "scale" of the landscape, that is the range of spatial dimensions over which landscape structure is scale independent. Generally, sampling and analysis must fall within this range of scales to provide meaningful results. Foremost among these measures is the fractal dimension D, discussed above. Others include contagion, diversity, evenness, and patchiness indices. Riitters et al. (1995, 1996) provide calculation formulae and discuss applications for a variety of landscape fractal measures, and other scale-related landscape structural parameters. Li and Reynolds (1994) and Cullinan and Thomas (1992) also provide analyses of scale-dependent landscape structural parameters.

Resampling or "blocking" techniques are used to frame or nest samples or landscape structures within a hierarchical framework to assess how scale and sequence of assembly affect landscape characteristics. "Moving windows" are samples grouped along a spatial or other axis to assess scale-dependence or spatial discontinuities (Cornelius and Reynolds 1991). Another set of techniques employs measures of spatial dependence or spatial covariance, such as variogram plots, spatial autocorrelation, or a "variance stairway" (after Turner et al. 1991; see also Dutilleul and Legrende 1993; Qi and Wu 1996). Palmer (1988) gives simulated variogram plots based on alternative scale-dependent landscape structures.

Geostatistical techniques such as kriging employ knowledge of the spatial covariance (as contained in the variogram plot) to produce a spatial model (Journal and Huijbregts 1978; Cressie 1993). Oliver and Webster (1990) have developed programs for multivariate kriging (cokriging) within a GIS framework. Example applications of some of these techniques include assessment of soil heterogeneity using a variogram plot (Jackson and Caldwell 1993a); spatial autocorrelation to assess spatial and temporal variability in pollen deposition (Graumlich and Davis 1993); sample plot "blocking" procedures to assess the effect of grain size and extent on plant community characteristics (Busing and White 1993; Palmer and White 1994); spectral analysis to identify contributions to heat flux at different spatial scales (Pielke et al. 1991); and moving window clustering to identify the scale of spatial discontinuity in Australian woodlands (Ludwig and Tongway 1995).

Simulation Models

Virtually any phenomenon can be simulated, and thus the diversity and breadth of simulation models in landscape ecology are great. Such models vary in complexity, from relatively simple one-dimensional diffusion models to compartmentalized models incorporating dynamic feedback loops within each of dozens of individual process inputs and outputs. The authors recognize three main conceptual groups of models: (1) neutral models; (2) "simple" simulation models; and (3) dynamic simulation models. Issues of scale vary depending on the complexity of the model and the intended purpose.

Neutral models in landscape ecology attempt to produce the simplest model possible to "explain" spatial patterns in the absence of any explanatory mechanism (Gardner et al. 1987; Gardner and O'Neill 1991). Typically, neutral models are generated to simulate alternative treatments, assumptions, and scenarios, relative to some "neutral" baseline. Using neutral models, it is possible to generate maps of a wide range of landscape structures, from random to highly structured. In landscape ecology, neutral models are frequently developed and manipulated to test for scale effects resulting from changes in grain size, extent, and spatial structure (e.g., after Caswell 1976; Milne 1992; O'Neill et al. 1992).

Here the term "simple" simulation models is used for those that do not specify a significant temporal dimension, time steps, or complex feedback mechanisms. These are often purely statistical or rule-based simulations of a limited number of phenomena. Many such simulations could also be called spatially explicit models (SEMs) because they are tied to or referenced to map coordinates in some way. One common element of interest in landscape simulation is the consequence of changing grain size and levels of aggregation on a hypothetical landscape. Thus the simulation involves manipulation of resolution and information content. Many of these employ neutral models as a baseline for comparison. Typical applications include assessment of scale effects on landscape structural characteristics (O'Neill et al. 1992; Li and Reynolds 1994); disturbance regimes (Baker 1993; Devasconcelos et al. 1993); species coexistence and habitat breadth (Palmer 1992; Lavorel et al. 1994b); species movement rules (Crist et al. 1992; Gross et al. 1995); and habitat use (Turner et al. 1993).

Dynamic simulation models take into consideration landscape changes through time, and fluxes of organisms, energy, water, and other matter. Fluxes have rates and are akin to the notion of the "geography of rates." Such dynamic models in landscape ecology span a wide range of spatial extent, from global vegetation and biome models, to site-specific models tied to sites or plots. Scaling issues place severe constraints on models throughout the range of extent covered.

Hierarchy Theory/Hierarchical Research Design

Much research in landscape ecology employs hierarchy theory, hierarchical research designs, or hierarchical structuring to address issues of spatial scale. Hierarchies, starting with the cell, seem a natural concept for biologists. According to O'Neill et al. (1989) "Hierarchical structuring simply means that, at a given level of resolution, a biological system is composed of interacting components (i.e., lower-level entities) and is itself a component of a larger system (i.e., higher level entity)." For landscapes the spatial dimension is critical, and hence the focus on "spatial hierarchies." Hierarchy theory predicts that each level in a hierarchy functions at rather distinct temporal and spatial scales (Allen and Starr 1982; O'Neill et al. 1986). Fine-scaled spatial phenomena appear to possess high rates of change while broad-scaled phenomena appear to change over long time scales (e.g., Meentemeyer 1989; and review by Deboer 1992). Although many authors recognize spatial and hierarchical relationships, relatively few employ true hierarchical structuring in field research design. Nevertheless, spatial hierarchies represent useful heuristic devices, as an aid in perspective and perhaps to help identify constraints on the landscape elements being studied.

How Scale Is Treated Within the Different Research Foci

The following paragraphs present a survey of the broad research themes in landscape ecology identified in Table 11.1. Here a sample of the literature both from *Landscape Ecology* and broader sources is synthesized to assess the manner and degree to which scale is considered, and to identify progress either in methodology, theory, or understanding of processes in that subdiscipline. The authors also attempt to identify what obstacles, if any, may render some research problems scale intractable.

Landscape Structure

Landscape structure refers to the spatial pattern of distinct landscape elements, such as their size, shape, configuration, number, and type (Turner and Gardner 1991b). Landscape structure is one of the areas in landscape ecology in which approaches to spatial scale are best developed. This is not surprising, because landscape ecology emerged as a discipline in the United States in part to explore "landscape scale" (Forman and Godron 1981), and because patch size, orientation, and shape are all readily amenable to study at multiple scales. A plethora of quantitative measures and indices have been developed to elucidate landscape structure, and most of these have been assessed at multiple scales. The effect of changing spatial grain and extent have been assessed for many landscape structural parameters (O'Neill et al. 1996), including measures of fractal dimension (Leduc et al. 1994; Ritters et al. 1995), contagion (Turner et al. 1989a, 1989b; Li and Reynolds 1994), evenness (Li and Reynolds 1994; Ritters et al. 1995), and landscape diversity (Turner et al. 1989a, 1989b; Ritters et al. 1995). O'Neill et al. (1996) report the effects of grain and sampling unit size on a number of landscape metrics, and suggest a grain size no larger than one half to one fifth the size of the smallest landscape feature of interest.

Simulation models and spatial statistics are used frequently to assess scale-dependent attributes of landscape structure (e.g., Turner et al. 1989b; Moloney et al. 1991; Cullinan and Thomas 1992; Bell et al. 1993). O'Neill et al. (1992) developed a hierarchical neutral model to compare structural characteristics of hierarchically scaled landscapes to patterns on random, unstructured landscapes. Landscape ecologists also have assessed scale-dependent interactions between landscape structure and ecological problems, such as disturbance regimes (e.g., Bergeron 1991; Turner et al. 1994); habitat structure (Milne 1992; With 1994a, 1994b); food webs (Martinez and Lawton 1995); structure and diversity of ecotones (Gosz 1993; Bretschko 1995) and hedge rows (Burel and Baudry 1990); floristic assemblages (McLaughlin 1994); old growth forest distribution (Mladenoff et al. 1993); and species diversity (McColin 1993; Huston 1994; Rescia et al. 1994), to name a few. Several studies have examined the characterization of landscape structure with changing grain size. Bretschko (1995) found that riverine ecotones become more numerous and diverse with decreasing grain size. Benson and Mackenzie (1995) examine landscape texture using differing levels of resolution for different remotely sensed images and found that some landscape structures are more sensitive to changes in grain size than others.

Recognizing that landscape structure varies with scale, landscape ecologists have strived for scale-invariant measures or indices. The fractal dimension D is the most commonly employed such measure. The range of spatial extent over which D is a constant is said to represent the "scale" of the landscape, or the scale over which the landscape is "self-similar" (after Burrough 1981; Mandelbrot 1982). This is taken to mean that over that range of scales, landscape units display similar behavior, appear structurally similar and are, presumably, affected by the same process and controls (Lathrop and Peterson 1992; Vedyushkin 1994). Plotnik et al. (1993) have introduced an alternative measure they designate the lacunarity index.

An alternative approach is to identify the scale of discontinuity in landscape structure, or assess the variability or similarity between landscape types or patches. These measures include variograms, spatial autocorrelation, and geostatistical methods. Loehle and Wien (1994) use information theory (ordination axis distances) to determine the degree to which locations are similar. Cornelius and Reynolds (1991) outline both scale-dependent and scale-independent approaches to identify spatial discontinuities using a moving split-window approach. Qi and Wu (1996) assess the sensitivity of spatial autocorrelation measures to changes in grain size. Conceptually these studies can be broken down into one group that seeks to find the natural scales of landscape structure/heterogeneity and another that attempts to find difference in how structure is revealed using different resolution/grain sizes.

Outside of ecology, physical geographers and geomorphologists have also contributed much to assessment of scale-dependent processes in landscape formation. These include assessments of scale effects in tectonics and mountain building (Lifton and Chase 1992; Schmidt and Montgomery 1995), stream channel and drainage network formation (Montgomery and Dietrich 1992; Tarboton et al. 1992; Willgoose et al. 1992), and glaciation (Sugden et al. 1992). These results yield considerable insight into processes that shape the physical matrix in which ecological processes operate.

From these and other studies it is possible to propose a number of general findings, that could be put forward as testable hypotheses. First, and most obviously, changes in grain size and/or study extent lead to changes in measured landscape structural properties. Second, increased resolution, or a smaller grain, leads to increased measures of landscape diversity, number of patches, patch perimeter and edge length. Third, landscapes have an inherent "scale," that is a range of spatial extent over which they are self-similar, and over which common measures of the fractal dimension are scale invariant. Less certain, but worthy of further inquiry, is the hypothesis that variability in the physical environment increases indefinitely with distance.

Land-Use/Land-Cover Classification

Land-use and land-cover classification has a long history in landscape ecology (e.g., Zonneveld 1989). Land managers, planners, and conservationists all develop and use maps, and these maps are based on some a priori scheme to classify or group land-cover types. There appears to be a convergence in this subdiscipline which involves remote sensing, hierarchical approaches and hierarchy theory, geographical information systems and scale issues per se.

Scale issues are critical components of these studies, which appear to converge around four main themes. First is the nature of the classification scheme itself, and foremost in question is the intended use or application. Many classification systems involve nested classes spanning a hierarchy of spatial scales (Bailey 1985, 1996; Zonneveld 1989; Dumanski et al. 1993; Klijn and de Haes 1994; Cooper 1995). Klijn and de Haes (1994) outline a generalized hierarchical approach to ecological land classification. Hains-Young (1992) developed a two-tiered land classification scheme in Great Britain, involving broad-scale cover types and landscape types. Cherrill (1994) compared classifications obtained using different input data and found that similar land classes are recognizable at different levels of environmental and ecological organization.

A second theme involves the actual process of classification. Issues include numerical methods and algorithms of classification (e.g., Michaelson et al. 1994), the nature of and scale compatibility in the underlying data (Edmonds et al. 1985; Malingreau and Belward 1992; Lathrop et al. 1995); sources of and the magnitude of the underlying error (Davis and Goetz 1990; Cherrill 1994); and the relationship of the classification to the underlying landscape (Cooper 1995). Data sources commonly include any combination of aerial photos, remote sensing imagery, standardized data bases, and original field work. A major research focus remains the proper scaling and interpretation of remotely sensed data (Malingreau and Belward 1992), and cross-scale compatibility of the various input data sets. Moody and Woodcock (1995) examined the effect of landscape structure on scale-dependent changes in land-cover classification. Despite an enormous body of literature, few methods of classification using remote sensing imagery are available that can be standardized from one location, or one phenomenon, to another. Rather, each application requires independent scaling, calibration, verification, and error analysis.

A third theme revolves around issues of output, typically a land-use/land-cover map, data base, or as input for other analyses. Scaling issues, especially rescaling mapped output for the desired application, are a major concern. Considerable effort has been directed toward developing methods for scaling up or scaling down, and for evaluating errors associated with the cross-scale applications. As should be expected, scale and classification issues are common to landscape variables for which much data already exists, such as elevation and soils (Edmonds et al. 1985; Lathrop et al. 1995). Bleecker et al. (1995) developed a classification for pesticide leaching potential from agricultural land and found that reliable results could not be obtained at a scale finer than 1:250,000 because of limitations of the input data.

A fourth theme includes applications of land cover classification, distribution, and abundance of vegetation types (Davis and Goetz 1990); mapping for landscape and land-use planning (Canters et al. 1991); identifying rare or priority conservation targets (Wright et al. 1994); and indentification of continental-scale floristic (McLaughlin 1994) and biome (Neilson et al. 1992) units.

Despite this effort, and rigorous quantitative work, scale appears to be treated in a relatively unsophisticated way in many land-use and land-cover classification projects. In many cases, approaches are limited to algorithms to extrapolate between scales, and statistical analyses of error rates and sources. Few general findings related to scale have been forthcoming, nor testable, theoretically based hypotheses offered.

Vegetation Analysis and Plant Ecology

This section refers to work on the ecology, physiognomy, composition, and mapping of vegetation and plant communities, other than that covered under other headings. Much of this work could be considered traditional plant ecology and biogeography, and therefore covers a very broad range of inquiry. The distinction from land cover classification becomes somewhat vague when whole communities are assessed, but here the emphasis is on the plants themselves, rather than the units on a landscape.

Plant ecologists have taken great effort to incorporate problems of scale in their research design and analysis in recent years, as evidenced by the number of citations identified in our literature scan. The results in Table 11.4 may therefore be somewhat misleading: Within the broad scope of plant ecology and vegetation science, treatment of scale is well advanced in many areas. However, relatively few authors in the journal *Landscape Ecology* address questions of traditional plant community ecology, such as species composition, physiognomy, or community ecology, and thus recent advances in treatment of scale-related issues are poorly represented there.

Within the broader context of plant ecology, treatment of scale appears best developed in studies that assess plant community composition, species turnover, diversity (discussed later), and plant community change over time (discussed later). A large body of work has addressed the influence of spatial and temporal scale on vegetation and plant community structure. Holling (1992) has proposed, convincingly, that processes that shape communities operate over a hierarchy of temporal and spatial scales. Many authors thus have attempted to determine the processes shaping different ecosystems over different spatial scales. For example, Coughenour and Ellis (1993) assessed vegetation composition in arid northern Kenya and found vegetation and plant community composition to be hierarchically constrained by physical factors: By climate at regional to continental scales; by topographic effects on rainfall and landscape water redistribution, and geomorphic effects on soil and plant available water at the landscape to regional scales; and finally by water redistribution and disturbance at local and patch scales. Such influences have been assessed for a number of other community types, including mid-latitude Kansas prairie grassland (Glenn et al. 1992; Gibson et al. 1993), temperate forest (Busing and White 1993; Cowell 1995), boreal forest (Schaefer 1993), tropical lowland forest (Colinvaux 1987; Mackey 1994), montane grasslands (Acosta et al. 1992; Herben et al. 1993), tropical savannah (Vetaas 1992), riparian woodland (Bendix 1994), riverine wetlands (Jean and Bouchard 1993); and others.

Scale-dependent factors have also been explored for many other structural and functional relationships in plant ecology, which are too lengthy to elucidate here. A few of these are scale-dependent assessments of species diversity (e.g., Palmer and White 1994); competition (Kenkel 1993; Bruns 1995); within-species genetic variability (Young et al. 1993); paleovegetational history (Delcourt and Delcourt 1988; Graumlich and Davis 1993; McLaughlan 1995) and even percentage serotiny in lodegepole pine cones (Tinker et al. 1994). Aarssen (1992) has discussed the scales of species interactions, and the implications for competition theory and coexistence. Collins et al. (1993) present evidence that species ranges are polymodal on various temporal and spatial scales, and propose a hierarchical continuum concept to account for community structure. Delcourt and Delcourt (1996) propose spatial unit areas appropriate for vegetation reconstruction using General Land Office Survey data. Phillips (1995) offers five conceptual approaches to deal with the contrasting temporal and spatial scales over which geomorphic versus biological processes operate.

Landscape Dynamics: Succession, Disturbance, Stability

Disturbance regimes and vegetation change are a special case of plant ecology, but are such a pervasive theme in landscape ecology that it has been given it its own category. Landscape dynamics is among the most extensively studied topics in landscape ecology, among which treatment of scale is most advanced. Ecologists have long recognized scale dependence of disturbance regimes, and as early as the 1970s began to build simulation models of landscape dynamics that accounted for scales of disturbance and recovery (Botkin et al. 1972) and which led to the shifting mosaic-steady state model (Bormann and Likens 1979). Today, most research on landscape dynamics includes scale as a concern, and more than half of the studies build multiscale analysis into the design (Table 11.4).

Too, landscape ecologists have applied the full suite of techniques discussed earlier in an effort to understand better spatial patterns of landscape change and disturbance. Simulation models (e.g., Turner et al. 1993), remote sensing and GIS (e.g., Knight et al. 1994), systems hierarchy (e.g., Devasconcelos et al. 1993), spatial statistics (e.g., Lundquist 1995), and cross-scale extrapolation (e.g., Kienast 1993; Boose et al. 1994; Simpson et al. 1994) have all been applied with good result to elucidate scale dependency in disturbance regimes.

Assessment of scale-dependent processes in landscape dynamics have taken a number of perspectives, including those that focus on landscape structure, on timing and recurrence of disturbance, on the nature of the disturbance, and on the ecosystem or biome type. Boose et al. (1994) combined models for the broad-scale patterns of winds generated by hurricanes with fine-scale models of topography and microclimate to predict landscape-scale patterns of forest disturbance based on topography, aspect, and exposure. Turner et al. (1993) developed a generalized scheme of landscape stability based on simulated disturbance at continuously varying ratios of disturbance size to landscape size and disturbance interval to recovery interval. They found varying combinations of size and interval ratios that lead to stable landscape composition, while others produced continuously variable landscapes with no stable endpoint. In a similar vein, Baker (1993) simulated disturbance regimes for boreal forest in northern

Minnesota and found both the timing and spatial scale of landscape effects to vary with fire intervals and return periods. Devasconcelos et al. (1993) constructed a hierarchical model of landscape dynamics in wet sclerophyllous forest, coupling stand-level population models with broader-scale, stochastic disturbance regimes. Frelich and Reich (1995) assessed the successional trajectory of boreal forest at different scales under different disturbance regimes. Glenn et al. (1992) assessed the effects of alternative artificial disturbance mechanisms (mowing and controlled burn) at two sales in Oklahoma tall grass prairie. A number of studies have focused on patch-scale disturbance and its influence on broad-scale patterns (Lavorel et al. 1994a; Lundquist 1995).

This focus has produced an advanced framework to assess landscape change, especially disturbance regimes, in a scale-dependent fashion. Increasingly, disturbance is seen driven by different processes at different scales, and researchers are actively tring to sort out the dominant factors at each spatial scale relevant to the analysis. Within this framework, spatially explicit models have been developed and in some cases verified by field data to map or predict landscape response to disturbance regimes at different spatial scales. Gap-phase regeneration models have become increasingly sophisticated, and many now include hierarchical subroutines to incorporate individual plant population and/or life-history strategies at the patch scale (e.g., Prentice and Leemans 1990; Lavorel et al. 1994b). Similar hierarchical models have been developed for specialized habitats, such as on serpentine soils subject to reccurrent fine-scale disturbance (e.g., Wu and Levin 1994).

Habitat Analysis: Fragmentation, Corridors, Foraging, and Movement Rules

Here, studies are included that address landscape in the context of its function as habitat for one or more species. These tend to cluster in three areas: (1) the influence of landscape structure, such as patch size, fragmentation, and connectivity, on the species that use it; (2) species range and species habitat preferences; and (3) patterns of species movement within landscapes, including foraging behavior and movement rules. Only movement and patterns of usage within habitat patches and mosaics are discussed. The more general aspects of animal ecology will be covered in the next section.

Habitat fragmentation and its potential impact on community structure and diversity emerged as one ramification of island biogeography theory (e.g., sensu MacArthur and Wilson 1967; Lovejoy et al. 1986) and is today a major focus of landscape ecology. Such problems are a direct question of spatial scale, including factors of size (area), proximity, and distance. One area of concern focuses on how habitat fragmentation affects one or more species, such as spotted owl (Carey et al. 1992) or black grouse (Kurki and Linden 1995); taxonomic groups (Beckon 1993; Dale et al. 1994; Hinsley et al. 1995; Wiens 1995); or diversity overall (e.g., Barrett and Peles 1994). These analyses often include scale variables, such as fragment size, distance separation, or heterogeneity, explicitly in sampling design or analysis. These analyses commonly address between-patch versus within-patch effects (e.g., Carey et al. 1992; Bowers and Dooley 1993; McCollin 1993), and the importance of corridors as habitat or conduits (e.g., Merriam and Lanoue 1990; Benninger-Truax et al. 1992; Demers 1993).

A second category of habitat analyses focus on factors associated with the ranges or habitats of species. GIS is frequently employed in such analyses (e.g., Johnson 1990; Duinker et al. 1991; Gagliuso 1991; Holt 1991; Aspinall 1992; Bagby and Bian, 1992; Herr and Queen 1993; Aspinall and Matthews, 1994; Kamradt 1995). One approach involves mapping the range of some target species together with environmental characteristics of the species' known habitat in a raster-based GIS and using cartographic modeling to produce predicted species ranges or evaluation of relative habitat favorability (Johnson 1990). These studies typically require extensive data, and often intensive fieldwork. Cross-scale extrapolation, raster averaging, or some other procedure is usually necessary for scale compatibility in input data. Duinker et al. (1991) outlined five principles to guide development of such habitat models. Cross-scale comparisons are common in habitat analyses, as for example in assessment of habitat use by bison (Pearson et al. 1995), various small mammals (Bennett 1993; Bowers and Dooley 1993), cattle (Devries and Schippers 1994), and moose (Forbes and Theberge 1993).

The authors also include here studies related to species movement rules, home range, and scale-dependence of habitat requirements. Many authors have noted that species movement is hierarchical, depending not only on distribution of individual home ranges, but also movement within the home range (Morris 1992). Loehle (1990) used fractal geometry to model species movement within the home range in heterogeneous territory, although this has been criticized (Gautestad and Mysterud 1994). Diffendorfer et al. (1995) assessed the influence of patch size and isolation on movement of three small mammals in heterogeneous landscapes and found no relationship of landscape structure to population sources and sinks. Several authors have suggested that landscape-scale habitat movement for large herbivores is random, or nearly so, but patch-scale movement is clumped, or selective (Pearson et al. 1995). Others have suggested that nearest-neighbor rules, rather than random walk models, are most consistent with within-habitat movement by large herbivores (Gross et al. 1995). Winter studies with moose in the boreal ecosystem suggest that moose select areas with disturbed canopy disproportionally at a broad scale, but select closed canopy at a local ($<100\,km^2$). Fractal analysis of habitat movement revealed that small species of grasshoppers interact with patch structure at a finer scale of resolution than larger species (With 1994b), and young individuals of a single species move at slower rates and resolve small-scale details of the landscape, while adults move at faster rates and operate at a greater spatial extent than nymphs (With 1994a).

Animal Ecology: Population Biology, Food Webs, Species Interactions

This section discusses scale-related analysis in topics of animal ecology, such as population biology, food web structure, competition, coexistence, and so forth, with the caveat that the organism(s) (and not the landscape) is (are) the primary unit of analysis. Thus many studies categorized above under "habitat analysis" address individual animal species or taxonomic groups, but the landscape, and its structure, is the main focus of study.

In the journal *Landscape Ecology*, again, relatively few studies focus primarily on animal species, but rather on some element of the landscape. Those that do tend not to treat scale in a very sophisticated way. Thus studies of population biology and habitat preferences of, for example, blue tits (*Parus caesuleus*) (Blondel et al. 1992), kirtland's warbler (*Dendroica kirtlandii*) (Zao et al. 1992), and cotton rats (*Sigmodon hispidus*) (Lidicker et al. 1992) all work at a single (generally landscape) scale, and none attempts to address methodologically the implications of scaling issues. This finding is the result, in part, of how the studies were categorized. It may also reflect the fact that landscape ecologists who address a single animal taxon or taxa may be concerned with organisms within a given landscape, and scale may be peripheral.

Within a broader context, however, animal ecologists have developed and applied many methods to address problems of spatial scale, and these have been applied to a wide variety of issues in animal ecology. These include population ecology (Johnson et al. 1992; Koricheva and Haukioja 1994; Rey 1995; Hamazaki 1996), especially density-dependent mortality (Stiling et al. 1991; Hails and Crawley 1992; Tuda 1993); competition and predation (Gascon and Travis 1992; O'Neill 1992; Oksanen et al. 1992; Ives et al. 1993); community composition of birds (Bersier and Meyer 1994), spiders (Ehmann 1994), beetles (Fuisz and Moskat 1992), grasshoppers (Kemp 1992), wireworms (Penev 1992), and peccaries (Taber et al. 1994); body size hierarchies (Brown and Nicoletto 1991; Holling 1992); and food web structure (Martinez 1994; Martinez and Lawton 1995).

One set of tools commonly employed to study animal population ecology is spatially explicit population models (SEPMs), which are often constructed within a GIS. There have been used to track, analyze, or simulate population dynamics, presence/absence, habitat selection, density-dependent processes, and habitat change for subject species, typically endangered or rare species (e.g., Dunning et al. 1995). They have also been used as management tools (Turner et al. 1995) and a logical extension would be to link dynamic animal population models to models of vegetation dynamics (Holt et al. 1995). The underlying data or model structures in SEPMs provide a ready opportunity to manipulate, and evaluate, effects at multiple scales (Conroy et al. 1995). Such studies have produced valuable insight into scale dependency in factors affecting the ecology and population biology for individual species, but appear to have produced relatively few general principles. One limitation to addressing problems of scale for individual animal species is the availability of requisite data for multiscale analysis (Turner et al. 1995).

Controlled experiments or simulation models may offer a more promising outlook for general findings and testable hypotheses. Using controlled experiments, Zhang and Sanderson (1993) demonstrated that specialist predators (mites) search selectively at fine spatial scale but randomly at broad scale, whereas generalists search randomly at fine scale but aggregate at broad scale.

Authors have also reported scale dependence in the variability of genetic differentiation (Mitchellolds 1992; Faith 1994; Joseph and Moritz 1994; Sandoval 1994), which has been used to study evolutionary mechanisms, elucidate historical biogeography (Joseph and Moritz 1994), and generate conservation plans (Faith 1994). Effects of scale are commonly incorporated into design of research on endangered species, such as the northern spotted owl (Carey et al. 1992), or pest species, such as the European starling (Clergeau 1995). Others have tested for but found no scale dependence in habitat factors affecting population biology of *Algyroides mancha*, a small endemic lizard of the Inberian Peninsula.

These results, together with those presented earlier, lend strong support to Holling's (1992) hypothesis that processes that shape landforms operate on spatial and temporal scales that structure ecosystems and thus entrain animal communities into hierarchies of spatial distribution and body size.

Biodiversity

Study of diversity has long been concerned with problems of scale. Among the earliest progress occurred in studies of species–area relationships (e.g., Gleason 1922, 1925), which were based on the recognition that species richness depended on the extent of the study area. This concept formed the basis for Whittaker's recognition of three *scales* of diversity, namely alpha-, beta-, and gamma-diversity, which represented diversity not only at the site (alpha) and landscape (gamma) spatial scales, but also the relationship, or "differentiation diversity," between the two (beta) (Whittaker 1972, 1977). Although conceptually, these two approaches appear to represent two sides of the same coin, there appears to be an important distinction. Species–area studies tend to focus on diversity as a property of the landscape, whereas the alpha-, beta-, and gamma-diversity approach treats diversity as a property of the biota. The plethora of indices developed to represent beta diversity (e.g., Patil and Taillie 1979; Magurran 1988) suggests that the change of species diversity with changing scale remains enigmatic.

A recent approach is a strategy that seeks to identify areas of high biological diversity or concentrations of rare or unique species that are vulnerable to habitat degradation (e.g., Scott et al. 1993; Bojorqueztapia et al. 1995). Various data, such as vegetation type, species composition, and rare species distributions, are input as layers into a GIS. These data are then overlaid by boundaries of property ownership and management status to identify areas with significant value for biological diversity that are unprotected from development or habitat degradation. The result is a spatial assessment of priority targets for acquisition or improved conservation management.

Alternative approaches seek to identify areas of specific environmental characteristics (potential natural vegetation, existing vegetation communities, presence of endangered species, etc.) that are underrepresented in nature reserve or other protected status (Davis et al. 1990). Miller et al. (1989) used this conceptual approach to analyze rare bird species richness in Tanzania, but noted serious shortcomings due to a paucity of data for meaningful explanatory variables, and the large grid size (0.86 km^2) of the available environmental data. These analyses all suffer the limitations of grain size and scale compatibility of the underlying GIS data base.

Until relatively recently, much of the "scale-literate" work on diversity focused on the functional nature of the species–area curve and variation in the underlying coefficients (sensu MacArthur and Wilson 1967). Connor and McCoy (1979) recognized various mechanisms in landscapes that contribute, in varying degrees, to the species–area relationship. Many authors have noted, of course, a relationship between species diversity and "habitat diversity" (e.g., Kohn and Walsh 1994), and others have called for a hierarchical approach to conservation based on ecosystem types and landscape units, which would each maintain its own suite of species (Norton and Ulanowicz 1992).

Recent work employing a landscape ecology framework has attempted to parcel diversity into different landscape units (Harris 1984; Diamond 1986; Forman and Godron 1986; and Zonneveld 1989). These issues are of obvious importance not only in landscape ecology and biogeography, but also in applied conservation biology (e.g., Diamond 1975; Margules et al. 1988; Zimmerman and Bierregard 1986). Several authors have begun to address variation in diversity associated with differences in successional and disturbance regimes over multiple scales (Huston 1994) and in different environments, such as tundra (Price 1971), desert bajadas (Withers 1992), fluvial floodplains (Hupp and Osterkamp 1985), coniferous forests (Romme and Knight 1981), and mixed hardwood forests (Witney 1982; Grimm 1984; Host et al. 1987; Cowell 1995). Others have addressed environmental gradients in diversity (e.g., Gentry 1988; Aronson and Schmida 1992; Wylie and Currie 1993).

Palmer and White (1994) extended these ideas by recognizing that species richness depends not only on grain and extent of the study area, but also on the starting point. Thus, any location possesses not one but a family of species–area curves. Similarly, familiar relationships between species diversity and, for example, latitude, elevation, and productivity, all depend on the spatial resolution and extent (Rosenzweig 1992; Huston 1994). And Stoms (1994) found that the alpha and beta components of diversity contribute to regional diversity to different degrees in different environments.

Considering the intensity of its study in general ecology, diversity is relatively little reported in *Landscape Ecology* ($n = 5$, Table 11.4) and scale issues remain peripheral. Within landscape ecology, attention appears to focus on the relationship between diversity and the size and complexity of habitat patches (e.g., McIntyre 1995). It appears that the landscape ecology approach may offer insight into the species–area problem through investigation of the relative contribution to regional diversity by individual landscape units (after Zonneveld 1989; Stoms 1994; Stohlgren et al. 1996).

Biogeochemical Cycles and Energy Flow

Analysis of ecosystem effects on hydrologic and nutrient cycles became a major focus in ecology as part of the International Biological Program (IBP). In landscape ecology, research in this area appears to focus on landscape and catchment-scale processes in soil formation and soil erosion, nutrient retention in plants, and nutrient and water allocation among competitors. Scaling fine-scale biogeochemical data for input to global-scale models is also a subject of intensive research, and will be addressed in the next section.

Studies on biogeochemical cycles may employ field data, simulation models, or GIS mapping, and spatial statistical methods are widely applied. Field studies are, for the most part, limited to landscape or, at greatest extent, catchment scale. Scale issues involve mainly the scale of spatial dependence (e.g. Jackson and Caldwell 1993a, 1993b), scale dependence in distribution pattern (Johnston 1993), or extrapolation across scale for regional-scale assessment (Groffman et al. 1992; Pierson et al. 1994; Bleecker et al. 1995). There is an enormous literature on estimating net primary productivity (e.g., Mooney and Field 1989; Running 1990; Begue et al. 1994; White and Running 1994) and on ecosystem-level carbon fluxes (e.g., Raupach et al. 1993; Klinger et al. 1994; see Post 1993 for a review), and the principal issue is how to extrapolate field measures at a site to the large grid scale required for regional and global modeling.

Simulation models provide an attractive alternative to field measurement of material and energy fluxes from complete landscapes and ecosystems, which presents severe methodological and logistic limitations. Such simulation models have been developed for a great number of biogeochemical processes, including soil decomposition in tropical soils (Lavelle et al. 1993), methane flux from wetlands (Klinger et al. 1994), and soil water storage capacity (Zheng et al. 1996). Lavelle et al. (1993) present a conceptual model of decomposition in tropical soils, outlining a hierarchy of processes operating at increasingly fine spatial scales.

Assessment of sources for chemical (Comeleo et al. 1996) and agricultural (Bleecker et al. 1995) contaminants is another important application of landscape ecology. Bleecker et al. (1995) produced regional estimates of atrazine leaching potential by combining an empirical model of leaching potential with a regional environmental data base mapped through a GIS. Comeleo et al. (1996) used a hierarchical sampling design to identify point and areal sources of chemical contaminants to Chesapeake Bay. Outside of ecology, there is a vast literature on biogeochemical cycles in the fields of agronomy, soil science, geomorphology, and hydrology that has also contributed greatly to our understanding of scale-dependent properties in material budgets.

Global Change: Climatic Models, Ecosystem Models, Climate Impacts

A flood of research has been generated in response to anticipated global warming, which has spawned a larger interest in broad issues of global change. Much of this research involves coupling a numerical climate model to ecosystem process models (see Smith et al. 1992; Solomon and Shugart 1993). In many of these studies, climate is seen as a forcing factor for phenomena such as net biomass or primary productivity (e.g., Pierce and Running 1995). Others include ecosystem feedback effects on the climate (Walker 1994). A major problem in these types of studies is the very large grid size ($1^\circ \times 1^\circ$ or larger) employed by climatic models (Henderson-Sellers 1987). Thus issues of scale in these studies focus on resolution, and on cross-scale extrapolation (Ehleringer and Field 1993).

For example, Pielke et al. (1991) examine the impact of landscape variability on the atmosphere, and Klaassen and Claussen (1995) looked at feedback by landscape variability on the surface drag coefficients in a climate model. As should be expected, the scales used in these studies are broad and beyond the extent of most landscape studies. Furthermore, the methods looking at impacts of climatic change are generally deductive in that specific phenomena are predicted from general empirical relations between the atmosphere and landscapes. Hence, much of the predicted outcomes of climatic change cannot be tested. Rosswall et al. (1988) and Clark (1985) provide excellent background on these issues. Goodess and Palatikoff (1992) examined the ecological effects of specific regional climatic scenarios, while Walker (1994) examined feedbacks between vegetation and the atmosphere at different scales. The large grid size of global climatic models generally precludes their use for assessment at a landscape scale. Such analyses to date have been largely deductive, and these have pushed as far as possible pending development of finer scale climatic models and essential empirical knowledge of climate/landscape interactions.

As such, one of the most active areas of research that involves questions of scale in this area falls into two broad categories: cross-scale extrapolation and resolution. And this extrapolation problem is bidirectional. That is, on the one hand, how can landscape-based experimental results, with an extent on the order of less than one to a few hectares, be incorporated into a broad-scale model, with a resolution on the order of 1 to 4° degrees of latitude/longitude (see e.g., Pierce and Rastetter et al. 1992; Raupach et al. 1992; Running 1995)? On the other hand, how can the results of global and regional climatic models be used to assess ecological and other environmental effects at the landscape level (see Smith et al. 1992; Hostetler and Giorgi 1993; White and Running 1994)? Efforts to model feedbacks of ecosystem perturbation at landscape scale back to the global climatic system are in their infancy, and these feedbacks at present are not understood even qualitatively, much less with any confidence for quantitative prediction (Walker 1994). Furthermore, about half of the studies on global change impacts in *Landscape Ecology* ignore issues of scale altogether (Table 11.4). Progress in these areas may allow progress beyond the largely speculative nature of much of the work on ecosystem effects of global change.

Land Planning, Design, and Management

Practical applications of their science to planning and conservation have provided much of the incentive for ecologists and others to pursue the landscape ecology approach. "It is crucial that landscape ecologists develop ways to convincingly express our understanding of the biosphere and then effectively apply this understanding to problem solving" (Golley 1987). Likewise, land planners and managers increasingly are incorporating landscape ecology and its methodology into management plans. Thus Norton and Ulanowicz (1992) have called for biodiversity policy that recognizes the landscape as the unit of conservation priority. Governmental agencies have also employed GIS models for habitat management directed at individual species or land resources in general. Assessing the effect of land-use change, and especially human developments, on species habitats is a particularly common theme. These have included assessment of human impact on the habitat of cougars (Gagliuso 1991); the black bear (Holt 1991); bobcat (Kamradt 1995); white deer (Bagby and Bian 1992); and moose and marten (Duinker et al. 1991). Commonly, the resulting data bases are used as a management tool to monitor the continuing influence of human land-use changes on population change, habitat use, and travel corridors of wildlife. Such analyses are typically of fixed raster size and require input data to be scaled for compatibility within the GIS. Often these employ relatively unsophisticated scaling techniques and are vulnerable to fallacious results (Ball 1994).

Other researchers have assessed scale-dependent issues surrounding many other aspects of land planning and natural resources management, including forest logging plans (Harris 1984; Li et al. 1993), nature reserve design (Price et al. 1995), and fragmentation effects on landscape planning (Gustafson and Parker 1994).

Central to many of these plans are issues of scale, and the parallel hierarchical structuring of landscapes and human management systems (Fox 1992). Freemark (1995) has outlined a hierarchical approach to assessment of agricultural impacts on native biota, and highlighted the need for institutional structures that are compatible with a hierarchical arrangement of taxa and the broad hierarchy of spatiotemporal scales in which populations, communities, and landscapes interact. Others too have noted the conflicts between national or regional conservation and economic objectives on the one hand, and objectives of local communities on the other (Barrett 1992; Wolf and Allen 1995). Schwartz (1994) has gone even further, noting potential cross-scale conflicts in underlying conservation goals. For example, introducing an endangered species outside its native range may enhance its survival prospects, but pose the risks of an exotic species in the new home range. Others have extended the concept of endangered species to threatened biotopes (Blab et al. 1995) or landscapes (Naveh 1993). They propose a national Red Data Book of threatened landscape units, modeled on the IUCN lists of endangered species. The slogan, "think globally, act locally," is thus not only a popular rallying cry for conservationists, but may reflect an increased awareness of scale dependence and hierarchical structuring in landscapes and human institutions.

Methodology

The vast majority of papers in *Landscape Ecology* and elsewhere provide useful information on the sampling and analysis methods employed. Many of these are particularly relevant to issues of scale, and have been cited throughout this chapter. Among useful general references are *Quantitative Methods in Landscape Ecology* (Turner and Gardner 1991a); *Scaling Ecophysiological Processes: From Leaf to Globe* (Ehleringer and Field 1993); *Landscape Ecology and GIS* (Haines-Young et al. 1993); and *Geographical Information Systems. Principles and Applications* (Maguire et al. 1991).

A few studies can be classified as strictly methodological, and these include methods to address issues of scale. Hunsaker et al. (1994) provide guidelines for landscape sampling procedures to characterize land cover and landscape structure and assess effects of sample area and study extent on landscape properties. Delcourt and Delcourt (1996) provide guidance on using land survey records at the proper scale. Hargrove and Pickering (1992) introduce the concept of pseudoreplication and its relevance to landscape structural measures.

A detailed discussion and an extensive list of references on methods relevant to scaling issues have been given earlier. Methods applicable to specific subdisciplines are discussed in the relevant paragraphs of this section.

Scale Per Se

A number of researchers have addressed directly the problem of scale in landscape ecology. Not surprisingly, these are the studies most focused on developing general findings and testable hypotheses regarding scale (Table 11.4). Many of these involve the effect of scale on measures of landscape structure (e.g., Turner 1990; O'Neill et al. 1991, 1996), the problems of scaling up or scaling down analyses at one scale to a different scale of interest (Simmons et al. 1992; Costanza and Maxwell 1994), or development of specific methods to address problems of spatial scale (Cullinan and Thomas 1992).

The common thread in most of these studies appears to revolve around the recognition that processes underlying ecological patterns occur on different scales (Meentemeyer 1989) and that to understand a given ecological system the researcher must examine properly matched spatial scales of process and pattern (Holling 1992; Levin 1992). Toward that end, Wiens et al. (1993) have outlined a framework for experimental model systems to integrate processes at multiple spatial scales in landscape ecology research.

A Proposed Standard Vocabulary for Issues of Scale in Landscape Ecology

As Norton and Lord (1990) have noted, science in general, and ecology in particular, are rife with jargon used to represent multiple phenomena. Below the authors identify some terms related to spatial scale that are used widely, but often ambiguously, not only in ecological but also in other scientific literature. A standardized useage is then proposed for these terms to deal with issues of spatial scale in ecology and landscape ecology. The authors have no illusion that this will be a final authority, but intend instead that the following be regarded as a first step in generating a dialogue toward development of a consistent, specific, and unambiguous vocabulary of scale.

In particular we beseech ecologists and other scientists to stop using the terms *large scale* and *small scale*, except when explicitly clear that they refer to a map.

Scale — As noted above, the term *scale* represents many phenomena, both conceptual and tangible, including resolution or grain size; sample size and density; area of study; range of taxa; landscape hierarchies; and other relationships in space and time. We therefore propose that the term *scale* be used only with a defining modifier. Thus, *spatial scale* to represent variation in sampling unit size, resolution, or extent; *taxonomic scale* to represent the level on the taxonomic hierarchy; and so on.

Grain size — As noted by Norton and Lord (1990), grain size has been invoked to define minimum sampling unit, as well as patch size, and response to habitat patchiness (Wiens, 1990). We propose the term "minimum sampling unit" to refer, unambiguously, to what Wiens (1989), Palmer and White (1994), and others have called *grain*. More universally grain can be equated with resolution.

Extent — Extent represents the boundary of the study area under consideration, and appears unambiguous.

Spatial scale — Dimensions expressed in distance and area, or for a map, a fraction.

Large-scale — Large fractional scale, small area, short distances, large detail: best expressed as "Large map-scales."

Samll-scale — Small fractional scale; large area, great distances less detail: best expressed as "small map-scales."

Broad scale — To replace "large-scale" (i.e., large extent).

Fine scale — To replace "small-scale" (i.e., small extent), generally meaning fine textured, or with much detail.

Summary

There can be no doubt that an explosion in research and interest in scale, primarily spatial scale, started in the mid-1980s. Perhaps it has always been known that size matters, but the new emphases go beyond just size. They also go beyond scaling and extrapolation. What caused this bloom? We speculate that some environmental biologists became less interested in their plots and one-dimensional models and returned to a form of natural history, but emerging tools and paradigm shifts occurred concurrently. The availability of remotely sensed images, geographic information systems, global satellite positioning, image storage and manipulation, and fractals are all implicated, and all are driven by technological advances and digital technologies. Inclusion of the spatial dimension in studies since the 1980s appeared possible with a minimal loss of cherished mechanistic detail and explanation.

When leading atmospheric chemists sounded the alarm on global warming, a justification for studies at broader spatial scales seemed to appear. By then, most environmental biologists were aware of hierarchy theory (Allen and Starr 1982; O'Neill 1988). Also in the 1980s Frank Golley started the landscape ecology initiative as part of his administrative role in the National Science Foundation (USA), as well as his creation of the journal, *Landscape Ecology*. If, however, Landscape Ecology should integrate at levels above the community or ecosystem, broader spatial scales of analyses would be needed. It became apparent that the models, explanation, and relevant variables at one spatial scale could often (usually?) not be applied at other levels. So then, the concurrent development of the digital technologies (remote sensing, GIS, desktop computers, etc.), the global warming initiative, hierarchy theory, the emergence of Landscape Ecology (e.g., the text by Forman and Godron 1986), and perhaps recognition that studies should go beyond the one meter plot, all conspired to put in place that "rare creature of the intellectual bestiary," the paradigm shift. The emergence of the "spatial revolution" then aided, if not caused, the emergence of scale.

It has now been a little over 10 years since appeals were made for a science of scale and the development of scale theory. Although hundreds of studies have now been directed to scale and much more is known about the consequences of changing scale, no "theory" has been forthcoming. Perhaps none should be expected—perhaps scale is just another variable, as it is now in the discipline of geography: Perhaps scale is simply being integrated or embedded within other theoretical constructs in ecology and landscape ecology. Scale studies reviewed in this chapter also seem to show that scale has acted as a bridge to the spatial paradigm. The answer to the question, "What happens to what we know about ecology and ecosystems when we broaden spatial scales?" was needed. What are we really looking at?

For reasons we cannot determine, landscape ecologists have not extended the spatial revolution and scale studies toward the physical environment, with the possible exception of soils. A rich literature on scale-dependent geomorphologic and tectonic processes that shape landforms, drainage basins, and mountain ranges may lend much support to landscape ecologists knowledge of the biological environment. There appears to be an opportunity for studies of the feedbacks among biotic and abiotic parts of systems, and how these feedbacks change with scale. Perhaps the relative scarcity of such studies indicates the academic training and background of professionals in landscape ecology.

The hundreds of paper reviewed also show that scale has value as an interpretative tool, although there is little consensus as to whether scale should be considered first or after a study has been initiated. Scale has heuristic value. It can be a way of organizing knowledge and a way of helping to communicate that knowledge to others especially students. One author (Meentemeyer) uses scale as a heuristic device to organize a course in biometeorology. Students start at fine scales and then become broader and broader. Scale is an interpretive tool. We found scale being mentioned or studied in nearly all parts of science from DNA to geomorphology and even political science, but the terminology was also confusing and inconsistent. The four terms—broad scale, fine scale, grain, and extent—show much promise for a basic and consistent terminology. We found many excellent reviews, but above all the discipline of landscape ecology currently is at the forefront in resolving issues of scale.

Such analyses have taken assessment of landscape patterns, of landscape structure and landscape dynamics, beyond the merely descriptive science that Wiens (1992) described, to the level of generating testable hypotheses and attempting to predict the behavior of real landscapes under alternative scenarios. In other words, landscape ecology may be emerging as a mature science—able to predict the unobserved based on theory and understanding of process.

References

Aarssen, L.W. 1992. Causes and consequences of variation in competitive ability in plant-communities. Journal of Vegetation Science 3(2):165–174.

Acosta, A., S. Diaz, M. Menghi, and M. Cabido. 1992. Community patterns at different spatial scales in the grasslands of Sierras De Cordoba, Argentina. Revista Chilena De Historia Natural 65(2):195–207.

Allen, T.F.H., and T.B. Starr. 1982. Hierarchy: Perspectives for Ecological Complexity. University of Chicago Press, Chicago.

Aronson, J., and A. Schmida. 1992. Plant species diversity along a Mediterranean-desert gradient and its correlation with interannual rainfall fluctuations. Journal of Arid Environments 23:235–247.

Aspinall, R.J. 1992. Bioclimatic mapping—Extracting ecological hypotheses from wildlife distribution data and climatic maps through spatial analysis in GIS. In GIS/LIS Proceedings, 10–12 November, 1992, Vol. 1, pp. 30–39.

Aspinall, R., and K. Matthews. 1994. Climate change impact on the distribution and abundance of wildlife species: an analytical approach using GIS. Environmental Pollution 86:217–223.

Avissar, R. and F. Chen. 1993. Development and analysis of prognostic equations for mesoscale kinetic-energy and mesoscale (subgrid scale) fluxes for large-scale atmospheric models. Journal of the Atmospheric Sciences 50(22):3751–3774.

Bagby, J.E., and L. Bian. 1992. Urbanization and its impact on white-tailed deer (Odocoileus virginianus) habitat: A GIS/remote sensing approach. In GIS/LIS Proceedings, 10–12 November, 1992, Vol. 1, pp. 42–50.

Bailey, R.G. 1985. The factor of scale in ecosystem mapping. Environmental Management 9:271–276.

Bailey, R.G. 1996. Ecosystem Geography. Springer-Verlag, New York.

Baker, W.L. 1993. Spatially heterogeneous multiscale response of landscapes to fire suppression. Oikos 66(1):66–71.

Ball, G.L. 1994. Ecosystem modeling with GIS. Environmental Management 18:345–349.

Band, L.E., and I.D. Moore. 1995. Scale–landscape attributes and geographical information systems. Hydrological Processes 9(3–4):401–422.

Barrett, G.W. 1992. Landscape ecology—designing sustainable agricultural landscapes. Journal of Sustainable Agriculture 2(3):83–103.

Barrett, G.W. and J.D. Peles. 1994. Optimizing habitat fragmentation—an agrolandscape perspective. Landscape and Urban Planning 28(1):99–105.

Beckon, W.N. 1993. The effect of insularity on the diversity of land birds. Oecologia 94(3):318–329.

Begue, A., N.P. Hanan, and S.D. Prince. 1994. Radiative-transfer in shrub savanna sites in Niger-preliminary results from hapex-sahel. 2. photosynthetically active radiation interception of the woody layer. Agricultural and Forest Meteorology 69(3–4):247–266.

Bell, G., M.J. Lechowicz, A. Appenzeller, M. Chandler, E. Deblois, L. Jackson, B. Mackenzie, R. Preziosi, M. Schallenberg, and N. Tinker. 1993. The spatial structure of the physical-environment. Oecologia 96(1):114–121.

Bendix, J. 1994. Scale, direction, and pattern in riparian vegetation environment relationships. Annals of the Association of American Geographers 84(4): 652–665.

Bennett, A.F. 1993. Microhabitat use by the long-nosed potoroo, potorous-tridactylus, and other small mammals in remnant forest vegetation of south-western Victoria. Wildlife Research 20(3):267–285.

Bennett, R.J. 1979. Spatial Time Series. Pion, London.

Benninger-Truax, M., J.L. Vankat, and R.L. Schaefer. 1992. Trail corridors as habitat and conduits for movement of plant species in Rocky Mountain National Park, Colorado, U.S.A. Landscape Ecology 6(4):269–278.

Benson, B.J., and M.D. Mackenzie. 1995. Effects of sensor spatial-resolution on landscape structure parameters. Landscape Ecology 10(2):113–120.

Bergeron, Y. 1991. The influence of island and mainland lakeshore landscapes on boreal forest-fire regimes. Ecology 72(6):1980–1992.

Bersier, L.F., and D.R. Meyer. 1994. Bird assemblages in mosaic forests—the relative importance of vegetation structure and floristic composition along the successional gradient. Acta Oecologica-International Journal of Ecology 15(5):561–576.

Blab, J., U. Riecken, and A. Ssymank. 1995. Proposal on a criteria system for a National Red Data Book of Biotypes. Landscape Ecology 10(1):41–50.

Bleecker, M., S.D. Degloria, J.L. Hutson, R.B. Bryant, and R.J. Wagenet. 1995. Mapping atrazine leaching potential with integrated environmental databases and simulation-models. Journal of Soil And Water Conservation 50(4):388–394.

Blondel, J., P. Perret, M. Maistre, and P.C. Dias. 1992. Do harlequin Mediterranean environments function as source sink for blue tits (*Parus caeruleus* L.)? Landscape Ecology 6(3):213–219.

Bojorqueztapia, L.A., I. Azuara, E. Ezcurra, and O. Floresvillela. 1995. Identifying conservation priorities in Mexico through geographic information-systems and modeling. Ecological Applications 5(1):215–231.

Boose, E.R., D.R. Foster, and M. Fluet. 1994. Hurricane impacts to tropical and temperate forest landscapes. Ecological Monographs 64(4):369–400.

Bormann, F.H., and G.H. Likens. 1979. Pattern and Process in a Forested Ecosystem. Springer-Verlag, New York.
Botkin, D.B., J.F. Janak, and J.R. Wallis. 1972. Some ecological consequences of a computer model of forest growth. Journal of Ecology 60:849–873.
Bowers, M.A., and J.L. Dooley. 1993. Predation hazard and seed removal by small mammals—microhabitat versus patch scale effects. Oecologia 94(2):247–254.
Bretschko, G. 1995. Riverland ecotones—scales and patterns. Hydrobiologia 303 (1–3):83–91.
Brown, J.H., and P.F. Nicoletto. 1991. Spatial scaling of species composition—body masses of North-American land mammals. American Naturalist 138(6): 1478–1512.
Bruns, T.D. 1995. Thoughts on the processes that maintain local species-diversity of ectomycorrhizal fungi. Plant and Soil 170(1):63–73.
Burel, F., and J. Baudry. 1990. Structural dynamics of a hedgerow network landscape in Brittany, France. Landscape Ecology 4(4):197–210.
Burrough, P.A. 1981. Fractal dimensions of landscapes and other environmental data. Nature 294:240–242.
Busing, R.T., and P.S. White. 1993. Effects of area on old-growth forest attributes —implications for the equilibrium landscape concept. Landscape Ecology 8(2): 119–126.
Calder, W.A. 1983. Ecological scaleing: Mammals and birds. Annual Review of Ecology and Systematics 14:213–230.
Canters, K.J., C.P. den Herder, A.A. de Veer, W.M. Veelenturf, and R.W. de Waal. 1991. Landscape-ecological mapping of the Netherlands. Landscape Ecology 5(3):145–162.
Carey, A.B., S.P. Horton, and B.L. Biswell. 1992. Northern spotted owls—influence of prey base and landscape character. Ecological Monographs 62(2):223–250.
Caswell, H. 1976. Community structure: A neutral model analysis. Ecological Monographs 46:327–354.
Cherrill, A. 1994. A comparison of 3 landscape classifications and investigation of the potential for using remotely-sensed land-cover data for landscape classification. Journal of Rural Studies 10(3):275–289.
Cherrill. A.J., C. Mcclean, P. Watson, K. Tucker, S.P. Rushton, and R. Sanderson. 1995. Predicting the distributions of plant-species at the regional-scale—a hierarchical matrix model. Landscape Ecology 10(4):197–207.
Clark, J.S. 1990. Integration of ecological levels: Individual plant growth, population mortality, and ecosystem processes. Journal of Ecology 78:275–299.
Clark, K.C. 1986. Computation of the fractal dimension of topographic surfaces using the triangular prism surface area method. Computational Geosciences 12:713–722.
Clark, W.C. 1985. Scales of climatic impacts. Climatic Change 7:5–27.
Clergeau, P. 1995. Importance of multiple scale analysis for understanding distribution and for management of an agricultural bird pest. Landscape and Urban Planning 32(1–3):281–289.
Colinvaux, P.A. 1987. Amazon diversity in light of the paleoecological record. Quaternary Science Reviews 6:93–114.
Collins, S.L., S.M. Glenn, and D.W. Roberts. 1993. The hierarchical continuum concept. Journal of Vegetation Science 4(2):149–156.

Comeleo, R.L., J.F. Paul, P.V. August, J. Copeland, S.S. Baker, and R.W. Latimer. 1996. Relationship between watershed stressors and sediment contamination in Chesapeake Bay estuaries. Landscape Ecology 11(5):307–319.

Connor, E.F., and E.D. McCoy. 1979. The statistics and biology of the species-area relationship. American Naturalist 113:791–833.

Conroy, M.J., Y. Cohen, F.C. James, Y.G. Matsinos, and B.A. Maurier. 1995. Parameter estimation, reliability, and model improvement for spatially explicit models of animal populations. Ecological Applications 5(1):17–19.

Cooper, A. 1995. Multivariate land class and land-cover correlations in northern-Ireland. Landscape and Urban Ecology 31(1–3):11–19.

Cornelius, J.M. and J.F. Reynolds. 1991. On determining the statistical significance of discontinuities within ordered ecological data. Ecology 72(6):2057–2070.

Costanza, R., and T. Maxwell. 1994. Resolution and predictability: an approach to the scaling problem. Landscape Ecology 9(1):47–57.

Coughenour, M.B., and J.E. Ellis. 1993. Landscape and climatic control of woody vegetation in a dry tropical ecosystem—Turkana district, Kenya. Journal Of Biogeography 20(4):383–398.

Cowell, C.M. 1995. Presettlement Piedmont forests—patterns of composition and disturbance in central Georgia. Annals of the Association of American Geographers 85(1):65–83.

Cressie, N. 1993. Statistics for Spatial Data. John Wiley & Sons, New York.

Crist, T.O., D.S. Guertin, J.A. Wiens, and B.T. Milne. 1992. Animal movement in heterogeneous landscapes—an experiment with eleodes beetles in shortgrass prairie. Functional Ecology 6(5):536–544.

Cullinan, V.I., and J.M. Thomas. 1992. A comparison of quantitative methods for examining landscape pattern and scale. Landscape Ecology 7(3):211–227.

Dale, V.H., S.M. Pearson, H.L. Offerman, and R.V. O'Neill. 1994. Relating patterns of land-use change to faunal biodiversity in the central Amazon. Conservation Biology 8(4):1027–1036.

Davis. F.W., and S. Goetz. 1990. Modeling vegetation pattern using digital terrain data. Landscape Ecology 4(1):69–80.

Davis, F.W., D.M. Stoms, J.E. Estes, J. Scepan, and J.M. Scott. 1990. An information systems approach to the preservation of biological diversity. International Journal of Geographic Information Systems 4:55–78.

Deboer, D.H. 1992. Hierarchies and spatial scale in process geomorphology—a review. Geomorphology 4(5):303–318.

Delcourt, H.R., and P.A. Delcourt. 1988. Quaternary landscape ecology: relevant scales in space and time. Landscape Ecology 2(1):23–44.

Delcourt, H.R., and P.A. Delcourt. 1996. Presettlement landscape heterogeneity: evaluating grain of resolution using general land office survey data. Landscape Ecology 11(6):363–381.

Demers, M.N. 1993. Roadside ditches as corridors for range expansion of the western harvester ant (*pogonomyrmex-occidentalis cresson*). Landscape Ecology 8(2):93–102.

Detilleul, P., and P. Legendre. 1993. Spatial heterogeneity versus heteroscedasticity: an ecological paradigm versus a statistical concept. Oikos 66:152–171.

Devasconcelos, M.J.P., B.P. Zeigler, and L.A. Graham. 1993. Modeling multiscale spatial ecological processes under the discrete-event systems paradigm. Landscape Ecology 8(4):273–286.

Devries, M.F.W., and P. Schippers. 1994. Foraging in a landscape mosaic—selection for energy and minerals in free-ranging cattle. Oecologia 100(1–2):107–117.

Diamond, J. 1975. The island dilemma: lessons of modern biogeographic studies for the design of natural reserves. Biological Conservation 7:129–146.

Diamond, J.M. 1986. The design of a nature reserve system for Indonesian New Guinea. In: Conservation Biology: The Science of Scarcity and Diversity, pp. 485–503. M.E. Soulé (ed.). Sinauer, Sunderland, MA.

Diffendorfer, J.E., M.S. Gaines, and R.D. Holt. 1995. Habitat fragmentation and movements of 3 small mammals (*Sigmodon, Microtus, and Peromyscus*). Ecology 76(3):827–839.

Duinker, P., P. Higgelke, and S. Koppikar. 1991. GIS-based habitat supply modelling in northwestern Ontario: moose and marten. In: Applications in a Changing World, GIS 91 Symposium Proceedings, pp. 271–275. Forestry Canada, Vancouver, British Columbia.

Dumanski, J., W.W. Pettapiece, D.F. Acton, and P.P. Claude. 1993. Application of agroecological concepts and hierarchy theory in the design of databases for spatial and temporal characterization of land and soil. Geoderma 60(1–4): 343–358.

Duncan, J., D. Stow, J. Franklin, and A. Hope. 1994. Assessing the relationship between spectral vegetation indexes and shrub cover in the Jornada Basin, New-Mexico. International Journal of Remote Sensing 14(18):3395–3416.

Dunning, J.B., D.J. Stewart, B.J. Danielson, B.R. Noon, T.L. Root, R.H. Lamberson, and E.E. Stevens. 1995. Spatially explicit population-models—current forms and future uses. Ecological Applications 5(1):3–11.

Edmonds, W.J., J.C. Baker, and T.W. Simpson. 1985. Variance and scale influences on classifying and interpreting soil maps. Soil Science Society of America Journal 49(4):957–961.

Ehleringer, J.R., and C.B. Field (eds.). 1993. Scaling Physiological Processes: Leaf to Globe. Physiological Ecology Series, Academic Press, San Diego.

Ehmann, W.J. 1994. Organization of spider assemblages on shrubs—an assessment of the role of dispersal mode in colonization. American Midland Naturalist 131(2):301–310.

Elston, D.A., and S.T. Buckland. 1993. Statistical modelling of regional GIS data: an overview. Ecological Modelling 67:81–102.

Faith, D.P. 1994. Phylogenetic pattern and the quantification of organismal biodiversity. Philosophical Transactions of the Royal Society of London, Series B-Biological Sciences 345(1311):45–58.

Fernandes, J.P. 1993. Ecogis-ecosad—a methodology for the biophysical environmental assessment within the planning process. Computers Environment and Urban Systems 17(4):347–354.

Fisher, P.F. 1991. Spatial data sources and data problems. In: Geographical Information Systems: Principles and Applications, Vol. 1, pp. 175–189. D.J. Maguire, M.F. Goodchild, and D.W. Rhind (eds.). Longman, Essex.

Forbes, G.J., and J.B. Theberge. 1993. Multiple landscape scales and winter distribution of moose (*Alces alces*) in a forest ecotone. Canadian Field-Naturalist 107(2):201–207.

Forman, R.T.T., and M. Godron. 1981. Patches and structural components for a landscape ecology. Bioscience 31:733–740.

Forman, R.T.T., and M. Godron. 1986. Landscape Ecology. John Wiley & Sons, New York.

Fox, J. 1992. The problem of scale in community resource-management. Environmental Management 16(3):289–297.

Freemark, K. 1995. Assessing effects of agriculture on terrestrial wildlife—developing a hierarchical approach for the United-States-EPA. Landscape and Urban Planning 31(1–3):99–115.

Frelich, L.E., and P.B. Reich. 1995. Spatial patterns and succession in a Minnesota southern-boreal forest. Ecological Monographs 65(3):325–346.

Friedman, S.K., and E.H. Zube. 1992. Assessing landscape dynamics in a protected area. Environmental Management 16:363–370.

Fuisz, T., and C. Moskat. 1992. The importance of scale in studying beetle communities—hierarchical sampling or sampling the hierarchy. Acta Zoologica Hungarica 38(3–4): 183–197.

Gagliuso, R.A. 1991. Remote sensing and GIS technologies: an example of integration in the analysis of cougar habitat utilization in southwest Oregon. In: GIS Applications in Natural Resources, pp. 323–329. M. Heit and A. Shortreid (eds.). GIS World, Fort Collins, CO.

Gallant, J.C., I.D. Moore, and P. Gessler. 1993. Estimating fractal dimension: a comparison of methods. Mathematical Geology 26(4):455–481.

Gardner, R.H., and R.V. O'Neill. 1991. Pattern, process, and predictability: the use of neutral models for landscape analysis. In: Quantitative Methods in Landscape Ecology, pp. 289–307. M.G. Turner and R.H. Gardner (eds.). Springer-Verlag, New York.

Gardner, R.H., B.T. Milne, M.G. Turner, and R.V. O'Neill. 1987. Neutral models for the analysis of broad-scale landscape patterns. Landscape Ecology 1:19–28.

Gascon, C., and J. Travis. 1992. Does the spatial scale of experimentation matter—a test with tadpoles and dragonflies. Ecology 73(6):2237–2243.

Gautestad, A.O., and I. Mysterud. 1994. Are home ranges fractals—comment. Landscape Ecology 9(2):143–146.

Gentry, A.H. 1988. Changes in plant community diversity and floristic composition on environmental and geographical gradients. Annals of the Missouri Botanical Garden 75:1–34.

Gibson, D.J., T.R. Seastedt, and J.M. Briggs. 1993. Management-practices in tallgrass prairie—large-scale and small-scale experimental effects on species composition. Journal of Applied Ecology 30(2):247–255.

Gleason, H.A. 1922. On the relation between species and area. Ecology 3:158–162.

Gleason, H.A. 1925. Species and area. Ecology 6:66–74.

Glenn, S.M., S.L. Collins, and D.J. Gibson. 1992. Disturbances in tallgrass prairie—local and regional effects on community heterogeneity. Landscape Ecology 7(4):243–251.

Golley, F.B. 1987. Introducing landscape ecology. Comments of the editor. Landscape Ecology 1(1):1–3.

Golley, F.B. 1989. Paradigm shift. Editorial comment. Landscape Ecology 3(2): 65–66.

Goodchild, M.F., and D.M. Mark. 1987. Review Article—The fractal nature of geographic phenomena. Annals of the Association of American Geographers 77:265–278.

Goodess, C.M., and J.P. Palutikof. 1992. The development of regional climate scenarios and the ecological impact of greenhouse-gas warming. Advances in Ecological Research 22:33–62.

Gosz, J.R. 1993. Ecotone hierarchies. Ecological Applications 3(3):369–376.

Gosz, J.R., and P.J.H. Sharpe. 1989. Broad-scale concepts for interaction of climate, topography, and biota at biome transitions. Landscape Ecology 3:229–243.

Gould, S.J. 1966. Allometry and size in ontogeny and phylogeny. Biological Review 41:587–640.

Graumlich, L.J., and M.B. Davis. 1993. Holocene variation in spatial scales of vegetation pattern in the upper great lakes. Ecology 74(3):826–839.

Greig-Smith, P. 1952. The use of random and contiguous quadrats in the study of the structure of plant communities. Annals of Botany 16:293–316.

Grimm, E.C. 1984. Fire and other factors controlling the big woods vegetation of Minnesota in the mid-nineteenth century. Ecological Monographs 54:291–311.

Groffman, P.M., J.M. Tiedje, D.L. Mokma, and S. Simkins. 1992. Regional scale analysis of denitrification in north temperate forest soils. Landscape Ecology 7(1):45–53.

Gross, J.E., C. Zank, N.T. Hobbs, and D.E. Spalinger. 1995. Movement rules for herbivores in spatially heterogeneous environments—responses to small-scale pattern. Landscape Ecology 10(4):209–217.

Gustafson, E.J., and G.R. Parker. 1994. Using an index of habitat patch proximity for landscape design. Landscape and Urban Planning 29(2–3):117–130.

Hails, R.S., and M.J. Crawley. 1992. Spatial density dependence in populations of a cynipid gall-former Andricus-quercuscalicis. Journal of Animal Ecology 61(3):567–583.

Hains-Young, R.H. 1992. The use of remotely sensed satellite imagery for landscape classification in Wales (UK). Landscape Ecology 7(4):253–274.

Hamazaki, T. 1996. Effects of patch shape on the number of organisms. Landscape Ecology 11(5):299–306.

Hargrove, W.W., and J. Pickering. 1992. Pseudoreplication: a sine qua non for regional ecology. Landscape Ecology 6(4):251–258.

Harris, L.D. 1984. The Fragmented Forest. University of Chicago Press, Chicago.

Harvey, D.W. 1968. Pattern, process, and the scale problem in geographical research. Transactions of the Institute of British Geographers 45:71–78.

Henderson-Sellers, A. 1987. A Climate Modelling Primer. Wiley, Chichester.

Herben, T., F. Krahulec, V. Hadincova, and H. Skalova. 1993. Small-scale variability as a mechanism for large-scale stability in mountain grasslands. Journal of Vegetation Science 4(2):163–170.

Herr, A.M., and L.P. Queen. 1993. Crane habitat evaluation using GIS and remote sensing. Photogrammetric Engineering & Remote Sensing 59:1531–1538.

Hinsley, S.A., P.E. Bellamy, I. Newton, and T.H. Sparks. 1995. Habitat and landscape factors influencing the presence of individual breeding bird species in woodland fragments. Journal of Avian Biology 26(2):94–104.

Holling, C.S. 1992. Cross-scale morphology, geometry, and dynamics of ecosystems. Ecological Monographs 62(4):447–502.

Holt, R.D., S.W. Pacala, T.W. Smith, and J.G. Liu. 1995. Linking contemporary vegetation models with spatially explicit animal population-models. Ecological Applications 5(1):20–27.

Holt, S. 1991. Human encroachment on bear habitat. In: GIS Applications in Natural Resources, pp. 319–321. M. Heit and A. Shortreid (eds.). GIS World, Fort Collins, CO.

Host, G.E., K.S. Pregitzer, C.W. Ramm, J.B. Hart, and D.T. Cleland. 1987. Landform-meditated differences in successional pathways among upland forest ecosystems in northwestern lower Michigan. Forest Science 33:445–457.

Hostetler, S.W., and F. Giorgi. 1993. Use of output from high-resolution atmospheric models in landscape-scale hydrologic-models—an assessment. Water Resources Research 29(6):1685–1695.

Hunsaker, C.T., R.V. O'Neill, B.L. Jackson, S.P. Timmins, D.A. Levine, and D.J. Norton. 1994. Sampling to characterize landscape pattern. Landscape Ecology 9(3):207–226.

Hupp, C.R., and W.R. Osterkamp. 1985. Bottomland vegetation distribution along Passage Creek, Virginia, in relation to fluvial landforms. Ecology 66:670–681.

Huston, M.A. 1994. Biological Diversity: The Coexistence of Species on Changing Landscapes. Cambridge University Press, Cambridge.

Hutchinson, G.E. 1953. The concept of pattern in ecology. Proceedings of the Academy of Natural Sciences of Philadelphia 105:1–12.

Hutchinson, G.E. 1971. Banquet address: Scale effects in ecology. In: Spatial Patterns and Statistical Distribution, pp. xvii–xxii. Statistical Ecology series, Vol 1. G.P. Patil, E.C. Pielou, and W.E. Waters (eds.). Pennsylvania State University Press, University Park.

Iverson, L.R., E.A. Cook, and R.L. Graham. 1994. Regional forest cover estimation via remote sensing—the calibration center concept. Landscape Ecology 9(3):159–174.

Ives, A.R., P. Kareiva, and R. Perry. 1993. Response of a predator to variation in prey density at 3 hierarchical scales—lady beetles feeding on aphids. Ecology 74(7):1929–1938.

Jackson, R.B., and M.M. Caldwell. 1993a. Geostatistical patterns of soil heterogeneity around individual perennial plants. Journal of Ecology 81(4):683–692.

Jackson, R.B., and M.M. Caldwell. 1993b. The scale of nutrient heterogeneity around individual plants and its quantification with geostatistics. Ecology 74(2):612–614.

Jean, M., and A. Bouchard. 1993. Riverine wetland vegetation—importance of small-scale and large-scale environmental variation. Journal of Vegetation Science 4(5):609–620.

Johnson, A.R., J.A. Wiens, B.T. Milne, and T.O. Crist. 1992. Animal movements and population-dynamics in heterogeneous landscapes. Landscape Ecology 7(1):63–75.

Johnson, L.B. 1990. Analyzing spatial and temporal phenomena using geographic information systems: A review of ecological applications. Landscape Ecology 4:31–44.

Johnston, C.A. 1993. Material fluxes across wetland ecotones in northern landscapes. Ecological Applications 3(3):424–440.

Johnston, K.B. 1992. Using regression analysis to build three prototype GIS wildlife models. In: GIS/LIS Proceedings, 10–12 November, 1992, Vol. 1, pp. 374–386.

Joseph, L., and Moritz. C. 1994. Mitochondrial-DNA phylogeography of birds in eastern Australian rain-forests—first fragments. Australian Journal of Zoology 42(3):385–403.

Journel, A.G., and C.J. Huijbregts. 1978. Mining Geostatistics. Academic Press, London.

Kamradt, D.A. 1995. Evaluating bobcat viability in the Santa Monica Mountains, California. Paper presented at the Annual Meeting of the Association of American Geographers, 14–18 March, Chicago.

Kemp, W.P. 1992. Rangeland grasshopper (*Orthoptera, Acrididae*) community structure—a working hypothesis. Environmental Entomology 21(3):461–470.

Kenkel, N.C. 1993. Modeling Markovian dependence in populations of Aralia-nudicaulis. Ecology 74(6):1700–1706.

Kennedy, B.A. 1977. A question of scale. Progress in Physical Geography 1:154–157.

Kesner, B.T., and V. Meentemeyer. 1989. A regonal analysis of total nitrogen in an agricultural landscape. Landscape Ecology 2(3):151–163.

Kienast, F. 1993. Analysis of historic landscape patterns with a geographical information system—a methodological outline. Landscape Ecology 8(2):103–118.

King, A.W. 1991. Translating models across scales in the landscape. In: Quantitative Methods in Landscape Ecology: The Analysis and Interpretation of Landscape Heterogeneity, pp. 479–517. M.G. Turner and R.H. Gardner (eds.). Springer-Verlag, New York.

Klaassen, W., and M. Claussen. 1995. Landscape variability and surface flux parameterization in climate models. Agricultural and Forest Meteorology 73(3–4):181–188.

Klijn, F., and H.A.U. de Haes. 1994. A hierarchical approach to ecosystems and its implications for ecological land classification. Landscape Ecology 9(2):89–104.

Klinger, L.F., P.R. Zimmerman, J.P. Greenberg, L.E. Heidt, and A.B. Guenther. 1994. Carbon trace gas fluxes along a successional gradient in the Hudson-bay lowland. Journal of Geophysical Research-Atmospheres 99(1):1469–1494.

Klinkenberg, B., and M.F. Goodchild. 1992. The fractal properties of topography: A comparison of measures. Earth Surface Processes and Landforms 12:217–234.

Knight, C.L., J.M. Briggs, and M.D. Nellis. 1994. Expansion of gallery forest on Konza Prairie Research Natural Area, Kansas, U.S.A. Landscape Ecology 9(2):117–125.

Kohn, D.D., and D.M. Walsh. 1994. Plant-species richness—the effect of island size and habitat diversity. Journal of Ecology 82(2):367–377.

Kolejka, J. 1992. Local GIS application in the planning of ecological landscape stability systems. Computers Environment and Urban Systems 16(4):329–335.

Koricheva, J., and E. Haukioja. 1994. The relationship between abundance and performance of eriocrania miners in the field—effects of the scale and larval traits studied. Journal of Animal Ecology 63(3):714–726.

Kurki, S., and H. Linden. 1995. Forest fragmentation due to agriculture affects the reproductive success of the ground-nesting black grouse Tetrao tetrix. Ecography 18(2):109–113.

Lathrop, R.G. Jr., and D.L. Peterson. 1992. Identifying structural self-similarity in mountainous landscapes. Landscape Ecology 6(4):233–238.

Lathrop, R.G., J.D. Aber, and J.A. Bognar. 1995. Spatial variability of digital soil maps and its impact on regional ecosystem modeling. Ecological Modelling 82(1):1–10.

Lavelle, P., E. Blanchart, A. Martin, S. Martin, A. Spain, F. Touta, I. Barois, and R. Schaefer. 1993. A hierarchical model for decomposition in terrestrial ecosystems—application to soils of the humid tropics. Biotropica 25(2):130–150.

Lavorel, S., J. Lepart, M. Debussche, J.D. Lebreton, and J.L. Beffy. 1994a. Small-scale disturbances and the maintenance of species-diversity in Mediterranean old fields. Oikos 70(3):455-473.

Lavorel, S., R.V. Oneill, and R.H. Gardner. 1994b. Spatiotemporal dispersal strategies and annual plant-species coexistence in a structured landscape. Oikos 71(1):75–88.

Leduc, A., Y.T. Prairie, and Y. Bergeron. 1994. Fractal dimension estimates of a fragmented landscape—sources of variability. Landscape Ecology 9(4):279–286.

Levin, S.A. 1992. The problem of pattern and scale in ecology. Ecology 73(6):1943–1967.

Li, H., J.F. Franklin, F.J. Swanson, and T.A. Spies. 1993. Developing alternative forest cutting patterns: a simulation approach. Landscape Ecology 8(1): 63–75.

Li, H.B., and J.F. Reynolds. 1994. A simulation experiment to quantify spatial heterogeneity in categorical maps. Ecology 75(8):2446–2455.

Lidicker, W.Z. Jr., J.O. Wolff, L.N. Lidicker, and M.H. Smith. 1992. Utilization of a habitat mosaic by cotton rats during a population decline. Landscape Ecology 6(4):259–268.

Lifton, N.A., and C.G. Chase. 1992. Tectonic, climatic and lithologic influences on landscape fractal dimension and hypsometry—implications for landscape evolution in the San-Gabriel Mountains, California. Geomorphology 5(1–2): 77–114.

Loehle, C. 1990. Home range: a fractal approach. Landscape Ecology 5:39–52.

Loehle, C., and G. Wein. 1994. Landscape habitat diversity—a multiscale information-theory approach. Ecological Modelling 73(3–4):311–329.

Lovejoy, S., and D. Schertzer (eds.). 1991. Scaling, Fractals, and Non-linear Variability in Geophysics. Kluwer Academic, Norwell, MA.

Lovejoy, T.E., R.O. Bierregaard Jr., A.B. Rylands, J.R. Malcolm, C.E. Quintela, L.H. Harper, K.S. Brown Jr., A.H. Powell, G.V.N.K. Powell, H.O.R. Schubart, and M.B. Hays. 1986. Edge and other effects of isloation on Amazon forest fragments. In: Conservation Biology: The Science of Scarcity and Diversity, pp. 257–285. M.E. Soulé (ed.). Sinauer, Sunderland, MA.

Ludwig, J.A., and D.J. Tongway. 1995. Spatial-organization of landscapes and its function in semiarid woodlands, Australia. Landscape Ecology 10(1):51–63.

Lundquist, J.E. 1995. Disturbance profile—a measure of small-scale disturbance patterns in ponderosa pine stands. Forest Ecology and Management 74(1–3):49–59.

MacArthur, R.H., and E.O. Wilson. 1967. The Theory of Island Biogeography. Princeton University Press, Princeton, NJ.

Mackey, B.G. 1994. Predicting the potential distribution of rain-forest structural characteristics. Journal of Vegetation Science 5(1):43–54.

Maguire, D.J., M.F. Goodchild, and D.W. Rhind (eds.). 1991. Geographical Information Systems. Principles and Applications, 2 vol. Longman, Essex.

Magurran, A.E. 1988. Ecological Diversity and Its Measurement. Princeton University Press, Princeton.

Malingreau, J.P., and A.S. Belward. 1992. Scale considerations in vegetation monitoring using AVHRR data. International Journal of Remote Sensing 13(12): 2289–2307.

Mandelbrot, B.B. 1982. The Fractal Geometry of Nature. W.H. Freeman, San Francisco.
Margules, C.R., A.O. Nicholls, and R.L. Pressey. 1988. Selecting networks of reserves to maximize biological diversity. Biological Conservation 43:63–76.
Martinez, N.D. 1994. Scale-dependent constraints on food-web structure. American Naturalist 144(6):935–953.
Martinez, N.D., and J.H. Lawton. 1995. Scale and food-web structure—from local to global. Oikos 73(2):148–154.
McCollin, D. 1993. Avian distribution patterns in a fragmented wooded landscape (North Humberside, UK)—the role of between-patch and within-patch structure. Global Ecology and Biogeography Letters 3(2):48–62.
McIntyre. N.E. 1995. Effects of forest patch size on avian diversity. Landscape Ecology 10(2):85–99.
McKendry, J.E., and G.E. Machlis. 1993. The role of geography in extending biodiversity gap analysis. Applied Geography 13:135–152.
McLaughlin, S.P. 1994. Floristic plant geography—the classification of floristic areas and floristic elements. Progress in Physical Geography 18(2):185–208.
Meentemeyer, V. 1989. Geographical Perspectives of space, time, and scale. Landscape Ecology 3(3–4):163–173.
Meentemeyer, V., and E.O. Box. 1987. Scale effects in landscape studies. In: Landscape Heterogeneity and Disturbance, pp. 15–34. M.G. Turner (ed.). Springer-Verlag, Berlin.
Merriam, G., and A. Lanoue. 1990. Corridor use by small mammals: field measurement for three experimental types of *Peromyscus leucopus*. Landscape Ecology 4(2/3):123–131.
Michaelsen, J., D.S. Schimel, M.A. Friedl, F.W. Davis, and R.C. Dubayah. 1994. Regression tree analysis of satellite and terrain data to guide vegetation sampling and surveys. Journal of Vegetation Science 5(5):673–686.
Miller, R.I., S.N. Stuart, and K.M. Howell. 1989. A methodology for analyzing rare species distribution patterns utilizing GIS technology: the rare birds of Tanzania. Landscape Ecology 2:173–190.
Milne, B.T. 1988. Measuring the fractal geometry of landscapes. Applied Mathematics and Computation 27:67–79.
Milne, B.T. 1991. Lessons from applying fractal models to landscape patterns. In: Quantitative Methods in Landscape Ecology: The Analysis and Interpretation of Landscape Heterogeneity, pp. 199–235. M.G. Turner and R.H. Gardner (eds.). Springer-Verlag, New York.
Milne, B.T. 1992. Spatial aggregation and neutral models in fractal landscapes. American Naturalist 139(1):32–57.
Mitchellolds, T. 1992. Does environmental variation maintain genetic-variation—a question of scale. Trends in Ecology and Evolution 7(12):397–398.
Mladenoff, D.J., M.A. White, J. Pastor, and T.R. Crow. 1993. Comparing spatial pattern in unaltered old-growth and disturbed forest landscapes. Ecological Applications 3(2):294–306.
Moloney, K.A., A. Morin, and S.A. Levin. 1991. Interpreting ecological patterns generated through simple stochastic processes. Landscape Ecology 5(3): 163–174.
Montgomery, D.R., and W.E. Dietrich. 1992. Channel initiation and the problem of landscape scale. Science 255(5046):826–830.

Moody, A., and C.E. Woodcock. 1995. The influence of scale and the spatial characteristics of landscapes on land-cover mapping using remote sensing. Landscape Ecology 10(6):363–379.

Mooney, H.A., and C.B. Field. 1989. Photosynthesis and plant productivity: scaling to the biosphere. In: Photosynthesis, pp. 19–44. W.R. Briggs (ed.). A.R. Liss, New York.

Morris, D.W. 1992. Scales and costs of habitat selection in heterogeneous landscapes. Evolutionary Ecology 6(5):412–432.

Naveh, Z. 1993. Red books for threatened Mediterranean landscapes as an innovative tool for holistic landscape conservation—introduction to the western Crete-red-book case-study. Landscape and Urban Planning 24(1–4):241–247.

Neilson, R.P., G.A. King, and G. Koerper. 1992. Toward a rule-based biome model. Landscape Ecology 7(1):27–43.

Nir, D. 1987. Regional geography considered from the systems perspective. Geoforum 18(2):187–202.

Norton, B.G., and R.E. Ulanowicz. 1992. Scale and biodiversity policy—a hierarchical approach. Ambio 21(3):244–249.

Norton, D.A., and J.M. Lord. 1990. On the use of "grain size" in ecology. Functional Ecology 4:719.

Oksanen, T., L. Oksanen, and M. Gyllenberg. 1992. Exploitation ecosystems in heterogeneous habitat complexes 2. impact of small-scale heterogeneity on predator prey dynamics. Evolutionary Ecology 6(5):383–398.

Oliver, M.A., and R. Webster. 1990. Kriging: A method of interpolation for geographical information systems. International Journal of Geographical Information Systems 4:313–332.

O'Neill, K.M. 1992. Temporal and spatial dynamics of predation in a robber fly (*efferia-staminea*) population (*diptera, asilidae*). Canadian Journal of Zoology (Revue Canadienne de Zoologie) 70(8):1546–1552.

O'Neill, R.V. 1988. Hierarchy theory and global change. In: Scales and Global Change: Spatial and Temporal Variability in Biospheric and Geospheric Processes, pp. 29–46. T. Rosswall, R.G. Woodmansee, and P.G. Risser (eds.). SCOPE, Vol. 35. John Wiley & Sons, Chicester.

O'Neill, R.V., D.L. DeAngelis, J.B. Waide, and T.F. Allen. 1986. A hierarchical concept of ecosystems. Monographs in Population Biology, Vol. 23. R.M. May (ed.). Princeton University Press, Princeton, NJ.

O'Neill, R.V., A.R. Johnson, and A.W. King. 1989. A hierarchical framework for the analysis of scale. Landscape Ecology 3(3–4):193–205.

O'Neill, R.V., M.G. Turner, V.I. Cullinan, D.P. Coffin, T. Cook, W. Conley, J. Brunt, J.M. Thomas, M.R. Conley, and J. Gosz. 1991. Multiple landscape scales: an intersite comparison. Landscape Ecology 5:137–144.

O'Neill, R.V., R.H. Gardner, and M.G. Turner. 1992. A hierarchical neutral model for landscape analysis. Landscape Ecology 7(1):55–61.

O'Neill, R.V., C.T. Hunsaker, S.P. Timmins, B.L. Jackson, K.B. Jones, K.H. Riitters, and J.D. Wickham. 1996. Scale problems in reporting landscape pattern at the regional scale. Landscape Ecology 11(3):169–180.

Palmer, M.W. 1988. Fractal geometry: a tool for describing spatial patterns of plant communities. Vegetatio 75:91–102.

Palmer, M.W. 1992. The coexistence of species in fractal landscapes. American Naturalist 139(2):375–397.

Palmer, M.W., and P.S. White, 1994. Scale dependence and the species-area relationship. American Naturalist 144(5):717–740.

Patil, G.P., and C. Taillie. 1979. An overview of diversity. In: Biological Diversity in Theory and Practice, pp. 3–27. J.F. Grassle, G.P. Patil, W. Smith, and C. Taillie (eds.). International Co-operative Publishing House, Fairland, MD.

Pearson, S.M., M.G. Turner, L.L. Wallace, and W.H. Romme. 1995. Winter habitat use by large ungulates following fire in Northern Yellowstone National Park. Ecological Applications 5(3):744–755.

Penev, L.D. 1992. Qualitative and quantitative spatial variation in soil wire-worm assemblages in relation to climatic and habitat factors. Oikos 63(2):180–192.

Peters, R.H. 1983. The Ecological Implications of Body Size. Cambridge University Press, Cambridge.

Phillips, J.D. 1995. Biogeomorphology and landscape evolution—the problem of scale. Geomorphology 13(1–4):337–347.

Pielke, R.A., G.A. Dalu, J.S. Snook, T.J. Lee, and T.G.F. Kittel. 1991. Nonlinear influence of mesoscale land-use on weather and climate. Journal of Climate 4(11):1053–1069.

Pierce, L.L., and S.W. Running. 1995. The effects of aggregating subgrid land-surface variation on large-scale estimates of net primary production. Landscape Ecology 10(4):239–253.

Pierson, F.B., W.H. Blackburn, S.S. Vanvactor, and J.C. Wood. 1994. Partitioning small-scale spatial variability of runoff and erosion on sagebrush rangeland. Water Resources Bulletin 30(6):1081–1089.

Plotnik, R.E., R.H. Gardner, and R.V. O'Neill. 1993. Lacunarity indices as measures of landscape texture. Landscape Ecology 8(3):201–211.

Post, W.M. 1993. Uncertainties in the global carbon cycle. In: Vegetation Dynamics and Global Change, pp. 116–132. A.M. Solomon and H.H. Shugart (eds.). Chapman & Hall, New York.

Prentice, I.C., and R. Leemans. 1990. Pattern and process and the dynamics of forest structure: a simulation approach. Journal of Ecology 78:340–355.

Price, L.W. 1971. Vegetation, microtopography, and depth of active layer on different exposures in subarctic alpine tundra. Ecology 52:638–647.

Price, O., J.C.Z. Woinarski, D.L. Liddle, and J. Russellsmith. 1995. Patterns of species composition and reserve design for a fragmented estate—monsoon rain-forests in the northern-territory, Australia. Biological Conservation 74(1):9–19.

Qi, Y., and J.G. Wu. 1996. Effects of changing spatial-resolution on the results of landscape pattern-analysis using spatial autocorrelation indexes. Landscape Ecology 11:39–49.

Rastetter, E.B., A.W. King, B.J. Cosby, G.M. Hornberger, R.V. O'Neill, and J.E. Hobbie. 1992. Aggregating fine-scale ecological knowledge to model coarser-scale attributes of ecosystems. Ecological Applications 2(1):55–70.

Raupach, M.R., O.T. Denmead, and F.X. Dunin. 1993. Challenges in linking atmospheric CO_2 concentrations to fluxes at local and regional scales. Australian Journal of Botany 40(4–5):697–716.

Rescia, A.J., M.F. Schmitz, P.M. Deager, C.L. Depablo, J.A. Atauri, and F.D. Pineda. 1994. Influence of landscape complexity and land management on woody plant diversity in northern Spain. Journal of Vegetation Science 5(4):505–516.

Rey, P.J. 1995. Spatiotemporal variation in fruit and frugivorous bird abundance in olive orchards. Ecology 76(5):1625–1635.

Riiters, K.H., R.V. O'Neill, C.T. Hunsaker, J.D. Wickham, D.H. Yankee, S.P. Timmins, K.B. Jones, and B.L. Jackson. 1995. A factor analysis of landscape pattern and structure metrics. Landscape Ecology 10(1):23–29.

Riiters, K.H., R.V. O'Neill, J.D. Wickham, and K.B. Jones. 1996. A note on cantagion indices for landscape analysis. Landscape Ecology 11(4):197–202.

Romme, W.H., and D.H. Knight. 1981. Fire frequency and subalpine forest succession along a topographic gradient in Wyoming. Ecology 62:319–326.

Rosenzweig, M.L. 1992. Species-diversity gradients—we know more and less than we thought. Journal of Mammalogy 73(4):715–730.

Rosswall, T., R.G. Woodmansee, and P.G. Risser (eds.). 1988. Scales and Global Change: Spatial and Temporal Variability in Biospheric and Geospheric Processes, SCOPE, Vol. 35. John Wiley & Sons, Chicester.

Running, S.W. 1990. Estimating terrestrial primary productivity by combining remote sensing and ecosystem simulation. In: Remote Sensing of Biosphere Functioning, pp. 65–86. R.J. Hobbs and H.A. Mooney (eds.). Springer-Verlag, Berlin.

Sandoval, C.P. 1994. The effects of the relative geographic scales of gene flow and selection on morph frequencies in the walking-stick *Timema christinae*. Evolution 48(6):1866–1879.

Schaefer, J.A. 1993. Spatial patterns in taiga plant-communities following fire. Canadian Journal of Botany (Revue Canadienne De Botanique) 71(12):1568–1573.

Schmidt, K.M., and D.R. Montgomery. 1995. Limits to relief. Science 270(5236): 617–620.

Schneider, D.C. 1994. Quantitative Ecology. Spatial and Temporal Scaling. Academic Press, San Diego.

Schwartz, M.W. 1994. Conflicting goals for conserving biodiversity—issues of scale and value. Natural Areas Journal 14(3):213–216.

Scott, J.M., F. Davis, B. Csuti, R. Noss, B. Butterfield, C. Groves, H. Aderson, S. Caicco, F. Derchia, T.C. Edwards, J. Ulliman, and R.G. Wright. 1993. Gap analysis: a geographical approach to protection of biological diversity. Wildlife Monographs 123:1–41.

Simmons, M.A., V.I. Cullinan, and J.M. Thomas. 1992. Satellite imagery as a tool to evaluate ecological scale. Landscape Ecology 7(2):77–85.

Simpson, J.W., R.E.J. Boerner, M.N. Demers, L.A. Berns, F.J. Artigas, and A. Silva. 1994. 48 Years of landscape change on 2 contiguous Ohio landscapes. Landscape Ecology 9(4):261–270.

Smith, T.M., H.H. Shugart, G.B. Bonan, and J.B. Smith. 1992. Modeling the potential response of vegetation to global climate-change. Advances in Ecological Research 22:93–116.

Snow, R.S., and L. Mayer (eds.). 1992. Fractals in Geomorphology, Geomorphology 5: Special issue.

Solomon, A.M., and H.H. Shugart (eds.). 1993. Vegetation Dynamics and Global Change. Chapman & Hall, New York.

Stevens, S.S. 1946. On the theory of scales of measurement. Science 103:677–680.

Stiling, P., A. Throckmorton, J. Silvanima, and D.R. Strong. 1991. Does spatial scale affect the incidence of density dependence—a field-test with insect parasitoids. Ecology 72(6):2143–2154.

Stohlgren, T.J., G.W. Chong, M.A. Kalkhan, and L.D. Schell. 1996. Effects of map resolution on landscape-scale studies of plant diversity: Tests of 100 ha, 50 ha, 2 ha, and 0.09 ha minimum mapping units. Presented at the 11th Annual Landscape Ecology Symposium, 26–30 March 1996, Galveston, TX.

Stoms, D.M. 1994. Scale dependence of species richness maps. Professional Geographer 46(3):346–358.

Stoms, D.M., and J.E. Estes. 1993. A remote-sensing research agenda for mapping and monitoring biodiversity. International Journal of Remote Sensing 14(10): 1839–1860.

Stone, K.H. 1968. Scale, Scale, Scale. Economic Geography 44:94.

Sugden, D.E., N. Glasser, and C.M. Clapperton. 1992. Evolution of large roches-moutonnees. Geografiska Annaler Series A-Physical Geography 74(2–3): 253–264.

Sugihara, G., and R.M. May. 1990. Applications of fractals in ecology. Trends in Ecology and Evolution 5:79–86.

Taber, A.B., C.P. Doncaster, N.N. Neris, and F. Colman. 1994. Ranging behavior and activity patterns of 2 sympatric peccaries, *Catagonus-wagneri* and *Tayassu-tajacu*, in the Paraguayan Chaco. Mammalia 58(1):61–71.

Tarboton, D.G., R.L. Bras, and I. Rodriguez-Iturbe. 1992. A physical basis for drainage density. Geomorphology 5(1–2):59–76

Tinker, D.B., W.H. Romme, W.W. Hargrove, R.H. Gardner, and M.G. Turner, 1994. Landscape-scale heterogeneity in lodgepole pine serotiny. Canadian Journal of Forest Research (Journal Canadien de la Recherche Forestière) 24(5):897–903.

Tomlin, C.D. 1991. Cartographic modelling. In: Geographical Information Systems: Principles and Applications, Vol. 1, pp. 361–374. D.J. Maguire, M.F. Goodchild, and D.W. Rhind (eds.). Longman, Essex.

Tuda, M. 1993. Density-dependence depends on scale—at larval resource patch and at whole population. Researches on Population Ecology 35(2):261–271.

Turner, M.G. 1990. Spatial and temporal analysis of landscape pattern. Landscape Ecology 4(1):21–30.

Turner, M.G., and R.H. Gardner (eds.). 1991a. Quantitative Methods in Landscape Ecology. Springer-Verlag, New York.

Turner, M.G., and R.H. Gardner. 1991b. Quantitative methods in landscape ecology: an introduction. In: Quantitative Methods in Landscape Ecology, pp. 3–14. M.G. Turner and R.H. Gardner (eds.). Springer-Verlag, New York.

Turner, M.G., R. Costanza, and F.H. Sklar. 1989a. Methods to evaluate the performance of spatial simulation models. Ecological Modelling 48:1–18.

Turner, M.G., V.H. Dale, and R.H. Gardner. 1989b. Predicting across scales: theory development and testing. Landscape Ecology 3(3–4):245–252.

Turner, M.G., W.H. Romme, R.H. Gardner, R.V. O'Neill, and T.K. Kratz. 1993. A revised concept of landscape equilibrium: disturbance and stability on scaled landscapes. Landscape Ecology 8(3):213–227.

Turner, M.G., W.W. Hargrove, R.H. Gardner, and W.H. Romme. 1994. Effects of fire on landscape heterogeneity in Yellowstone-National-Park, Wyoming. Journal of Vegetation Science 5(5):731–742.

Turner, M.G., G.J. Arthaud, R.T. Engstrom, S.J. Hejl, J.G. Liu, S. Loeb, and K. Mckelvey. 1995. Usefulness of spatially explicit population-models in land management. Ecological Applications 5(1):12–16.
Turner, S.J., R.V. O'Neill, W. Conley, M.R. Conley, and H.C. Humphires. 1991. Pattern and scale: statistics for landscape ecology. In: Quantitative Methods in Landscape Ecology, pp. 3–14. M.G. Turner and R.H. Gardner (eds.). Springer-Verlag, New York.
Vedyushkin, M.A. 1994. Fractal properties of forest spatial structure. Vegetatio 113(1):65–70.
Vetaas, O.R. 1992. Micro-site effects of trees and shrubs in dry Savanna. Journal of Vegetation Science 3(3):337–344.
Vitousek, P.M. 1993. Global dynamics and ecosystem processes: scaling up or scaling down? In: Scaling Physiological Processes. Leaf to Globe, pp. 169–177. J.R. Ehleringer and C.B. Field (eds.). Academic Press, San Diego.
Walker, B.H. 1994. Landscape to regional-scale responses of terrestrial ecosystems to global change. Ambio 23(1):67–73.
White, J.D., and S.W. Running. 1994. Testing scale-dependent assumptions in regional ecosystem simulations. Journal of Vegetation Science 5(5):687–702.
Whittaker, R.H. 1972. Evolution and measurement of species diversity. Taxon 21:213–251.
Whittaker, R.H. 1977. Evolution of species diversity in land communities. Evolutionary Biology 10:1–67.
Wiens, J. 1989. Spatial scaling in ecology. Functional Ecology 3:385–397.
Wiens, J. 1990. On the use of "grain" and "grain size" in ecology. Functional Ecology 4:720.
Wiens, J.A. 1992. What is landscape ecology, really? Landscape Ecology 7(3): 149–150.
Wiens, J.A. 1995. Habitat fragmentation—island versus landscape perspectives on bird conservation. Ibis 137(1):97–104.
Wiens, J.A., and B.T. Miline. 1989. Scaling of "landscapes" in landscape ecology, or, landscape ecology from a beetle's perspective. Landscape Ecology 3:87–96.
Wiens, J.A., N.C. Stenseth, B. Vanhorne, and R.A. Ims. 1993. Ecological mechanisms and landscape ecology. Oikos 66:369–380.
Willgoose, G., R.L. Bras, and I. Rodriguez-Iturbe. 1992. The relationship between catchment and hillslope properties—implications of a catchment evolution model. Geomorphology 5(1–2):21–37.
With, K.A. 1994a. Ontogenic shifts in how grasshoppers interact with landscape structure—an analysis of movement patterns. Functional Ecology 8(4):477–485.
With, K.A. 1994b. Using fractal analysis to assess how species perceive landscape structure. Landscape Ecology 9(1):25–36.
Withers, M.A. 1992. An empirical test of the peninsula effect in Baja California, Mexico. M.A. Thesis, University of California, Los Angeles.
Witney, G.G. 1982. Vegetation-site relationships in the presettlement forests of northeastern Ohio. Botanical Gazette 143:225–237.
Wolf, S.A., and T.F.H. Allen. 1995. Recasting alternative agriculture as a management model—the value of adept scaling. Ecological Economics 12(1):5–12.
Wright, R.G., J.G. Maccracken, and J. Hall. 1994. An ecological evaluation of proposed new conservation areas in Idaho—evaluating proposed Idaho national-parks. Conservation Biology 8(1):207–216.

Wu, J.G., and S.A. Levin. 1994. A spatial patch dynamic modeling approach to pattern and process in an annual grassland. Ecological Monographs 64(4): 447–464

Wylie, J.L., and D.J. Currie. 1993. Species-energy theory and patterns of species richness. 1. patterns of bird, angiosperm, and mammal species richness on islands. Biological Conservation 63(2):137–144.

Xu, T.B., I.D. Moore, and J.C. Gallant. 1993. Fractals, fractal dimensions and landscapes—a review. Geomorphology 8(4):245–262.

Young, A.G., S.I. Warwick, and H.G. Merriam. 1993. Genetic-variation and structure at 3 spatial scales for *Acer saccharum* (sugar maple) in Canada and the implications for conservation. Canadian Journal of Forest Research (Journal Canadien de la Recherche Forestière) 23(12):2568–2578.

Zao, X., C. Theiss, and B.V. Barnes. 1992. Pattern of Kirtlands's warbler occurrence in relation to the landscape structure of its summer habitat in northern Lower Michigan. Landscape Ecology 6(4):221–231.

Zhang, Z.Q., and J.P. Sanderson. 1993. Spatial scale of aggregation in 3 acarine predator species with different degrees of polyphagy. Oecologia 96(1):24–31.

Zheng, D.L., E.R. Hunt, and S.W. Running. 1996. Comparison of available soil-water capacity estimated from topography and soil series information. Landscape Ecology 11:3–14.

Zimmerman, B.L., and R.O. Bierregaard. 1986. Relevance of the equilibrium theory of island biogeography and species-area relations to conservation with a case from Amazonia. Journal of Biogeography 13:133–143.

Zonneveld, I.S. 1989. The land unit—a fundamental concept in landscape ecology, and its applications. Landscape Ecology 3:67–89.

12 Spatial Statistics in Landscape Ecology

Marie-Josée Fortin

In ecology, especially in plant ecology, it is commonplace to identify and quantify spatial patterns (Greig-Smith 1952, 1964; Kershaw 1964). In fact, most ecological data are inherently composed of several levels of spatial structure: large-scale trends (species responses to climate conditions, to migration process, etc.), small scale patterns, patchiness (weather conditions, physical conditions, dispersal mechanisms, predation, competition, etc.), and local random noise. Therefore, the notion of spatial autocorrelation implies that "Everything is related to everything else, but near things are more related than distant things" (Tobler 1970) and "with spatial data, there is a better than random chance that one can predict attribute values from a given areal unit from those taken on by its juxtaposed areal units . . ." (Haining 1980). Griffith (1992) provides several other definitions of spatial autocorrelation, which include "a diagnostic tool for spatial model misspecification; a surrogate for unobserved geographical variables; a nuisance in applying conventional statistical methodology to spatial data series; an indicator of the appropriateness of, and possible artifact of, areal unit demarcation."

The first methods developed to assess the spatial structure of data were primary first-order statistics (Table 12.1), which test whether the large scale spatial trend (i.e., the mean spatial trend in the data), is significantly different or not from a random pattern (variance-to-mean ratio; clumping index, Green's index, Lloyd's index, Morisita's index, etc.). Such first-order statistics identify whether a study area shows an overall spatial structure (clump or regular) or not (random). These indices are however limited in their ability to distinguish among various spatial patterns, for example a trend from a patch. This drawback explains why second-order statistics (Ripley's K, Moran's I, Geary's c, semivariance, etc.) were developed. They allow the quantification of small-scale spatial pattern intensity (i.e., magnitude, degree) and scale (i.e., spatial extent) (Table 12.1). All these second-order methods measure the square deviations to the mean (Bailey and Gatrell 1995), so they are sensitive to outlier data.

Two families of second-order spatial statistics were developed in parallel by human geographers (spatial statistics: Cliff and Ord 1981), and by mining

TABLE 12.1. First- and second-order statistics.

First-order	Variance-to-mean
	Clumping index
	Green's index
	Lloyd's index
	Morisita's index
	Nearest Neighbor
Second-order	Ripley's K, Ripley's K_{12}
	Moran's I, Geary's c
	Semivariance

engineers (geostatistics: Matheron 1970; Journel and Huijbregts 1978). Both types of techniques quantify the degree of spatial autocorrelation of the values of a variable (i.e., the intensity and the scale of the spatial structure of a variable), but the former was developed to also test the spatial pattern significance (Upton and Fingleton 1985; Legendre and Fortin 1989; Haining 1990; Cressie 1991), while the latter was developed to model the spatial pattern (Journel and Huijbregts 1978; Cressie 1991; Deutsch and Journel 1992; Rossi et al. 1992). The major drawback of this parallel development is that all of these methods are based on the same principles of computing local spatial deviations as spatial correlation or spatial covariance (Rossi et al. 1992) but using completely different notations (e.g., distance can be referred to as t, d, or h) and terminology (e.g., geostatistics do not interpolate but "Krige" named after the mining engineer who first developed the interpolation method that takes into account the degree of spatial covariance of the data).

These spatial statistics, although developed separately, rest on the same statistical assumption of stationarity. The stationarity assumption, also the root of time series analysis methods (Chatfield 1984), implies that the data have a normal distribution as well as the same mean, variance, and isotropy throughout the study plot. The term isotropy refers to the fact that a spatial pattern shows the same intensity in all directions. When one direction shows a spatial pattern for a longer distance or of various intensity according to the direction, the spatial pattern is said to be anisotropic.

This chapter presents how spatial statistics can be useful to landscape ecology studies first by stressing when spatial statistics should be used (Tables 12.1 and 12.2); second by highlighting their particular behaviors (Table 12.3); and finally by presenting how the relationship between spatially autocorrelated data should be statistically assessed. The main objective is not to summarize all of the previous reviews of spatial statistics (Cliff and Ord 1981; Upton and Fingleton 1985; Burrough 1987; Legendre and Fortin 1989; Haining 1990; Rossi et al. 1992; Bailey and Gatrell 1995) but rather to provide the missing links among the various spatial statistics and also to offer some guidelines about decisions to make in selecting and

TABLE 12.2. Spatial statistics for sampled data classified by objective.

1. Objective: Description of the spatial structure
   - Ripley's K, Ripley's K_{12}
   - Correlograms (Moran's I, Geary's c, Join-Count)
   - Variogram (semivariance)
   - Spatial clustering
   - Edge detection
2. Objective: Mapping (interpolation)
   - Trend surface analysis
   - Kriging
   - Spline
3. Objective: Testing for the presence of spatial autocorrelation
   - Ripley's K, Ripley's K_{12}
   - Correlograms (Moran's I, Geary's c, Join-Count)
   - Variograms (semivariance with a posteriori randomization test)
   - Mantel's test (variable and geographical coordinates)

→No significant spatial autocorrelation: use parametric statistical tests.

→Significant spatial autocorrelation: quantify pattern intensity and scale.

4. Objective: Correlation between spatially autocorrelated data
   - Clifford et al.'s correction
   - Partial Mantel's test
   - Partial CCA

computing these spatial statistics. Only spatial statistics for two-dimensional data will be presented; good reviews of spatial statistics for one-dimensional data can be found in Ludwig and Reynolds (1988), Dale (1990), Dale and Blundon (1990), Turner et al. (1991), and ver Hoef et al. (1993), among others.

Spatial Statistics versus Landscape Metrics

Spatial statistics allow the quantification of the spatial structure from sampled data, while landscape metrics characterize the geometric and spatial properties of mapped data (e.g., mosaic of patches). They describe the degree of spatial autocorrelation, that is, the spatial dependency of the values of a variable (or self-correlation) that has been sampled at various geographical coordinates. Usually, such samples are gathered to better understand the spatial heterogeneity of ecological data. Such quantitative knowledge about the spatial structure of the data can then be used to group samples into relatively spatially homogeneous clusters (i.e., patches)

TABLE 12.3. Spatial statistics for sampled data according to data types and measurements.

	Point Pattern	Surface Pattern
Qualitative, categorical data	**Ripley's *K*, Ripley's K_{12}** ? Intensity; Scale. − Significance: Monte Carlo simulations of CSR. + x, y (exhaustive mapping). → Isotropic (in term of distances). † Edge effect: ½ of the shorter edge. (Kenkel 1988)	**Join-Count** (two- or *k*-categories) ? Clustered, Uniform or Random; − Significance: parametric tests. + x, y and z = category. → Isotropic (in term of neighbor links). † Need links network. (Cliff and Ord 1981)
Quantitative, numerical data	**Correlogram** (univariate = Moran's *I*; Geary's *c*; multivariate = Mantel's test) ? Intensity; Scale; Omni- or Directional. − Significance: parametric statistics or randomization tests. + x, y and z = quantity; Stationarity. † Edge effect: interpret only first half of correlogram. † Reliable coefficient: more than 20 or 30 pairs. → Moran's I is sensitive to outliers which affect the mean. → Mantel's test estimates a linear relationship between distance measures rather than raw data. (Legendre and Fortin 1989)	
	Variogram/Kriging ? Intensity; Scale; Omni- or Directional; Interpolation. − Significance: randomization tests. + x, y and z = quantity; Stationarity, pseudo-stationarity. † Edge effect: interpret only first half of correlogram. † Reliable coefficient: more than 20 or 30 pairs. † Search neighborhood: range size and/or 12 to 25 locations. (Rossi et al. 1992)	
	Mantel's and Partial Mantel's Tests; Partial CCA ? Intensity; Scale; Omni- or Directional. − Significance: randomization tests. + Square distance matrices. † Partial CCA is sensitive to errors in environmental data. → Estimate linear relationships. (Smouse et al. 1986; Borcard et al. 1992; Palmer 1993)	

? indicates which spatial characteristics are computed; − indicates how to test for significance; + indicates what the spatial statistics requirements are; † indicates the statistics sensitivity and/or how to use them; → indicates the properties of the results.

(Fig. 12.1). In its simplest form, patch definition implies the delineation of an area showing at least one distinct characteristic that permits its differentiation from its surrounding (Kotliar and Wiens 1990). Patch delineation can also be achieved by using edge detection algorithms (for more details see, among others, Johnston et al. 1992; Fortin 1994; Fortin and Drapeau 1995). Hence, field data can be reclassified into a mosaic of patches. In doing so, quantitative–numerical fine-grain data are transformed into a qualitative–categorical coarse-grain patch map (Fig. 12.1). Therefore, patch mosaics constitute another perception of the spatial structure of the data in which data within patches are defined as spatially homogeneous entities and in which spatial arrangement among patches is of interest. A suite of characteristics can be computed on this new representation of the data. These measures are known as landscape metrics (among others, Baker and Cai 1992; Gustafson and Parker 1992; Li and Reynolds 1995; McGarigal and Marks 1995; Riitters et al. 1995) and they quantify, among others, the geometric properties of patches (area, perimeter, shape, etc.) and the spatial arrangement and diversity of the latter at the landscape scale. This new categorical representation of the data can also be analyzed using spatial statistics, such as join-count spatial autocorrelation coefficients (surface pattern methods; Table 12.3), in which patch centroids can be analyzed with point pattern methods (Table 12.3). Any other type of spatial analysis requires a resampling of this new data to obtain some quantitative–numerical values of these new variables (Anselin 1992). For example, patch diversity spatial structure could be assessed by first resampling the study area using

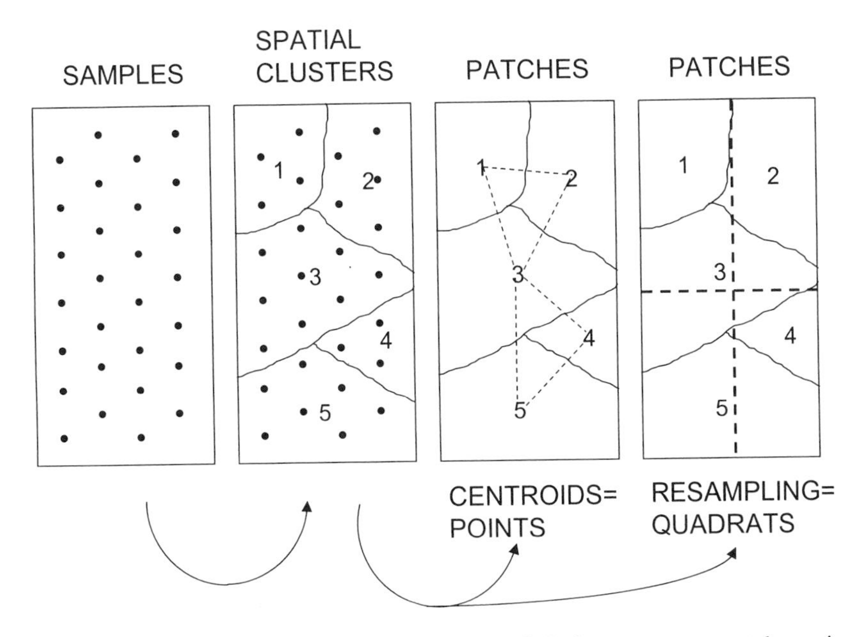

FIGURE 12.1. Field samples transformed into spatially homogeneous patches using spatial clustering. Patch mosaic used either as is, by using patch centroids as points for point pattern methods, or by resampling it to obtain quantitative landscape data for surface pattern methods.

quadrat sizes big enough to contain at least one patch (Fig. 12.1). With these new landscape samples, quantitative spatial autocorrelation coefficients can be used to estimate the spatial pattern of patch abundance or patch diversity at the landscape scale. It is important to keep in mind that spatial statistics are tools that estimate the spatial structure of the values of a sampled variable while landscape metrics are tools that characterize the properties of a patch (a spatially homogenous entity) or mosaic of patches.

Spatial Statistics

The study of the spatial dynamics of quantitative data (density, biomass, height) cannot be analyzed with the same tools as with qualitatively exhaustive data (exhaustive mapping, $x - y$ coordinates, of each plant in a plot). While surface pattern methods, such as those reviewed by Legendre and Fortin (1989) or Rossi et al. (1992), can be used for additive quantitative data gathered with areal sampling units (e.g., species abundance per quadrat), point pattern methods (Boots and Getis 1988) have to be used with exhaustive mapping of discrete individual data (e.g., mapping of $x - y$ coordinates of all stems of a species within a study area). Although all of these methods have the same goal of quantifying the intensity and scale of a variable's spatial structure, they must meet the same underlying assumption that the data are stationary over the entire area under study such that the variable has a positive, not null, probability of being found over the entire study area. Therefore, these methods are sensitive to the location of the study area and its size (known as the extent in landscape ecology literature), such that the bigger the area the more likely it is to incorporate environmental variability (Palmer and Dixon 1990; Fortin 1992). A consequence of this environmental variability is the presence of nonstationary data. Surface pattern methods are also sensitive to the size of the sampling unit, the quadrat or the grain in plant and landscape ecology terminology, respectively. This problem of quadrat size is well known in both plant ecology (Greig-Smith 1952) and in geostatistics (Journel and Huijbregts 1978), where it is referred to as either the Change of Support problem (Isaaks and Srivastava 1989; Deutsch and Journel 1992) or the Modifiable Area Unit Problem (MAUP; Haining 1990; Cressie 1991).

Point Pattern Methods

Point pattern methods quantify spatial pattern intensity and scale from point data, that is, points in space which correspond to the locations of discrete events such as individual plant stems. These methods require that all of the events on the study plot need to be mapped, that is, to locate the $x - y$ coordinates of each individual plant.

The simplest point-to-point method is the nearest neighbor distance technique (Clark and Evans 1954). This statistic measures the mean nearest distance among all points, $\bar{d}_i$, where $i = 1$ for the first neighbor. This technique has been extended to higher neighbors, when i is greater than 1, and is known as the refined nearest neighbor method (Boots and Getis 1988; Cressie 1991). To assess whether or not there is a significant spatial pattern, the observed mean nearest distance, $\bar{d}_i$, is tested against the expected mean nearest distance, $E(\bar{d}_i)$, under a complete spatial randomness (CSR) pattern having the same number of points over the same total area. Under such a CSR, the point pattern distribution follows a Poisson distribution where the point intensity, λ, is estimated as the density n/A, where A is the total area sampled and n the number of points mapped. The expected mean nearest distance is computed as:

$$E(d_i) = \gamma_i \sqrt{\frac{A}{n}}$$

where γ_i is a constant which varies as a function of the i-neighbor analyzed (see Boots and Getis 1988 for mathematical details). Given that the expected nearest neighbor values are a function of the total sampled area, they are very sensitive to the location and size of the study plot (the extent). This sensitivity can affect the power that these statistics have to detect a significant spatial pattern. Furthermore, the nearest neighbor statistics cannot distinguish among certain clustered patterns (Boots and Getis 1988). Finally, point-to-point spatial relationships with these statistics are computed in terms of i-neighbor and not actual Euclidean distance between points. For all of these reasons, Ripley's K second-order statistic (Ripley 1976, 1977, 1979, 1981, 1988) is currently the most widely used point pattern method.

Ripley's K Statistic

Ripley's K is a second-order statistic (Ripley 1976, 1977, 1979, 1981; Upton and Fingleton 1985; Getis and Franklin 1987; Boots and Getis 1988; Moeur 1993) that counts the number of events (e.g., points, plant stems) that lie in an area within a distance t of a randomly chosen event (Fig. 12.2):

$$K(t) = \lambda^{-1} E(\text{number of events within a circular area of radius } t),$$

where λ is n/A, n is the total number of events, A is the total study area, and the expected number of events under a Poisson process (CSR) is πt^2. The observed $K(t)$ value is calculated for different distances as follows:

$$K(t) = \lambda^{-1} \sum_{i=1}^{n} \sum_{j=1}^{n} I(e_i - e_j)/n, \quad \text{for } i \neq j \text{ and } t > 0,$$

where I is the number of events e_j within distance t of all events e_i. This statistic is an overall mean value of the number of points per given circle

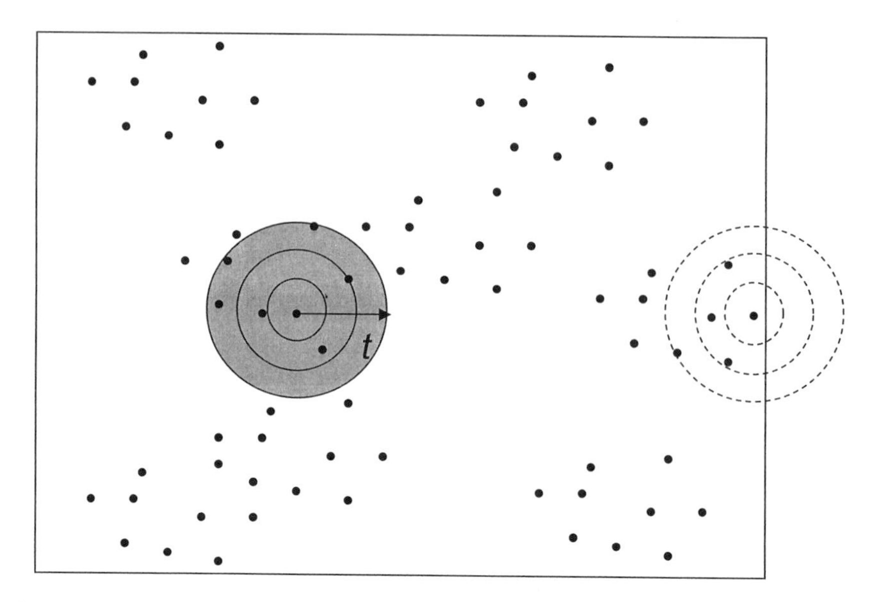

FIGURE 12.2. Point pattern method. Example of Ripley's K search area for a distance t which is the gray area from the target point up to t. Here $t = 3$ units (3 rings of 1 unit each). The dotted rings illustrate the edge effect problem that points near the edge have fewer neighbors than points in the center of the plot.

area of radius t. It is important to notice that Ripley's K statistic is a cumulative statistic since it is computed with all points from 0 up to the distance t (Fig. 12.2). This is different than surface pattern methods (described below) where spatial structure is computed for samples lying within an interval distance class (Fig. 12.3). By computing this statistic using a circular search area, Ripley's K statistic is an average isotropic measure of spatial pattern. To linearize the plot of $K(t)$ against t, as well as for stabilizing the variance, $L(t)$ rather than the $K(t)$ statistic is used (Diggle 1983) where $L(t)$ is:

$$L(t) = \sqrt{\frac{K(t)}{n}} - t.$$

The expected value of $L(t)$ under a Poisson process is 0. Positive values indicate spatial clustering, while negative ones spatial segregation. Ripley's K, and its derived L statistic, are not bounded such that comparisons between spatial patterns of different species or different study plot sizes should be based on the significant spatial scale rather than on significant spatial intensity per se. Significance is achieved by generating, using Monte Carlo simulation, n simulations of a Poisson point pattern process (CSR) that provide a confidence envelope. This confidence envelope is defined by the extreme maximum and minimum values of the simulation where 99 simulations correspond to a 0.01 significance level (Ripley 1979). This method of generating a confidence envelope is the most extreme case of absence of spatial pattern against which to test ecological data. Given the inherent spatial clustering of ecological data due to dispersal processes, other point pattern processes such as the Poisson cluster process can be used to generate a more realistic confidence envelope (Cressie 1991).

Ripley's K_{12} Statistic

The univariate Ripley's K can easily be extended to analyze the spatial relationship between points from two event types. This statistic is termed the Ripley's K_{12} statistic (Ripley 1976, 1979, 1981; Upton and Fingleton 1985; Cressie 1991):

$$K_{12}(t) = \{n_2\tilde{K}_{12}(t) + n_2\tilde{K}_{21}(t)\}/(n_1 + n_2)$$

where,

$$\tilde{K}_{12}(t) \frac{A}{n_1 n_2}\sum_{i=1}^{n_1}\sum_{j=1}^{n_2} I(e_i - e_j)$$

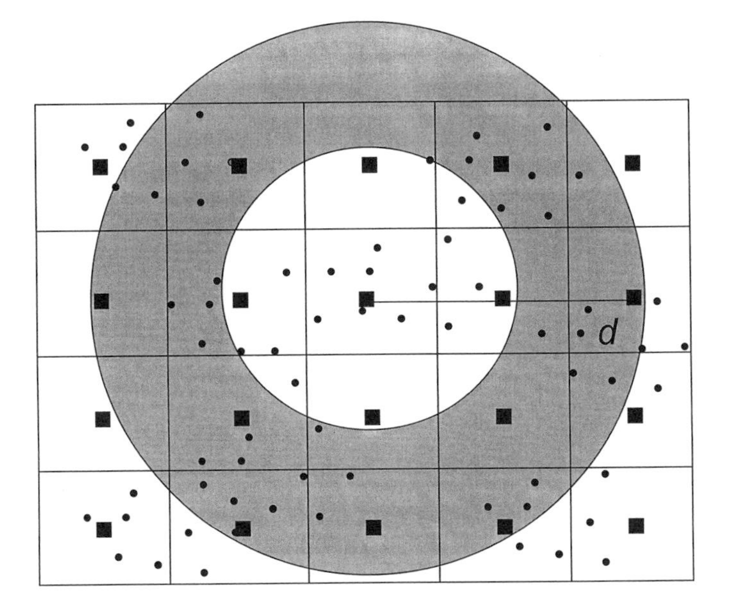

FIGURE 12.3. Surface pattern method. Point data sampled using quadrats where quadrat centroids are used as coordinates to compute spatial autocorrelation using the equidistance class d. Here $d = 2$ units (the distance between the target quadrat centroid to the distance d) and is only the gray ring area, therefore the interval between the $d > 1$ to $d = 2$.

and

$$\tilde{K}_{21}(t) = \frac{A}{n_1 n_2} \sum_{i=1}^{n_1} \sum_{j=1}^{n_2} I(e_j - e_i), \quad \text{for } i \neq j \text{ and } t > 0.$$

Here, I is the number of events of one state e_i within distance t of all events of the other state e_j. As with the univariate statistic, the K_{12} can be transformed into a L_{12} statistic which linearizes the plot and stabilizes the variance. Positive values indicate a positive interaction between the two variables, while negative values indicate spatial segregation or repulsion between them. This bivariate second-order statistic has been used in ecology by Kenkel (1988) to analyze the pattern of self-thinning in jack pine, and by Duncan (1991, 1993) to analyze the spatial interaction component between species.

Edge Effects

Although point pattern methods, such as the nearest neighbor and Ripley's *K* uni- and bivariate statistics, are not based on sampling units, they are not affected by sampling unit sizes as are the surface pattern methods. However, they are affected by the total study area (the extent). This is inherent to the fact that the expected values of these statistics are based on the CSR pattern estimated as the n/A ratio, λ. Furthermore, they are sensitive to edge effects (also known as boundary effects), which result from the fact that points near the edge boundary of the study plot have fewer nearby neighboring points than points located in the middle of the study area. This implies that computed short distances, such as Ripley's *K* values, are biased since they are calculated with potentially fewer points (Fig. 12.2). A similar bias occurs for Ripley's *K* values for longer distances. Several solutions have been proposed to correct for such edge effects (Fig. 12.4): (1) to sample a buffer zone around the study area and use these extra points to compute the statistics at short distances; (2) to not include the points near the edge of the plot in the computation; (3) to make a torus around the edges of the study area creating a donut shaped distance between points, such that points near one side find close neighbors from the opposite side; (4) to measure Ripley's *K* statistics only for half the smallest dimension of the study plot (Table 12.3); and (5) to use one of the various weight correction methods for points near the edge (Diggle 1983; Boots and Getis 1988; Cressie 1991; Haase 1995). Haase (1995), in a comparative study of these various edge effects correction methods, recommends using caution with the torus method and introduces a new weight correction which deals better with points next to corners, which implies a correction for double edge effects.

Ripley's *K* Plot

The plot of $L(t)$ values against t illustrates the number of points within a circular area of radius t. At very short distances, the points usually follow a

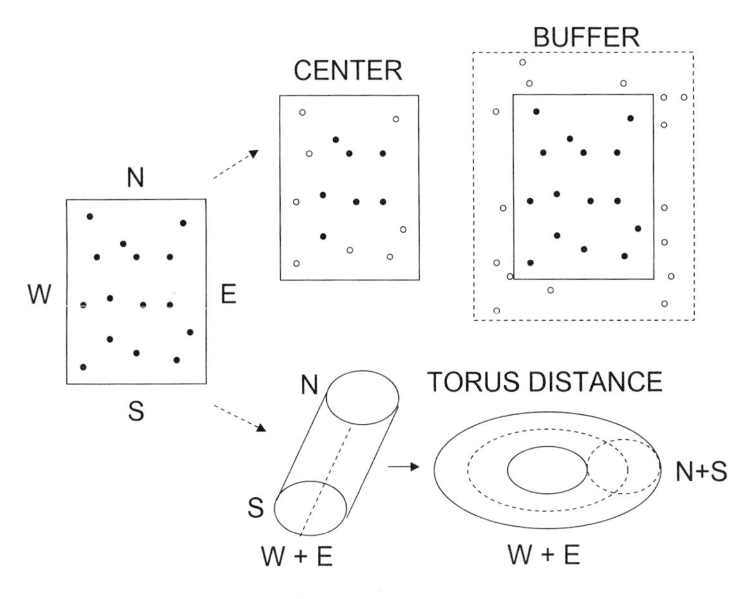

FIGURE 12.4. Edge effects correction alternatives: use only the points at the center of the plot (Center diagram); sample a buffered zone surrounding the plot (Buffer diagram); or compute torus distance such that points near the edge can have neighbors in points located near the opposite edge (Torus distance diagram).

Poisson pattern because of random dispersal and environmental fluctuations; hence $L(t)$ values are small and may lie within the confidence envelope. As distance increases, the $L(t)$ values increase as well since the $L(t)$ values include all of the points within the circle of t radius such that there are enough points to be able to detect a significant pattern when it is present (Fig. 12.2). Then, as the distance t increases, the $L(t)$ values can fluctuate, increase and decrease, as a function of the overall patchy spatial pattern of the variable over the study area as distance increases. A too large distance step t can affect our ability to detect significant spatial patterns since a large t will have tendency to smooth out spatial patterns given that the K statistic is a mean value over the entire circle of radius t. Too small a t implies more computer time which is less and less of an issue these days.

Surface Pattern Methods

Surface pattern methods require that the within study area data be sampled using areal units (quadrats). Since gathering data using quadrats is easier and faster to carry out in the field, it is the most common form of data collection used by field ecologists. With surface pattern methods, within quadrat data is assumed to be homogeneous and focus is placed on the spatial structure among quadrats. As previously mentioned, the inherent problem with areal data is the arbitrary choice of the sampling unit size which directly affects the quantification of spatial pattern intensity and scale (among others, Fortin 1992; Qi and Wu 1995).

Join-Count for Qualitative–Categorical Data

Significant spatial patterns for categorical data can be tested using the join-count statistics (Cliff and Ord 1981; Upton and Fingleton 1985). These spatial statistics were first developed to analyze binary data. They test whether neighboring areas are more likely to be of the same category, say 0 (black) or 1 (white), or not under a pattern of complete spatial randomness (CSR). These statistics count the number of join encounters of the same category in adjacent areas (e.g., *BB* and *WW* to assess positive spatial autocorrelation and *BW* for negative spatial autocorrelation). The significance of the join-count statistics is achieved by computing a Standard Normal Deviate (SND) and using a two-tailed test to detect positive or negative spatial autocorrelation. The join-count statistics have been extended to multicategorical data (Cliff and Ord 1981; Upton and Fingleton 1985). Spatial correlograms based on join-count statistics are computed in terms of neighbor link networks, such as those produced by the Delaunay triangulation algorithm (Upton and Fingleton 1985), and therefore correspond to nearest neighbor rather than Euclidean distances.

These join-count statistics were developed by geographers to assess significant spatial patterns of diseases or social behaviors among an a priori defined area, such as a county (Cliff and Ord 1981). In such cases, areas (patches) are already defined by external criteria rather than by the variable of interest (i.e., the one on which spatial autocorrelation is computed). With landscape ecology data, however, patches are usually defined by the variable under study (e.g., forest stands delineated by photointerpretation) (Beaulieu and Lowell 1994) or by spatial clustering (Fig. 12.1; Fortin and Drapeau 1995). By doing so, patch delineation, by definition, generates a spatially homogeneous entity different from its surrounding such that join-count statistics computed on such categorical maps are likely to find no significant spatial autocorrelation for the within category and significant positive or negative autocorrelation only among categories. Caution is recommended when interpreting spatial autcorrelation based on such defined patches since within patches are spatially homogeneous while adjacent patches are different by definition.

Spatial Autocorrelation Coefficients for Quantitative–Numerical Data

Spatial intensity and scale can be estimated for one quantitative variable using, among others, Moran's *I*, Geary's c or semivariance coefficients (Cliff and Ord 1981; Upton and Fingleton 1985; Haining 1990; Cressie 1991; Bailey and Gatrell 1995). Moran's *I* computes the degree of correlation between the values of a variable as a function of spatial locations. This coefficient is structurally comparable to a Pearson product-moment correlation coefficient, in that it represents the deviation between the values of the variable and its mean (see equation below). Moran's I values vary from -1 (negative autocorrelation) to 1 (positive autocorrelation), and has an expected value close to zero in the absence of spatial autocorrelation ($E(I) = -(n-1)^{-1}$). Geary's c, on the other hand, measures the difference (distance) among values of a variable at nearby locations (see equation below). It behaves somewhat like a distance measure and varies from 0, for perfect positive autocorrelation, to around 2 for a strong negative autocorrelation. The expected value $E(c)$ is 1 in the absence of spatial autocorrelation. Both coefficients are sensitive to outlier values but for different reasons: Moran's I is based on the comparisons of all values to the mean such that an under- or overestimated mean will bias all the estimated spatial autocorrelated values; while Geary's c is the squared difference among values such that outlier values will have more weight on the estimated spatial autocorrelated values. These techniques have been extended to measure the spatial autocorrelation of multivariate data (Wartenberg 1985).

Both Moran's I and Geary's c coefficients can be computed at several equidistance classes, d:

$$I(d) = \frac{n\sum_{i=1}^{n}\sum_{j=1}^{n} w_{ij}(z_i - \bar{z})(z_j - \bar{z})}{W_d \sum_{i=1}^{n}(z_i - \bar{z})^2}, \quad \text{for } i \neq j \text{ and}$$

$$c(d) = \frac{(n-1)\sum_{i=1}^{n}\sum_{j=1}^{n} w_{ij}(z_i - z_j)^2}{2W_d \sum_{i=1}^{n}(z_i - \bar{z})^2}, \quad \text{for } i \neq j,$$

where z_i and z_j are the values of the variable at locations i and j, respectively; $\bar{Z}$ is the variable mean; w_{ij} is a weight matrix where a value of 1 indicates a pair of locations i and j that are in the same distance class d and a value of 0 for all other case; W_d is the sum of the w_{ij}'s for the dth distance class. w_{ij} is mostly used as an indicator of the presence (1) or absence (0) of a connection between a pair of locations. Other weights, such as $1/d^2$, could be used to give more importance to nearby locations than further ones. The significance of individual Moran's I, or Geary's c, coefficients can be tested under two hypotheses. The first asserts that the values of the variable are a random sample taken from a normal distribution (Cliff and Ord 1981). The second assumes that any randomization of the observed surface values over the localities is equally likely.

A graph of Moran's I or Geary's c coefficients against the distance classes is called a spatial correlogram or simply a correlogram. The distance at which the value of spatial autocorrelation crosses the expected value, $E(I)$

or E(c), is considered the patch size (i.e., the spatial zone of influence of the variable), thus the spatial range in geostatistics terminology. Spatial correlograms have been used in ecology as tools to either identify or discriminate among various potential underlying processes responsible for the spatial patterns detected (Sokal 1979; Legendre and Fortin 1989; Duncan and Stewart 1991; Brodie et al. 1995). Before the shape of a spatial correlogram can be interpreted, however, one needs to test for the overall significance of the correlogram. To properly test this significance, a test needs to take into account the fact that the individual coefficients of a spatial correlogram are not independent from one another. Therefore the significance can be assessed by using a Bonferroni method (Oden 1984) that approximates the adjusted significance probability while having multiple testing, in this case multiple distance classes. Using the Bonferroni adjustment, the probability level, α', needed to test the entire spatial correlogram is determined by dividing the probability level, say $\alpha = 0.05$, by the number of distance classes, say $k = 10$ (e.g., $\alpha' = \alpha/k = 0.05/10 = 0.005$). A correlogram is deemed significant if the significance level of at least one individual coefficient is lower than this α' level.

These spatial autocorrelation coefficients provide an averaged isotropic estimation of the intensity of spatial autocorrelation at each distance class. Since in plant ecology, the dispersal of seeds by the wind can be largely influenced by dominant wind direction, most plant spatial structures are likely to show more or less some degree of spatial anisotropy (i.e., the degree to which the autocorrelation function is not the same in all directions). To identify such spatial anisotropy, spatial autocorrelation can be quantified by calculating the spatial autocorrelation for pairs of locations grouped not only by distance class but also by direction (Oden and Sokal 1986). The membership to the distance and direction classes is given by the weight matrix. Anisotropy is identified when patch sizes vary depending on direction, that is by comparing all-directional (omnidirection) and directional correlograms (Oden and Sokal 1986; Legendre and Fortin 1989; Rossi et al. 1992).

The intensity of spatial autocorrelation for a given distance class should be evaluated for all possible pairs of locations in the study area. This creates two problems: (1) if a given distance class contains fewer pairs of locations (e.g., less than 20), the estimation of the spatial autocorrelation is less reliable resulting in Moran's I values greater than 1, or lower than -1, and Geary's c values greater than 2; and (2) because the intensity of autocorrelation is evaluated as an average for the entire area, details about intensity fluctuations in some subregions of the study plot are lost. The first problem is often solved either by not trying to interpret the spatial autocorrelation values based on less than 20 to 30 pairs of locations; by interpreting only the first half to two thirds of the correlogram (Table 12.3); or by dividing the sample locations into equifrequent distance classes (Sokal and Wartenberg 1983). The second problem is inherent to all second-order statistics (Table 12.1, Table 12.3). This latter problem can be more pronounced when the study plot is large, where it is more likely to have additional environmental variability, therefore nonstationary data.

The estimation of spatial autocorrelation, intensity and scale, can be affected by the number of distance classes (Fortin 1992), by quadrat sizes and shapes (among others, Cohen et al. 1990; Fortin 1992; Qi and Wu 1995), and by the number and spatial arrangement of quadrats used (Fortin et al. 1989). Some of these effects are more important than others. First, the intensity of spatial autocorrelation varies for the first distance class, depending on the number of equidistant classes used: the larger the equidistance class, the more likely it is to include more environmental variability that results in a lower averaged spatial autocorrelation signal (Palmer and Dixon 1990; Fortin 1992). Because species abundance data depend on the quadrat size, spatial autocorrelation estimations based on such data will also vary depending on quadrat size. Therefore, when interpreting spatial correlogram from species abundance data, one should mainly be concerned with the overall correlogram shape and its significance, since they remain mainly consistent according to quadrat size, but not the intensity per se (Fortin 1992). Quadrat shape can also affect and bias estimates of spatial autocorrelation by generating artificial spatial anisotropy. To avoid spatial anisotropy artefact, isotropic quadrat shapes (squares, circles, or hexagons) should be used.

Furthermore, not only the number of samples (minimum 20 to 30) but their spatial arrangement also affect the ability of detecting a significant spatial pattern (Fortin et al. 1989): samples that are too close to one another tend to duplicate information, while samples too far apart may increase the environmental variability which in turn smoothes out the spatial signal. Hence, to detect a significant spatial pattern (Table 12.2), samples should be located within the spatial range of the variable. To identify the spatial range of the variable under study, prior knowledge such as a pilot study or aerial photographs could be used. On the other hand, when the goal of the study is to use parametric statistics to test some hypotheses (Table 12.2), spatially independent data can be obtained by sampling spatial steps that exceed the spatial range of the variable (Fortin et al. 1989).

Finally, environmental conditions can make the presence of some species unlikely such that study plots should be limited to the area where the species have an equal and positive probability of being found. This is the stationarity assumption on which spatial statistics are based. Having a study area (an extent) which covers a wide range of environmental conditions (e.g., a forest area and a recently burned or clear-cut area; a forest stand containing several water bodies, etc.) will not fulfill the stationarity requirement. Ecologists shoud therefore be carefull when selecting the location and total area covered by their study plot. For example, when the spatial distribution of a species is limited to one part of the study area, as is the case at an ecotone margin, the estimation of spatial autocorrelation should be established using only the area where the species has a positive probability of being found, thereby avoiding nonstationarity.

Geostatistics: Variogram and Kriging

Ecologists have shown an increasing interest in geostatistics (among others, Robertson 1987; Fortin et al. 1989; Rossi et al. 1992) because these techniques not only quantify (semivariance) but also permit be modeling of spatial patterns (krige, i.e., interpolate using known spatial autocorrelation functions). The first step consists in estimating spatial autocorrelation parameters using an experimental variogram and then using these parameters to estimate interpolated values of unsampled locations using kriging. While geostatistics are based on the stationarity assumption, recent developments in this field make these methods more flexible such that they can be used with a less strict stationarity assumption: local-, quasi-, pseudo-, weak-, and second-order stationarity all refer to the fact that stationarity is respected only at short distances (Cressie 1991). Such a relaxed stationarity assumption is very interesting for ecological data since they are rarely stationary (as mentioned above).

Spatial pattern intensity and scale are quantified using the spatial semivariance function which is computed at equidistance classes, h:

$$\gamma(h) = \frac{1}{2n_h} \sum_{i=1}^{n} w_{ij} (z_i - z_{i+h})^2$$

where n_h is the number of pairs of points located at distance h from one another. As Geary's c autocorrelation coefficient (above), the semivariance is a distance function; the difference between these two coefficients lies mainly in the lack of a denominator that standardizes the spatial autocorrelation estimation. Therefore, the semivariance is not bounded which makes comparisons among variables difficult. The latter drawback explains why recent studies in geostatistics have used standardized semivariance estimations rather than semivariance per se (Rossi et al. 1992). The spatial semivariance function is often referred to as either the semivariogram (because of the division by two) or simply the variogram for the plot of the semivariance against the distance h. Finally, since the variogram is an average of squared differences, as Geary's c, it is also sensitive to outlier. The semivariance coefficient has been extended to permit the computing of variograms for multivariate data (Bourgault and Marcotte 1991).

As for the spatial correlogram, the shape of the variogram obtained with sampled data, known as the experimental variogram, allows the description of the overall spatial pattern (Burrough 1987) and the estimation of spatial autocorrelation parameters needed to krige: (1) the spatial range, a, where the variable is spatially influenced by the same underlying process; (2) the sill, $C_0 + C_1$, that quantifies the spatial pattern intensity; and (3) the nugget effect, C_0, which is the estimate of the inherent error induced by measurement errors (sampling design and the sampling unit size) and by environmental variability (Journel and Huijbregts 1978; Isaaks and Srivastava 1989; Deutsch and Journel 1992; Griffith 1992).

According to the shape of the experimental variogram, different theoretical variogram models can be used to krige the spatial structure. The most commonly used models include the linear, the exponential, the spherical, and the Gaussian (Journel and Huijbregts 1978; Isaaks and Srivastava 1989; Deutsch and Journel 1992). Because most of the spatial autocorrelation signal is comprised of the first part of the variogram (i.e., usually up to the spatial range) good parameter fitting for short distance lags is very important. Although in the early years of geostastistics such model fitting was often guesstimated (Englund 1990), now generalized least-squares, maximum likelihood, and restricted maximum likelihood methods can be used which make the choice of theoritical variogram models more objective (Cressie 1991). A good estimation of these parameters is crucial for the subsequent kriging steps and the resulting interpolated map. Hence, a spatial range that is too short will result in a spiky map where spatial autocorrelation structure will be included only in interpolated locations near the sampling points and all the other locations will have the mean intensity value (the modeled sill value); while too large a spatial range will result in a very smoothed map.

Although geostatistics offer more flexibility for analyzing pseudostationary data, it is still sensitive to the way in which samples were gathered: the sampling unit sizes (MAUP), the number of samples and their spatial arrangement (Fortin et al. 1989; Cohen et al. 1990; Webster and Oliver 1990; Fortin 1992; Qi and Wu 1995). As for spatial correlograms, variograms are computed for distance classes, which implies that the number of pairs of points used in the computation decreases as distance increases. Thus, only values computed with more than 30 pairs of points should be taken into account when describing the spatial structure. This often corresponds to the first half to two thirds of the experimental variogram (Table 12.3).

The kriging procedure is known as BLUE (Best Linear Unbias Estimator) because it returns the observed values at sampling locations, it interpolates values using the intensity and shape of the spatial autocorrelation function of the data (using a neighborhood and/or distance search radius; Table 12.3), and it provides the standard errors of the interpolated values (Journel and Huijbregts 1978). These estimation errors have often been used to optimize sampling design (Webster and Oliver 1990) by identifying areas where sampling effort should be increased or decreased. However, these errors are a function of the theoretical variogram model used and not the raw data (Deutsch and Journel 1992). Caution should therefore be taken in interpreting the meaning of these estimation errors while optimzing sampling design. Finally, as mentioned above, kriging is like other interpolation techniques and produces smoothed maps. To obtain less smoothed map with comparable spatial autocorrelation functions, conditional simulation procedures can be used (Journel and Huijbregts 1978; Deustch and Journel 1992).

Spatial Statistics for Population Data

The spatial statistics and geostatistics mentioned above when used on remotely sensed data, or GIS raster layer data, to estimate spatial autocorrelation is overkill. Indeed, spatial statistics were developed to quantify spatial pattern from sampled data (n). When the data is exhaustive (i.e., the whole population, N) over the study plot, like it is the case with remotely sensed data, to quantify spatial structure one should directly use either two-dimensional spectral analysis for stationary data (Ford and Renshaw 1984; Renshaw and Ford 1984; Legendre and Fortin 1989) or two-dimensional wavelet analysis for nonstationary data (Bradshaw 1991). Furthermore, there is no reason to use kriging to interpolate missing sample points because data for the whole area are available.

Relationship Between Spatially Autocorrelated Variables

To compute the degree of relationship among variables, the first tendency is to use correlation coefficients such as Pearson's or Spearman's. With spatially autocorrelated data, however, the assumption of independence of the observations required by parametric tests (Cliff and Ord 1981) is not fulfilled. Indeed, in the presence of positive spatial autocorrelation, the relationship among spatially autocorrelated data is too often significant when in fact it is not (Cliff and Ord 1981; Legendre et al. 1990; Legendre 1993). Furthermore, it is possible that the relationship between two variables is a spurious one induced by a third ungathered variable or simply by their joint spatial cooccurrence (Legendre 1993). Two alternatives exist to test the relationship among spatially distributed variables: (1) by first removing the spatial structure in the data and then computing the relationship on the residuals (Cliff and Ord 1981) or (2) by using approaches that can correct for the spatially autocorrelated structure of the data (Clifford et al. 1989; Dutilleul 1993).

Detrending Large Spatial Pattern

Detrending time series data is relatively common and it is achieved by using autoregressive models (Chatfield 1984). Detrending large spatial pattern from two-dimensional data can be achieved by using either trend-surface analysis (Cliff and Ord 1981; Upton and Fingleton 1985; Haining 1990) or universal kriging (Cressie 1991; ver Hoef 1993) such that the relationship among variables is computed on the residuals.

Another way to assess the degree of relationship among spatially autocorrelated data is to control for the spatial pattern in controlling for it as in partial correlation (Sokal and Rohlf 1995). This can be achieved by using either a Partial Mantel's test (Smouse et al. 1986; Legendre and Troussellier 1988; Legendre and Fortin 1989; Leduc et al. 1992; Fortin and Gurevitch 1993; Brodie et al. 1995) or a Partial Canonical Correspondence Analysis (Partial CCA; ter Braak 1987, 1988; Borcard et al. 1992; Legendre 1993; Palmer 1993).

Partial Mantel's Test

Mantel's test (Mantel 1967) computes the relationship between two distance matrices, for example, dissimilarity among plant species and among the environmental variables obtained at the same sampling locations (Fig. 12.5). The Mantel's statistic, which is the sum of all of the products between corresponding elements of the distance matrices, can be normalized into a product moment correlation coefficient, r, that varies from -1 to 1. Hence, the Mantel's statistic is a linear estimate of the relationship between the distance or dissimilarity values between pairs of samples rather than between the raw data (Table 12.3). There is no restriction on the type of distance coefficients to use, except that the matrices have to be square. When the Mantel's statistic is calculated between a set of variables and a

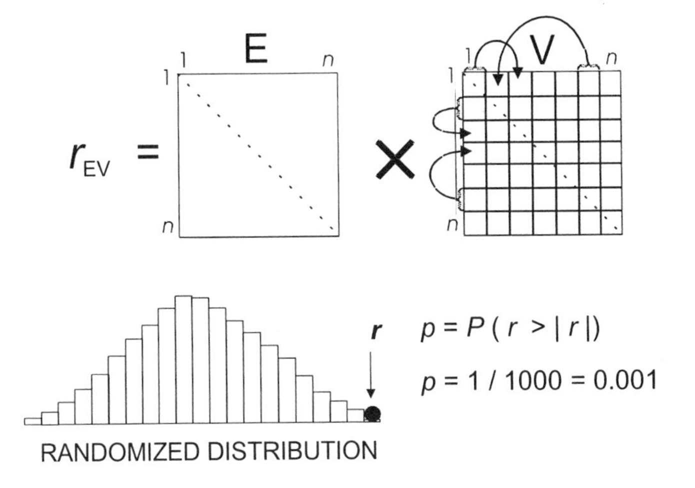

FIGURE 12.5. Mantel's test between two squared distance matrices. Significance is achieved by randomizing the rows and columns of at least one of the matrices. The probability level is computed by including the observed statistic to the randomization reference distribution of $n = 999$ such that $p(r) = 1/1000 = 0.001$.

geographical distance matrix (i.e., Euclidean distances among samples), the outcome corresponds to an averaged isotropic intensity of spatial autocorrelation of the variable for the entire study plot.

Significance is assessed either by using a randomization test to construct a reference distribution, or by using an asymptotic *t*-approximation (Mantel 1967). In the randomization test, the statistic which is calculated on the observed data is compared against a reference distribution generated by randomly shuffling the rows and columns of one of the matrices (Fig. 12.5). Under the null hypothesis of no relationship between the two distance matrices, the observed Mantel's statistic is expected to have a value located near the mode of the reference distribution obtained by randomization of the data. On the other hand, if there is a strong relationship, positive or negative, between the two matrices, the observed Mantel statistic is expected to be more extreme, either higher or lower, than most of the reference distribution values. The probability level is computed to include the observed statistic in the randomization reference distribution (Hope 1968). The higher the number of randomizations, the higher the probability resolution. Several authors recommend as many randomizations as possible (i.e., 1000–10,000), depending on the sample size (Manly 1991). Given that the relationship between the two matrices is based on distance values rather than on the raw data, the degree of relationship between the to sets of variables is often weaker than what would have been expected between the raw data. Therefore, one should not be too concerned about the intensity of the relationship but rather whether or not it is significant.

To identify potential spurious correlations, a Partial Mantel's test (Smouse et al. 1986) can be used. This statistic is a multivariate equivalent of a partial correlation coefficient (Sokal and Rohlf 1995) and it computes the degree of relationship between two distance matrices that are the regression residuals where the effects of a third matrix (i.e., geographic distances) has been controlled for (Fig. 12.6). Significance testing is achieved by randomization tests, such as the Mantel's test (Fig. 12.5). Usually Euclidean distances can be used but other distance measures can also be used such as $1/d^2$, which will take into consideration nonlinear relationships (e.g., patchy structure) among the variables or set of variables and geographic locations (Table 12.3). The third matrix, the one being factored out, can be a model (contrast) matrix to test the causality between spatially autocorrelated variables (Leduc et al. 1992; Fortin and Gurevitch 1993; Sokal et al. 1993) or a weight (connection) matrix to compute a multivariate correlogram (Oden and Sokal 1986; Legendre and Fortin 1989). Partial Mantel's test was also extended to deal with more than one covariables matrix (Manly 1991). The major problem with the Mantel's and the Partial Mantel's statistics is that multivariate data are summarized into a single distance, or dissimilarity, value such that the resulting relationship is a global outcome for all of the variables and it is not possible to identify which

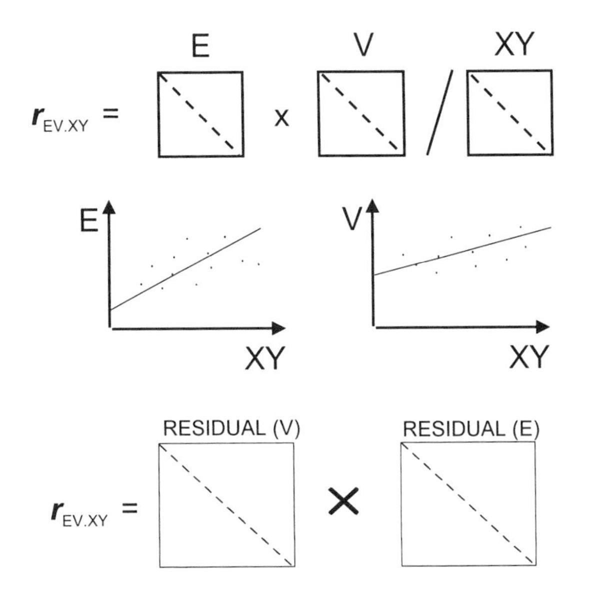

FIGURE 12.6. Partial Mantel's test between two squared matrices where the effects of a third one (e.g., covariables, geographic distances, or model), are factored out by linear regression. Mantel's test is then computed on the regression residuals.

variable(s) contributed the most to its intensity. To circumvent this problem, Partial CCA can be used.

Partial CCA

The relationship between two sets of variables, say species and environmental data, can be quantified using multivariate ordination methods such as Canonical Correspondence Analysis, CCA (ter Braak 1987). In this method, species ordination axes are constrained, using multiple regression, to be linear combinations of the environmental variables in maximizing species variance. This method is an iterative procedure which estimates the best linear fit between two data sets. Hence the outcome will be the best possible linear relationship between species and environmental ordination axes. The outcome of this method is however very sensitive to measurement errors in the environmental data and is limited to the environmental variables included in the analysis. Often the relationship between species and environmental data is induced by other underlying factors such as climate, topography or historical events. One way to control for the effects of these other variables would be to use Partial CCA (ter Braak 1987, 1988; Borcard et al. 1992; Legendre 1993; Palmer 1993).

Partial CCA, unlike Partial Mantel's test, allows us to disentangle the relative contributions of more than one set of factors species responses by partialling out the effects of a third data set considered as covariables.

Randomization tests are then used to assess the significance of the relationship (ter Braak 1990). The use of this third set of covariables allows the testing of specific hypotheses by coding it as a contrast matrix in ANOVA (Fortin and Gurevitch 1993; Sokal et al. 1993) or by using geographic locations as surrogate for nonsampled variables (Borcard et al. 1992; Legendre 1993; Harvey 1996). Furthermore, Partial CCA can also be used to quantify the relationship between wildlife and habitat characteristics such as the landscape ones obtained by FRAGSTATS (McGarigal and Marks 1995) while controlling for environmental components such as topography and space as covariables (McGarigal and McComb 1995; Harvey 1996).

Correcting for the Presence of Spatial Autocorrelation

The degree of relationship between spatially autocorrelated variables can be calculated using a correlation coefficient, but its significance must be assessed by reducing the degree of freedom proportionally to the amount of spatial autocorrelation in the data (Clifford et al. 1989; Haining 1990, 1991; Dutilleul 1993). Indeed, when there is positive spatial autocorrelation each sample point does not provide a full degree of freedom but only a part which determined by the degree of spatial autocorrelation (Legendre et al. 1990). Therefore, the effective sample size is less than the actual sample size, n, so the degree of freedom at which the significance of the correlation should be tested is usually less than the parametric one ($\upsilon = n - 2$) (Cliff and Ord 1981 Legendre et al. 1990). The estimate given in Clifford et al. (1989) is the effective sample size as a function of the intensity of spatial autocorrelation contained in both variables. When the sample size is less than 20, a modified estimated degree of freedom algorithm should be used (Dutilleul 1993; but see also Clifford et al. 1993). The significance of the correlation coefficient is then tested by a t-test with the latter estimated degree of freedom minus 2. The advantage of this approach is that the relationship is computed on the raw data rather than on distance measures as in Mantel's and Partial Mantel's tests.

Spatial Statistics in Landscape Ecology

The aim of landscape ecology is to analyze ecological processes in their spatial context. Firstly, this chapter glances over spatial statistics which describe spatial patterns (Table 12.2) and secondly it presents statistics which test causality between species spatial structures and potential underlying factors (e.g., environmental variables). As presented, there are several spatial statistics available, all having specific assumptions about the data (Table 12.3). It is important to keep in mind that these statistics were developed to analyze samples of data, and that when exhaustive data (the population) are available, other methods can be more appropriate.

Finally, although spatial description analysis is an important step, and can involve using several sophisticated methods (Table 12.3), it is not an end by itself; testing relationship among variables is subsequently needed to identify the underlying processes responsible of species spatial responses. In such relationship of analysis, it is crucial to realize that the presence of spatial autocorrelation in the data can mask potential relationships between variables (Legendre and Troussellier 1988; Fortin and Gurevitch 1993; Legendre 1993; Sokal et al. 1993). Hence, by detrending or controlling for spatial pattern, one can quantify and disentangle whether or not relationships between variables are due to a potential causality link between them, to the presence of spatial autocorrelation, or to both (Borcard et al. 1992; Palmer 1993; Sokal et al. 1993).

Acknowledgments. I am grateful to Dean Urban to have invited me to give a workshop at the 12th Annual US-IALE meeting at Duke University. I am thankful to F. Fournier for his helpful comments. This work was supported by a NSERC-individual grant to M.-J. Fortin.

References

Anselin, L. 1992. SpaceStat: A Program for the Analysis of Spatial Data. National Center for Geographic Information and Analysis. University of California, Santa Barbara.

Bailey, T.C., and A.C. Gatrell. 1995. Interactive Spatial Data Analysis. Longman Scientific & Technical. Essex.

Baker, W.L., and Y. Cai. 1992. The r.le programs for multiscale analysis of landscape structure using the GRASS geographical information system. Landscape Ecology 7:291–302.

Beaulieu, P., and K. Lowell. 1994. Spatial autocorrelation among forest stands identified from the interpretation of aerial photographs. Landscape and Urban Planning 29:161–169.

Boots, B.N., and A. Getis. 1988. Point Pattern Analysis. Sage University Scientific Geography Series, Vol. 8. Sage publications, Beverly Hills.

Borcard, D., P. Legendre, and P. Drapeau. 1992. Partialling out the spatial component of ecological variation. Ecology 73:1045–1055.

Bourgault, G., and D. Marcotte. 1991. Multivariable variogram and its application to the linear model of coregionalization. Mathematical Geology 23:899–928.

Bradshaw, G.A. 1991. Hierarchical analysis of pattern and processes in Douglas-fir forests using the wavelet transform. Ph.D. Dissertation, Oregon State University.

Brodie, C., G. Houle, and M.-J. Fortin. 1995. Development of a *P. balsamifera* clone in subarctic Québec reconstructed from spatial analyses. Journal of Ecology 83:309–320.

Burrough, P.A. 1987. Spatial aspects of ecological data. In: Community and Landscape Ecology, pp. 213–251. R.H.G. Jongman, C.J.F. ter Braak, and O.F.R. van Tongeran (eds.). Pudoc Wegeningen, Nertherlands.

Chatfield, C. 1984. The Analysis of Time Series: An Introduction, 3d ed. Chapman and Hall, London.

Clark, P.J., and F.C. Evans. 1954. Distance to nearest neighbour as a measure of spatial relationships in populations Ecology 35:445–453.

Cliff, A.D., and J.K. Ord. 1981. Spatial Processes: Models and Applications. Pion, London.

Clifford, P., S. Richardson, and D. Hémon. 1989. Assessing the significance of the correlation between two spatial processes. Biometrics 45:123–134.

Clifford, P., S. Richardson, and D. Hémon. 1993. Response: reader reaction. Modifying the t test for assessing the correlation between two spatial processes. Biometrics 49:305–314.

Cohen, W.B., T.A. Spies, and G.A. Bradshaw. 1990. Semivariograms of digital imagery for analysis of conifer canopy structure. Remote Sensing Environment 34:167–178.

Cressie, N.A.C. 1991. Statistics for Spatial Data. Wiley John & Sons, New York.

Dale, M.R.T. 1990. Two-dimensional analysis of spatial pattern in vegetation for site comparison. Canadian Journal of Botany 68:149–158.

Dale, M.R.T., and D.J. Blundon. 1990. Quadrat variance analysis and pattern development during primary succession. Journal of Vegetation Science 1:153–164.

Deutsch, C.V., and A.G. Journel. 1992. GSLIB: Geostatistical Software Library and User's Guide. Oxford University Press, New York.

Diggle, P.J. 1983. Statistical Analysis of Spatial Point Patterns. Academic Press, London.

Duncan, R.P. 1991. Competition and the coexistence of species in a mixed podocarp stand. Journal of Ecology 79:1073–1084.

Duncan, R.P. 1993. Flood disturbance and the coexistence of species in lowland podocarp forest, south Westland, New Zealand. Journal of Ecology 81:403–416.

Duncan, R.P., and G.H. Stewart. 1991. The temporal and spatial analysis of tree age distributions. Canadian Journal of Forostur Research 21:1703–1710.

Dutilleul, P. 1993. Reader Reaction. Modifying the *t* test for assessing the correlation between two spatial processes. Biometrics 49:305–314.

Englund, E.J. 1990. A variance of geostatisticians. Mathematical Geology 22:417–455.

Ford, E.D., and E. Renshaw. 1984. The interpretation of process from pattern using two-dimensional spectral analysis: modeling single species patterns in vegetation. Vegetatio 56:113–123.

Fortin, M.-J. 1992. Detection of ecotones: definition and scaling factors. Ph.D. Dissertation, State University of New York at Stony Brook.

Fortin, M.-J. 1994. Edge detection algorithms for two-dimensional ecological data. Ecology 75:956–965.

Fortin, M.-J., and P. Drapeau. 1995. Delineation of ecological boundaries: comparison of approaches and significance tests. Oikos 72:323–332.

Fortin, M.-J., and J. Gurevitch. 1993. Mantel tests: spatial structure in field experiments. In: Design and analysis of ecological experiments, pp. 342–359. S.M. Scheiner and J. Gurevitch (eds.). Chapman & Hall, New York.

Fortin, M.-J., P. Drapeau, and P. Legendre. 1989. Spatial autocorrelation and sampling design in plant ecology. Vegetatio 83:209–222.
Getis, A., and B. Boots. 1978. Models of Spatial Process: An Approach to the Study of Point, Line and Area Patterns. Cambridge Univ. Press, Cambridge.
Getis, A., and J. Franklin. 1987. Second-order neighborhood analysis of mapping point patterns. Ecology 68:473–477.
Greig-Smith, P. 1952. The use of random and contiguous quadrats in the study of the structure of plant communities. Annals of Botany. New Series 16:293–316.
Greig-Smith, P. 1964. Quantitative plant ecology, 2nd ed. Butterworth, London.
Griffith, D.A. 1992. What is spatial autocorrelation? Reflection on the past 25 years of spatial statistics. L'Espace géographique 3:265–280.
Gustafson, E.J., and G.R. Parker. 1992. Relationships between landcover proportion and indices of landscape spatial pattern. Landscape Ecology 7:101–110.
Haase, P. 1995. Spatial pattern analysis in ecology based on Ripley's *k*-function: Introduction and methods of edge correction. Journal of Vegetation Science 6:575–582.
Haining, R. 1980. Spatial autocorrelation problems. In: Geography and the Urban Environment, pp. 1–44. D. Herbert and R. Johnston (eds.). John Wiley & Sons, London.
Haining, R. 1990. Spatial Data Analysis in the Social and Environmental Sciences. Cambridge University Press, Cambridge.
Haining, R. 1991. Bivariate correlation with spatial data. Geographical Analysis 23:210–227.
Harvey, L.E. 1996. Macroecological studies of species composition, habitat and biodiversity using GIS and canonical correspondence analysis. In: Proceedings, Third International Conference/Workshop on Integrating GIS and Environmental Modeling, Sante Fe, NM, January 21–26, 1996, CD-ROM and WWW (http: / /WWW.ncgia.ucsb.edu/conf/SANTE_FECD_ROM/main.htlm).
Hope, A.C.A. 1968. A simplified Monte Carlo significance test procedure. Journal of the Royal Statistical Society Seres B 30:582–598.
Isaaks, E.H., and R.M. Srivastava. 1989. An Introduction to Applied Geostatistics. Oxford University Press, Oxford.
Johnston, C.A., J. Pastor, and G. Pinay. 1992. Quantitative methods for studying landscape boundaries. In: Landscape boundaries: Consequences for Biotic Diversity and Ecological Flows, pp. 107–128. A. Hansen and F. di Castri (eds.). Springer-Verlag, New York.
Journel, A.G., and C. Huijbregts. 1978. Mining Geostatistics. Academic Press, London.
Kenkel, N.C. 1988. Pattern of self-thinning in jack pine: testing the random mortality hypothesis. Ecology 69:1017–1024.
Kershaw, K.A. 1964. Quantitative and Dynamic Ecology. Edward Arnold, London.
Kotliar, N.B., and J.A. Wiens. 1990. Multiple scales of patchiness and patch structure: a hierarchical framework for the study of heterogeneity. Oikos 59:253–260.
Leduc, A., P. Drapeau, Y. Bergeron, and P. Legendre. 1992. Study of spatial components of forest cover using partial Mantel tests and path analysis. Journal of Vegetation Science 3:69–78.
Legendre, P. 1993. Spatial autocorrelation: trouble or new paradigm? Ecology 74:1659–1673.

Legendre, P., and M.-J. Fortin. 1989. Spatial pattern and ecological analysis. Vegetatio 80:107–138.

Legendre, P., R.R. Sokal, N.L. Oden, A. Vaudor, and J. Kim. 1990. Analysis of variance with spatial autocorrelation in both the variable and the classification criterion. Journal of Classification 7:53–75.

Legendre, P., and M. Troussellier. 1988. Aquatic heterotrophic bacteria: modeling in the presence of spatial autocorrelation. Limnology and Oceanography 33:1055–1067.

Li, H., and J.F. Reynolds. 1995. On definition and quantification of heterogeneity. Oikos 73:280–284.

Ludwig, J.A., and J.F. Reynolds. 1988. Statistical Ecology. John Wiley & Sons, New York.

Manly, B.F.J. 1991. Randomization and Monte Carlo methods in biology. Chapman & Hall, London.

Mantel, N. 1967. The detection of disease clustering and a generalized regression approach. Cancer Research 27:209–220.

Matheron, G. 1970. La théorie des variables régionalisées, et ses applications. Les Cahiers du Centre de Morphologie Mathématique de Fontainebleau, Fascicule 5, Fontainebleau.

McGarigal, K., and B.J. Marks. 1995. FRAGSTATS: spatial pattern analysis program for quantifying landscape structure. U.S. Forest Service General Technical Report PNW 351.

McGarigal, K., and W.C. McComb. 1995. Relationship between landscape structure and breeding birds in the Oregon Coast Range. Ecological Monographs 65:235–260.

Moeur, M. 1993. Characterizing spatial patterns of trees using stem-mapped data. Forest Science 39:756–775.

Oden, N.L. 1984. Assessing the significance of a spatial correlogram. Geographical Analysis 16:1–16.

Oden, N.L., and R.R. Sokal. 1986. Directional autocorrelation: an extension of spatial correlograms to two dimensions. Systematic Zoology 35:608–617.

Palmer, M.W. 1993. Putting things in even better order: the advantages of canonical correspondence analysis. Ecology 74:2215–2230.

Palmer, M.W., and P.M. Dixon. 1990. Small-scale environmental heterogeneity and the analysis of species distributions along gradients. Journal of Vegetation Science 1:57–65.

Qi, Y., and J. Wu. 1996. Effects of changing spatial resolution on the results of landscape pattern analysis using spatial autocorrelation indices. Landscape Ecology 11:39–49.

Renshaw, E., and E.D. Ford. 1984. The description of spatial pattern using two-dimensional spectral analysis. Vegetatio 56:75–85.

Riitters, K.H., R.V. O'Neill, C.T. Hunsaker, J.D. Wickham, D.H. Yankee, S.P. Timmins, K.B. Jones, and B.L. Jackson. 1995. A factor analysis of landscape pattern and structure metrics. Landscape Ecology 10:23–29.

Ripley, B.D. 1976. The second-order analysis of stationarity processes. Journal of Applied Problems 13:255–266.

Ripley, B.D. 1977. Modelling spatial patterns. Journal of the Royal Statistical Society Series B 39:172–212.

Ripley, B.D. 1979. Tests of randomness for spatial point patterns. Journal of the Royal Statistical Society Series B 41:368–374.

Ripley, B.D. 1981. Spatial Statistics. John Wiley & Sons, New York.

Ripley, B.D. 1988. Statistical Inference for Spatial Processes. Cambridge University Press, Cambridge.

Robertson, G.P. 1987. Geostatistics in ecology: interpolating with known variance. Ecology 68:744–748.

Rossi, R.E., D.J. Mulla, A.G. Journel, and E.H. Franz. 1992. Geostatistical tools for modeling and interpreting ecological spatial dependence. Ecological Monographs 62:277–314.

Sokal, R.R. 1979. Ecological parameters inferred from spatial correlograms. In: Contemporary Quantitative Ecology and Related Ecometrics, pp. 167–196. G.P. Patil and M.L. Rosenzweig (eds.). Statistical Ecology Series 12. International Cooperative Publishing House Fairland, MD.

Sokal, R.R., and F.J. Rohlf. 1995. Biometrics, 3rd ed. W.H. Freeman, New York.

Sokal, R.R., and D.E. Wartenberg. 1983. A test of spatial autocorrelation using an isolation-by-distance model. Genetics 105:219–237.

Sokal, R.R., N.L. Oden, B.A. Thomson, and J. Kim. 1993. Testing for regional differences in means: distinguishing inherent from spurious spatial autocorrelation by restricted randomization. Geographical Analysis 25:199–210.

Smouse, P.E., J.C. Long, and R.R. Sokal. 1986. Multiple regression and correlation extensions of the Mantel test of matrix correspondence. Systematic Zoology 35:627–632.

ter Braak, C.J.F. 1987. Ordination. In: Community and Landscape Ecology, pp. 91–173. R.H.G. Jongman, C.J.F. ter Braak, and O.F.R. van Tongeran (eds.). Pudoc Wageningen, Nertherlands.

ter Braak, C.J.F. 1988. Partial canonical correspondende analysis. In: Classification and Related Methods of Data Analysis, pp. 551–558. H.H. Bock (ed.). Elsevier, North Holland.

Tobler, W.R. 1970. A computer movie simulating urban growth in the Detroit region. Economic Geography Supplement 46:234–240.

Turner, S.J., R.V. O'Neill, W. Conley, M.R. Conley, and H.C. Humpries. 1991. Pattern and scale: statistics for landscape ecology. In: Quantitative Methods in Landscape Ecology, pp. 17–49. M.G. Turner and R.H. Gardner (eds.). Springer-Verlag, New York.

Upton, G.J.G., and B. Fingleton. 1985. Spatial data analysis by example, Vol. 1: Point pattern and quantitative data. John Wiley & Sons, New York.

ver Hoef, J. 1993. Universal kriging for ecological data. In: Environmental modeling with GIS, pp. 447–553. M.F. Goodchild, B.O. Parks, and L.T. Steyaert (eds.). Oxford University Press, New York.

ver Hoef, J., N.A.C. Cressie, and D.C. Glenn-Lewin. 1993. Spatial models for spatial statistics: some unification. Journal of Vegetation Science 4:441–452.

Wartenberg, D. 1985. Multivariate spatial autocorrelation: a method for explanatory geographical analysis. Geographical Analysis 17:263–283.

Webster, R. and M.A. Oliver. 1990. Statistical Methods in Soil and Land Resource Survey. Oxford University Press, Oxford.

13 RULE: Map Generation and a Spatial Analysis Program

Robert H. Gardner

The use of random maps as neutral landscape models was first introduced more than 10 years ago (Gardner et al. 1987). The concept is to investigate the effect of random processes on landscape patterns and use this information as a null hypothesis testing actual landscapes. Although the concept of a neutral model is quite simple, the issues have sometimes been confused in the literature (see With and King 1997 for a more complete discussion). Gardner and O'Neill (1991) have offered a prescription to avoid this confusion: (1) make a clear statement of the problem, (2) define the simplest model that allows each variable to be examined, (3) compare prediction with available data and observations, and (4) establish an objective measure of the adequacy of the results. Presented here are a set of simple methods for generating neutral landscape models that have been extensively used over the last 10 years.

The development of a single computer program for the generation of random maps and the analysis of spatial patterns was begun in 1989 by assembling a collection of programs, routines, and algorithms that our group at Oak Ridge National Laboratory had been using for a number of years (see the References section for a listing of citations using RULE). Along the way, a unique approach to the analysis of spatial patterns was developed which allowed the user to specify the neighborhood rule used to identify contiguous sites (i.e., patches of land cover). Methods usually employed for the identification of patches assume a "nearest neighbor rule" (i.e., patches identified by enumerating all sites connected via contact of adjacent vertical and horizontal surfaces). Sites that are diagonally adjacent, but not connected along the four cardinal directions, are not considered a part of the patch. The inclusion of these additional sites requires one to specify a "nearest-neighbor rule." The use of a "nearest-neighbor rule" can affect both the size and number of patches identified depending on the orientation of the grid. For instance, a long thin but connected land-cover type would be regarded as a single patch with the "nearest-neighbor rule" if the patch was oriented parallel to the axis of the map grid. If the patch was rotated by 45 degrees than the same area might be defined by many small, separate patches. RULE allows the user to specify the number and location of neighbors (i.e., the "contact rule") identifying habitat patches. In addition, the analysis of spatial patterns used in RULE includes indices not commonly available in other analysis packages. The output from RULE can be controlled by the user to create several useful data sets, including a scale-dependent measure of map texture called "lacunarity" (Plotnick et al. 1993). This chapter presents overview of the methods used by RULE and provides a general program guide for using the software. The user should refer to the references cited for detailed descriptions of the techniques used by RULE.

The original version of RULE was written in FORTRAN 77 for execution on UNIX workstations. The version presented here has been converted from FORTRAN 77 to FORTRAN 90 for execution under MS-DOS on the Intel Pentium family of personal computers (RULE can be executed in Windows, WindowsNT, or Windows95/98 by opening an MS-DOS window). Several routines have been completely rewritten to take advantage of the recursion possible in FORTRAN 90, while other routines have been modified to improve program flow and insure reliability. Even with these improvements, portions of RULE may seem awkward and inconvenient because RULE is essentially a dynamic research tool. Hopefully, this version of RULE will prove to be a reliable and useful addition to the tools used by landscape ecologists to characterize spatial patterns.

Description of RULE

RULE is designed to generate and analyze multiple sets of maps of the same dimension and number of habitat types. The results of the analysis can be statistically summarized, providing a useful basis for comparison against actual landscape patterns. For instance, one might be interested in the distribution of sizes of habitat clusters on an actual landscape. The generation of a series of random maps of the same dimension provides a basis for stating that the actual landscape does or does not differ from the random pattern (it is surprising that many attributes of actual landscapes *often do not differ* from random). RULE has been designed to make this process as efficient as possible by providing a large variety of methods for map generation and by using a set of analysis methods that are quantitatively robust.

Five categories of information are required by RULE to generate and analyze land-cover maps: (1) the size and type of map to be generated, (2) the initial value for the pseudorandom number generator, (3) the number of habitat types and the probabilities of occurrence associated with each habitat type; (4) the number of iterations of maps to be generated and analyzed, and (5) the analysis methods and output desired.

Generation of Random Maps

Four types of maps (Fig. 13.1A) can be generated by RULE. These are: (1) a simple random map; (2) a random map generated by sampling without replacement; (3) a hierarchically structured map; and (4) a multifractal map. Additionally actual maps, or maps generated by other programs, may be read into RULE for analysis of spatial patterns.

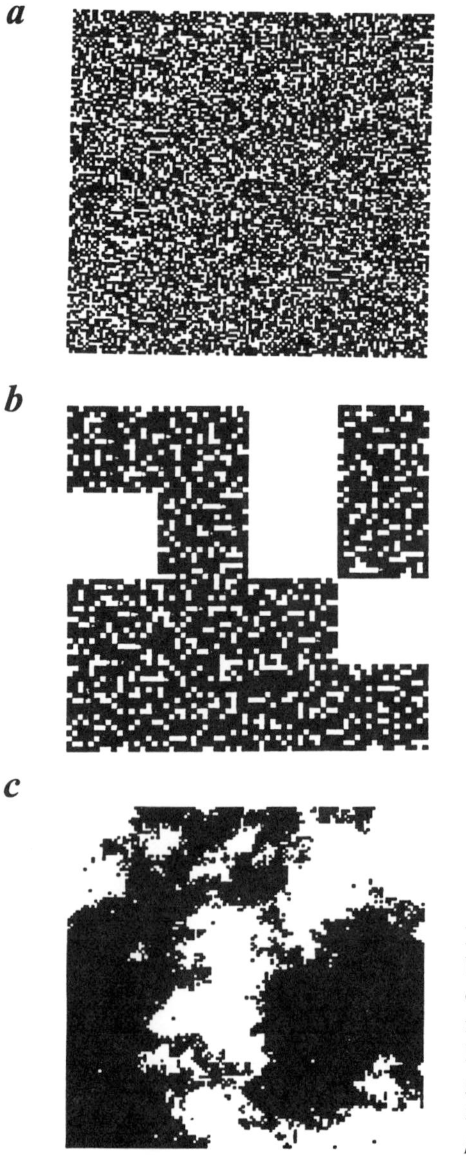

FIGURE 13.1. Three maps generated by RULE with number of rows and columns equal to 128. (A) A random map with p equal to 0.6; (B) A four level curdled map with the p_is as given in Table 13.3A; and (C) A multifractal map with H = 0.5 and p = 0.6.

TABLE 13.1A. Log file for a 16 by 16 simple random map ($N = 10$) with a single habitat type.

Neighborhood Analysis
Map choice: s
Rows × Columns = 16 × 16
Random number seed: −19187176
Rule choice is: 1
Map classes = 1
p (0) = 0.4000
p (1) = 1.0000
N_Reps = 10
Map output choice = S
Size map file: simple.map
Analysis method: R

STATISTICAL SUMMARY ($N = 10$)

Variable	Mean	St.Dev.	C. V.	Minimum	Maximum
Map Type 1					
L.C. size	91.800	29.903	32.574	32.000	135.000
L.C. edge	137.200	44.012	32.079	50.000	210.000
L.C. fractal	1.709	0.035	2.028	1.658	1.769
L.C._rms	5.935	1.024	17.255	3.685	6.958
TTL clusters	11.900	3.725	31.305	6.000	18.000
TTL edges	267.200	15.838	5.927	242.000	286.000
Sav size	66.671	25.999	38.997	22.082	117.218
S_Freq	156.100	4.748	3.042	146.000	164.000
p = 0.6098					
Cor_len	5.802	1.000	17.230	3.629	6.947
Perc/freq	0.400	0.516	129.099	0.000	1.000

Simple random maps are created by specifying the number of rows and columns in the map, the number of habitat types to be generated, and the probabilities associated with each habitat type—including the probability for areas lacking any habitat at all (Table 13.1). The process of map generation is, as the name implies, a simple one. A random number generator is placed within 2 nested ***do*** loops with the variables ***irow*** and ***jcol*** defining the number of rows and columns of the map, ***in_map***, to be generated. If ***pr*** is the probability that any given site in the map will be occupied by the habitat type of interest—then the code sufficient for generating a simple random map with a single habitat type is:

```
do i = 2, irow+1
   do j = 2, jcol+1
      y = ran1 (iseed)
      if (y .le. pr) then
         in_map(i,j) = 1
      else
         in_map (i,j) = 0
      endif
   enddo
enddo
```

TABLE 13.1B. A single realization of a cluster size map produced by the simple random map generation procedure shown in Table 13.1A. The map has 16 rows and columns with 7 cluster sizes identified. See text for a explanation of how this map was generated.

0	6	6	0	7	0	0	1	1	0	0	1	1	0	1	0
7	0	0	7	0	1	0	1	1	0	0	1	0	1	1	1
0	4	4	0	1	1	1	1	0	1	1	1	1	1	1	1
0	0	4	0	0	0	1	1	1	1	1	1	1	0	1	0
4	0	4	0	1	0	1	1	0	1	0	0	1	0	1	1
4	0	0	1	1	0	1	1	1	0	3	3	0	0	0	1
4	0	1	1	0	0	0	0	1	1	0	3	3	0	1	1
4	0	1	1	1	1	1	0	1	1	0	3	0	0	0	1
0	1	1	1	0	0	0	0	1	1	0	3	0	1	0	1
1	1	1	1	1	1	1	1	1	0	7	0	1	1	1	1
0	1	1	1	0	0	0	1	0	5	0	7	0	0	0	1
0	1	1	1	0	7	0	1	0	5	5	0	2	2	2	0
0	1	0	0	2	0	7	0	0	0	0	0	2	0	2	2
0	1	0	2	2	2	0	0	0	0	0	2	2	2	2	2
7	0	2	0	2	2	2	2	2	2	2	0	0	2	2	2
0	0	2	2	2	2	0	2	0	2	2	2	2	2	2	0

Because the generated map is surrounded by a border of −9s to indicate the map edge, the indices for the "do loops" listed above start at 2 and end at 1 plus the number of rows or columns. The pseudorandom number generator used in this example is ***ran1*** (see Press et al. 1992 for a description of this function). ***Ran1*** is used because it is a reliable method of generating random numbers and is easily transported from one computer type to another. ***Ran1*** requires a large, negative integer, ***iseed***, as an initial value. Rerunning the program with the same value for ***iseed*** will produce an identical copy of the results.

The procedure used to generate a simple random map "decides" on the number and location of habitat sites at random on a site by site basis. This procedure does not guarantee that the fraction of the map occupied by the habitat of interest will be exactly equal to the value ***pr*** that the user specified on input. The problem can be annoying when map sizes are small (***irow, jcol*** < 50), but a minor complication as map size increases. A second routine is available in RULE that will produce a map with *exactly* the probabilities specified (Table 13.2). A process of "sampling without replacement" is used to insure that the exact number of habitat sites are located on each map. First the number of sites to be generated is calculated and then these sites are randomly located on the map. Because the algorithm must search for unoccupied sites, the method is more time consuming than the method used to generate simple random maps. However, when comparing a large number of maps the variance in amount of habitat type between maps will be exactly zero (Table 13.2A). One caveat should be kept in mind—the precision of the map generation process is limited by the

TABLE 13.2A. Log file for a 16 by 16 random map ($N = 10$) with a single habitat type generated without replacement.

Neighborhood Analysis
Map choice: r
Rows × Columns = 16 × 16
Random number seed: −19187176
Rule choice is: 1
Map classes = 1
p (0) = 0.4000
p (1) = 1.0000
N_Reps = 10
Map output choice = S
Size map file: random.map
Analysis method: R

STATISTICAL SUMMARY ($N = 10$)

Variable	Mean	St.Dev.	C. V.	Minimum	Maximum
Map Type 1					
L.C. size	102.000	36.588	35.870	50.000	142.000
L.C. edge	156.200	57.290	36.678	70.000	224.000
L.C. fractal	1.727	0.043	2.490	1.640	1.787
L.C._rms	6.110	1.070	17.518	4.422	7.146
TTL clusters	12.300	2.830	23.011	8.000	18.000
TTL edges	271.400	9.336	3.440	262.000	288.000
Sav size	80.853	37.344	46.187	31.338	131.130
S_Freq	154.000	0.000	0.000	154.000	154.000
p = 0.6016					
Cor_len	5.986	1.206	20.144	4.126	7.141
Perc/freq	0.600	0.516	86.066	0.000	1.000

TABLE 13.2B. A single realization of a cluster size map produced by the generation of a random map without replacement shown in Table 13.2A. The map has 16 rows and columns with 10 cluster sizes identified. See text for a explanation of how this map was generated.

0	3	3	3	3	0	6	6	0	0	0	1	0	8	8	8
3	3	3	3	0	0	6	6	0	0	1	1	1	0	0	0
3	0	3	3	3	0	0	6	0	1	1	0	1	1	1	0
0	5	0	3	3	0	2	0	1	1	0	0	1	1	1	0
5	5	5	0	0	2	2	0	1	0	0	1	1	1	0	10
5	5	0	2	0	0	2	0	0	0	1	1	1	0	1	0
0	0	2	2	2	2	2	2	2	2	0	1	1	1	1	1
2	0	2	0	0	2	2	2	2	2	0	1	1	1	0	0
2	0	2	2	0	2	0	2	2	2	0	1	0	1	1	0
2	0	2	0	10	0	2	0	2	0	0	1	1	1	1	1
2	2	2	2	0	2	2	2	0	1	1	1	1	0	1	0
2	2	2	2	2	2	2	0	0	0	1	1	1	0	1	1
0	0	2	0	0	2	0	9	9	0	1	1	0	6	0	0
7	7	0	9	9	0	10	0	0	0	0	1	0	6	0	10
0	7	7	0	0	0	0	0	4	4	4	0	6	6	0	0
10	0	0	9	9	0	4	4	4	0	4	0	6	0	10	0

map dimension. For instance, the precision of the 16 by 16 map example shown in Table 13.2 is: $1/(16 \times 16) = 0.0039$. Although the value requested on input (Table 13.2A) was 0.6, the value obtained for all 10 iterations of the map was equal to 0.6016.

The third type of map that can be generated is a hierarchical, or curdled, map (Fig. 13.1B, Table 13.3). The algorithm used to generate curdled maps is derived from fractal geometry (Mandelbrot 1983). The generation of a curdled map requires that the number of levels (m) of the hierarchy (m is generally between 2–5), the size of each level (l_i, $i = 1, n$), and the probabilities (P_i, $i = 1, n$) that habitat is located within any given level be specified (see O'Neill et al. 1983; Lavorel et al. 1993; Lavorel et al. 1994 and Plotnick et al. 1993 for further details and illustrations of curdled maps). The final dimension of the map is the product of the sizes of all n levels, and the overall fraction of the map occupied by the habitat type of interest is the product of the probabilities for each level. The example given in Table 13.3 is for a 4-level hierarchical map. The size, l_i, of all levels is equal to 2. The top level will be composed of a total of $2 \times 2 = 4$ units, with a probability of occupancy of 0.75. Table 13.3B shows that the first quadrant of the map is unoccupied, but the remaining 3 quadrants are all occupied. Thus, at the coarsest level, the amount of habitat is equal to 0.75. At the second level ($l_i = 2$) the value of p_i is equal to 1.0—thus all subquadrants are fully occupied. At the third level the value of p_i is equal to 0.75, producing a pattern of 2 by 2 units of unoccupied habitat within the larger 4 by 4 units of occupied habitat. At level 4 the value of p_i is 1.0, so all subunits at this scale are occupied. Thus the total probability of map occupancy is equal to $0.75 \times 1.0 \times 0.75 \times 1.0 = 0.5625$. The results shown in Table 13.3A indicate that all 10 maps had exactly 144 sites occupied by the habitat type. The user should be aware that the precision of the results of the curdled map are constrained by the sizes of each level. For example, Table 13.3 shows that results are highly constrained by the small sizes of th l_is. Because each level has only 4 units, the precision is equal to 0.25 (i.e., $1/(2 \times 2)$). The possible dimensions of curdled maps that can be generated are limited by m (currently set to 6) and the maximum value for each level (set ot 25). Although it is theoretically possible to produce a map of dimensions 25^6, the maximum values specified in RULE for the number of rows and columns places an upper limit on map dimensions at 512 rows by 512 columns.

Multifractal maps (Fig. 13.1C) are generated by the midpoint displacement algorithm (MidPointFM2D; Saupe 1988:101) which creates a map of real numbers by successive division, interpolation, and random perturbation. Two parameters are used by this method: the variance associated with the random perturbations, and H, the parameter describing the correlation between points (Table 13.4A). Values of H range between 0.0 and 1.0. Adjustment of the value of H between 0.0 to 1.0 results in maps that range from extremely fragmented to highly aggregated. RULE converts the grid

Table 13.3A. Log file for a 16 by 16 curdled map ($N = 10$) with a single habitat type.

Neighborhood Analysis
Map choice: c
Rows × Columns = 16 × 16

Curds (levels = 4)	N	P
	2	0.750000
	2	1.00000
	2	0.750000
	2	1.00000
Geo. Totals	16	0.562500

Random number seed: −19187176
Rule choice is: 1
Map classes = 1
N_Reps = 10
Map output choice = S
Size map file: curd.map
Analysis method: R

STATISTICAL SUMMARY ($N = 10$)

Variable	Mean	St.Dev.	C. V.	Minimum	Maximum
Map Type 1					
L.C. size	136.800	15.179	11.096	96.000	144.000
L.C. edge	115.600	15.714	13.594	88.000	140.000
L.C. fractal	1.791	0.021	1.163	1.761	1.837
L.C._rms	6.910	0.428	6.193	5.856	7.277
TTL clusters	1.300	0.483	37.157	1.000	2.000
TTL edges	123.200	10.799	8.766	108.000	140.000
Sav size	133.200	20.810	15.623	80.000	144.000
S_Freq	144.000	0.000	0.000	144.000	144.000
p = 0.5625					
Cor_len	6.872	0.522	7.590	5.529	7.277
Perc/freq	1.000	0.000	0.000	1.000	1.000

Table 13.3B. A single realization of a cluster size map produced by the curdled map generation procedure shown in Table 13.3A. The map has 16 rows and columns with 2 cluster sizes identified. See text for a explanation of how this map was generated.

0	0	0	0	0	0	0	0	1	1	0	0	1	1	1	1
0	0	0	0	0	0	0	0	1	1	0	0	1	1	1	1
0	0	0	0	0	0	0	0	1	1	1	1	0	0	1	1
0	0	0	0	0	0	0	0	1	1	1	1	0	0	1	1
0	0	0	0	0	0	0	0	1	1	0	0	1	1	1	1
0	0	0	0	0	0	0	0	1	1	0	0	1	1	1	1
0	0	0	0	0	0	0	0	1	1	1	1	0	0	1	1
0	0	0	0	0	0	0	0	1	1	1	1	0	0	1	1
0	0	2	2	2	2	0	0	1	1	1	1	1	1	1	1
0	0	2	2	2	2	0	0	1	1	1	1	1	1	1	1
2	2	2	2	2	2	2	2	0	0	1	1	0	0	1	1
2	2	2	2	2	2	2	2	0	0	1	1	0	0	1	1
2	2	0	0	2	2	0	0	1	1	1	1	1	1	1	1
2	2	0	0	2	2	0	0	1	1	1	1	1	1	1	1
2	2	2	2	2	2	2	2	0	0	1	1	1	1	0	0
2	2	2	2	2	2	2	2	0	0	1	1	1	1	0	0

TABLE 13.4A. Log file for a 16 by 16 fractal map ($N = 10$) with a single habitat type.

Neighborhood Analysis
Map choice: m
Maxlevel 4 H = 0.500000
Rows × Columns = 16 × 16
n Wrap = F
Random number seed: −19187176
Rule choice is: 1
Map classes = 1
p (0) = 0.4000
p (1) = 1.0000
N_Reps = 10
Map output choice = S
Size map file: fractal.map
Analysis method: R

STATISTICAL SUMMARY ($N = 10$)

Variable	Mean	St.Dev.	C. V.	Minimum	Maximum
Map Type 1					
L.C. size	139.800	14.078	10.070	122.000	163.000
L.C. edge	101.800	22.240	21.847	72.000	130.000
L.C. fractal	1.813	0.044	2.410	1.741	1.862
L.C._rms	6.570	0.615	9.365	5.728	7.718
TTL clusters	6.100	2.234	36.616	1.000	9.000
TTL edges	136.800	23.649	17.287	108.000	174.000
Sav size	128.004	20.897	16.325	101.953	163.000
S_Freq	154.400	6.867	4.448	142.000	165.000
p = 0.6031					
Cor_len	6.550	0.614	9.374	5.681	7.714
Perc/freq	0.800	0.422	52.705	0.000	1.000

TABLE 13.4B. A single realization of a cluster size map produced by the fractal map generation procedure shown in Table 13.4A. The map has 16 rows and columns with 4 cluster sizes identified. See text for a explanation of how this map was generated.

1	1	1	1	0	0	0	0	0	0	0	0	0	0	0	3
1	1	1	1	0	0	0	0	0	0	0	0	0	0	0	3
1	1	1	1	1	1	0	0	0	0	0	0	0	0	0	0
1	1	1	1	1	1	0	0	0	0	0	0	0	0	4	0
1	1	0	1	1	1	1	0	0	0	0	0	0	0	0	0
1	1	1	1	1	1	0	1	1	0	4	0	0	0	0	0
1	1	1	1	1	1	0	0	1	0	0	0	4	0	2	2
1	1	1	1	1	1	1	1	1	0	1	0	0	0	2	2
1	1	1	1	1	1	1	1	1	1	1	0	0	0	0	2
1	1	1	1	1	1	1	1	1	1	0	0	0	0	0	2
1	1	1	1	1	1	1	1	1	1	0	0	0	0	0	0
1	1	1	1	1	1	1	1	1	1	1	0	0	0	0	0
1	1	1	1	1	1	1	1	1	1	1	1	1	0	0	0
1	1	1	1	1	1	1	1	1	1	1	0	0	0	0	0
1	1	1	1	1	1	1	1	1	1	1	0	0	0	0	0
1	1	1	1	1	1	1	1	1	1	1	0	0	4	0	0

of real numbers to integers representing the proportion of habitat types specified by the vector p (see description of probabilities below). Maps with gradients of habitat change and/or maps that are "wrapped" (i.e., opposite edges have similar habitat patterns) can be produced by specifying the appropriate options for the multifractal maps (Table 13.5).

Maps created by other programs, or produced from existing data sets, can be input into RULE for analysis of pattern. Rule assumes that these maps are ASCII integer values with a blank space delimiting adjacent sites.

Specification of the Neighborhood Rule

A unique feature of RULE is the ability of the user to specify the neighborhood rule used to identify clusters of habitat. The program uses the neigh-

TABLE 13.5A. Log file for a 16 by 16 fractal map with a gradient (N = 10) with a single habitat type.

Neighborhood Analysis
Map choice: g
Maxlevel 4 H = 0.500000
Rows × Columns = 16 × 16
n Wrap = F
−1.00000
−1.00000
1.00000
1.00000
Random number seed: −19187176
Rule choice is: 1
Map classes = 1
p (0) = 0.4000
p (1) = 1.0000
N_Reps = 10
Map output choice = S
Size map file: gradient.map
Analysis method: R

STATISTICAL SUMMARY (N = 10)

Variable	Mean	St.Dev.	C. V.	Minimum	Maximum
Map Type 1					
L.C. size	150.100	10.577	7.047	132.000	164.000
L.C. edge	92.400	20.844	22.559	62.000	126.000
L.C. fractal	1.849	0.048	2.614	1.761	1.912
L.C._rms	6.493	0.510	7.851	6.056	7.755
TTL clusters	5.400	1.897	35.136	2.000	9.000
TTL edges	119.600	26.796	22.405	78.000	162.000
Sav size	141.027	15.717	11.145	113.599	163.012
S_Freq	160.700	6.360	3.958	147.000	170.000
p = 0.6277					
Cor_len	6.482	0.508	7.832	6.050	7.752
Perc/freq	0.900	0.316	35.136	0.000	1.000

Table 13.5B. A single realization of a cluster size map produced by the fractal map with gradient procedure shown in Table 13.5A. The map has 16 rows and columns with 2 cluster sizes identified. See text for a explanation of how this map was generated.

1	1	1	1	1	1	0	0	0	0	0	0	0	0	0	0
1	1	1	1	1	1	0	0	0	0	0	0	0	0	0	2
1	1	1	1	1	1	0	0	0	0	0	0	0	0	0	0
1	1	1	1	1	1	1	0	0	0	0	0	0	0	0	0
1	1	1	1	1	1	1	1	0	0	0	0	0	0	0	0
1	1	1	1	1	1	1	1	1	0	2	0	0	0	0	0
1	1	1	1	1	1	0	1	1	1	0	0	2	0	2	0
1	1	1	1	1	1	1	1	1	1	1	0	0	0	0	0
1	1	1	1	1	1	1	1	1	1	1	0	0	0	0	0
1	1	1	1	1	1	1	1	1	1	1	0	0	0	0	0
1	1	1	1	1	1	1	1	1	1	1	0	0	0	0	0
1	1	1	1	1	1	1	1	1	1	1	0	0	0	0	0
1	1	1	1	1	1	1	1	1	1	1	1	1	0	0	0
1	1	1	1	1	1	1	1	1	1	1	0	0	0	0	0
1	1	1	1	1	1	1	1	1	1	1	0	0	0	0	0
1	1	1	1	1	1	1	1	1	1	1	0	0	0	0	0

borhood rule to recursively search the area around each site for other similar habitat sites. When similar habitat sites are found, these are labeled as cluster members and the process is iterated for these additional sites. The algorithm used to identify clusters has undergone extensive development, testing and optimization. It is extremely fast at identifying clusters when cluster sizes are small, but when clusters become very large the recursive method can be much slower than brute force methods. The slow execution speed will be most noticeable when p, the probability of site occupancy, exceeds the critical threshold for that neighborhood rule (see Plotnick and Gardner 1993 for a discussion of the relationship between the neighborhood rule and critical thresholds).

Three neighborhood rules are built into the program. Rule 1 searches the four neighbors located in the cardinal directions (the nearest neighbor case). Rule 2 searches the four neighbors located in the cardinal directions and the four neighbors in the diagonal directions (the next-nearest neighbor case for a total of 8 neighbors). Rule 3 investigates all neighbors of Rule 2, but also adds 4 neighbors in the cardinal direction that are one site removed from the nearest neighbors (a total of 12 neighbors). The user can choose "Rule 4" as an option, which causes the program to query the user for the number of neighbors and relative coordinates that define the rule. For instance, the input to duplicate the nearest neighbor rule would specify 4 neighbors with the four coordinates of (0, 1), (1, 0), (0, −1), and (−1, 0). Note that actual program input does not require parentheses or commas. The current version of RULE allows the specification of up to 50 neighbors.

Specifying Probabilities for Multiple Habitat Types

All maps, except curdled maps, allow multiple habitat types to be specified. The generation of maps with multiple habitat types requires the specification of probabilities for each habitat type, including a probability that a site will not be occupied by any habitat at all (i.e., an empty site). Confusion can be avoided by remembering that the vector of probabilities starts with p_0 which equals the probability of a site being unoccupied. Obviously the vector must sum to 1.0. The program actually converts the input values to a cumulative frequency distribution and normalizes to 1.0. Thus, to specify a simple random map with 2 habitat types that each occupy 1/3 of the map (leaving 1/3 of the map blank) we can type in the values 3, 3, 3 and the log file will show that the values have been normalized to 0.333, 0.666 and 1.0.

Input–Output of Maps

RULE can input maps from other programs for spatial analysis. Random maps generated by RULE can also be output for analysis or display by other programs. All maps are read and written as space delimited ASCII files. Several choices are offered for map output: (1) the random maps generated by RULE may be output with each habitat type labeled with consecutive integer values; (2) a second type of output allows patches of a single habitat type to be uniquely labeled without regard to size; or (3) output with clusters sequentially labeled by size from the largest (a label of 1) to smallest. (Map types 2 and 3 are generated as RULE analyzes maps and are primarily useful for display purposes only.) Tables 13.1B, 13.2B, 13.3B, and 13.4B show the result of labeling cluster sizes. A blank line is inserted between each map when multiple maps are generated by RULE. Input of multiple maps to RULE requires an insertion of a blank line between maps.

Visualization of Maps

A color display of the generated map and the clusters associated with each habitat type can be displayed by RULE. Two color maps are simultaneously shown on the screen. The map in the lower left corner shows the different habitat types within a single map while the map in the upper right corner iteratively displays the different patches within a each habitat type. The colors selected by RULE are intended to best distinguish between habitats and patches and are not under the control of the user.

The analysis and display of maps can be temporarily interrupted if the user strikes the "escape" key. A menu appears allowing the user to either continue the analysis and display, print the color image to a disk file, or abort the current program analysis. The viewed image may be saved in several different formats, including: 1 = HP-GL; 2 = postscript; 3 = Acorn Draw format; 4 = raster printer; 5 = Tektronix 4041; 6 = Raster image file or "bmp" format; 7 = Lotus PIC format; 8 = AutoCad DXF format; and 9 = CGM format. RULE will write up to 10 files that are consecutively labeled as "hardcp1.ps" through "hardcp10.ps." The user should rename files after each program execution or they will be written over the next time that RULE is run.

The graphical library of programs used by RULE for the display of generated maps was developed by INTERACTER (Interactive Software Services, Ltd., Huntington, UK). Although source code is given in the distribution file, distrib zip, the user will not be able to change and recompile the program unless the compilier has access to the INTERACTER libraries.

Data Files Produced by RULE

RULE uses a variety of data sets for input and output of maps and documentation of the analysis results. A log file, "rulelog.dat", is written each time RULE is executed. This file records the input-output stream, and gives a listing of the analysis results. Examples of this file are given in Tables 13.1 to 13.4. Two other data files, "cfd_scr.dat" and "lacun.dat" are written during each program execution. The purpose of the "cfd_scr.dat" file is to provide information on each cluster identified for each habitat type, including the size of the cluster, the amount of edge, the fractal dimension, and the mean-squared radius (**rms**) of the cluster. A computational trick is used to estimate the fractal based on the average length of a box needed to contain the cluster of. The code for calculation of the fractal is:

```
red = float(mxrw + mxcl)  /  2.0
rsz = float(size)
if (red .gt. 1 .and. rsz .ge. 1)      then
    fract = log(rsz)  /  log(red)
 else
   fract = 0.0
endif
```

where **red** is the average of the number of rows and columns (**mxrw** and **mxcl**, respectively) needed to enclose the box and **rsz** is the size of the cluster (the **float** function converts integer values to real numbers). The ratio of the logs of **rsz** and **red** provides an unbiased estimate of the fractal dimension. The fractal is set to zero if the cluster is composed of a single site. As we will see later, only the fractal of the largest cluster is reported in the statistical summary (Tables 13.1–13.4) because average values tend to be biased by the large number of samll clusters. An important rule of thumb is that the fractal dimension of any cluster that has fewer than 50 sites will be strongly influenced by the shape of the grid used to represent each site (a square grid has a fractal dimension of 2). The user can be convinced of this fact by using the results in "cfd_scr.dat" to plot the fractal dimension against cluster size.

The **rms** is equivalent to the radius of gyration estimated as: **rms** = $[\mathbf{sumr}^2 / \mathbf{size}^2]^{1/2}$, where $\mathbf{sumr}^2$ is one half the squared sum of the distance between sites within a cluster and $\mathbf{size}^2$ is the squared cluster size. The **rms** is a useful index of the average distance between sites within a cluster. Clusters that are highly fragmented will have a larger **rms** than when sites are more densely packed.

The results of a lacunarity analysis are written to a file "lacun.dat" (see the next section for a description of this file). The results of each run of the RULE program can be permanently saved by renaming the "rulelog.dat" and "lacun.dat" with unique names.

Map Analysis

The analysis of map pattern is statistically summarized by a listing to the screen and to "rulelog.dat" of the mean, standard deviation (St.Dev.), coefficient of variation (C. V.), minimum and maximum values of 10 indices (see Tables 13.1–13.4 for examples). These indices are: (1) The size of the largest cluster (L.C. size), (2) the amount of edge of the largest cluster (L.C. edge), (3) the fractal dimension of the largest cluster (L.C. fractal), (4) the mean-squared radius of the largest cluster (L.C._rms), (5) the total number of clusters (TTL clusters), (6) the total amount of edge on the map (TTL edge), (7) the area weighted average cluster size (SAV size), (8) the total number of sites of a given habitat type (S_Freq), (9) the correlation length (cor_len), and (10) the frequency of map percolation (Perc/freq).

The emphasis of the statistical output on characteristics of the largest cluster is important. When the fraction of the landscape occupied by a particular habitat type is above the critical threshold[1] the cumulative frequency distribution of cluster attributes is dominated by a single, large cluster. That is, the distribution of cluster sizes is highly skewed. Simple averages of skewed distributions may result in a biased characterization of the pattern. Therefore, it is useful to note the frequency of the maps that percolate (Gardner et al. 1989) and compare the relative size of the largest cluster against the mean values to check for possible bias in the results.

---

[1]The critical thresholds for the nearest-neighbor rule is 0.5928; the next-nearest neighbor rule is 0.4073, and the third-nearest neighbor rule is 0.292 (Plotnick and Gardner 1993).

SAV (also labeled "Sav size" in the RULE output) is calculated as the area weighted average cluster size: $SAV = \Sigma S_i^2/\Sigma S_i$, where S_i is the size of cluster ***i***, and the sums are taken over all clusters of the same habitat type. The correlation lengthe quals the square root of the ratio of two sums, $\Sigma RMS_i/\Sigma S^2$, where RMS_i is the mean-squared radius of the *i*th cluster and ΣS^2 is the squared sum of the cluster sizes. The correlation length provides a measure of the average within-cluster distances for the entire map. One may think of this statistic as the average distance an organism might move without leaving a habitat cluster.

One option in RULE is to perform (or not perform) a lacunarity analysis. This option is provided to the user because lacunarity analysis can be extremely time-consuming—especially if the map size is large. Because of the novelty of lacunarity analysis, an in-depth description of this technique is presented here. Mandelbrot (1983) first suggested a method, that he termed *lacunarity analysis*, for the analysis of distributions of "gaps" or "holes" within a geometric structure. More recently, Allain and Cloitre (1991) described a simple algorithm for calculating *lacunarity* that uses a series of "gliding boxes" of increasing size to sample the spatial pattern of gridded data. This method is used to analyze gridded data by placing a "gliding box" with linear dimension of r, and area of r^2, at the upper left corner of the grid and counting the number of sites occupied by the object of interest within that "gliding box". The box is then moved one column to the right and the count repeated. The process continues until all the columns of the top set of r rows have been sampled. The box is then returned to the left edge of the grid, but moved one row down from the starting position. Counting continues until all rows and columns have been sampled and the counts summarizes as a frequency distribution, Q. $N(r)$, the number of gliding boxes of size r that can be placed on a grid with linear dimension of M, will be equal to $(M - r + 1)^2$, and the intervals of Q will range from 1 to r. The frequencies for each interval, Q_i, $(i = 1,r)$, can range from 0 (no samples contained exactly i sites) to $N(r)$ (all samples contained exactly i sites). The frequency distribution, Q, is converted to a probability distribution, q, by dividing each frequency by $N(r)$ and the first and second moments ($Z^{(1)}$ and $Z^{(2)}$, respectively) are calculated as $Z^{(1)} = \sum_{i=1}^{r} iq_i$, and $Z^{(2)} = \sum_{i=1}^{r} i^2 q_i$. The value of lacunarity, $\Lambda(r)$, for box size r is estimated by the ratio $Z^{(2)}/(Z^{(1)})^2$. Lacunarity values are similar to the variance-mean ratio, which can be calculated as $[Z^{(2)} - (Z^{(1)})^2]/Z^{(1)}$. Additional details concerning the calculation of the lacunarity index can be found in Plotnick et al. (1993)

The gliding box sampling is the most intensive, nonredundant sampling possible and differs substantially from methods that strive for independent, non-overlapping samples of spatial data. For example, a grid with $M = 100$ and $r = 10$ will result in $(M - r + 1)^2 = 8{,}281$ gliding box samples. However, the TTLQV (Ludwig and Reynolds 1989) or quadrat variance methods (Dale and MacIsaac 1989) will produce only $(M/r)^2 = 100$ quadrat samples.

Another way to visualize the difference is that each gliding box samples a unique combination of contiguous rows and columns resulting in each site being sampled by r gliding boxes, while the TTLQV and quadrate variance methods are restricted to sampling each site a single time.

RULE samples a grid with box sizes, r, ranging from 1 to $M/2$. The lacunarity results are listed in "lacun.dat" and plots of Λ against r illustrate the changes in lacunarity with changes in scale. The "lacun.dat" is composed of 4 columns of data. The first column is the box size, r, the second column is the untransformed lacunarity value, Λ, the third column is the normalized lacunarity value (all values of Λ divided by the value of Λ when $r = 1$), and the fourth column is the variance/mean ratio. The characteristic changes in Λ with change in scale, and interpretation of the results, are given in Plotnick et al. (1993). It is possible to perform a lacunarity analysis on gridded data that are non-rectangular (i.e., the boundaries of the grid are irregular). This is accomplished by placing the irregular data within a larger grid with missing values (sites outside the irregular grid) labeled with negative values. During the lacunarity analysis, all quadrats with sites having negative values will be ignored.

Quantitative Issues and Pitfalls

A number of quantitative issues arise when analyzing landscape patterns. A particularly serious one is the truncation effect of the map boundary on the number, size and shape of habitat patches. Gardner et al. (1987) showed that for simple random maps the size of the largest cluster is severely impacted by the extent of the map—especially when p, the fraction of the map occupied by the habitat type of interest, is greater than the critical threshold. This is, of course, predicted from theory. What is not well appreciated is that the size of the largest cluster below the critical threshold is also impacted by boundary effects. The rule of thumb seems to be that the bigger the map the better the results will be. It is quite clear that results are unreliable when map size is less than 100 rows by 100 columns (10,000 sites).

It is very useful to generate many iterations of a map type to make quantitative comparisons against actual landscape patterns. It is probably true that landscape patterns will never be random. However, the statement that the attributes of a single landscape ($N = 1$) lies outside the 95% confidence limits of a set of random maps is much more useful than simply stating that the landscape demonstrates unique properties. This is, of course, the basis of statistical comparison—the testing of results against a null hypothesis. RULE was specifically designed to perform these kinds of tests.

It is also clear from running a variety of maps of different types that the most important variable in explaining pattern is p, the fraction of the map occupied by the habitat type of interest. Assuming that the total extent of the map is the same, the variance in pattern with change in p is clear

(Gardner et al. 1987). However, it is possible to produce dramatically different results by imposing a structure on the map. The curdled and fractal maps allow comparison against random maps while holding p constant. Unfortunately, many examples of analysis of landscape pattern have violated the 3 principles of comparison: (1) map grain and extent must be the same; (2) the maps must be of sufficient size so that boundary effects are minimized; and (3) comparisons must account for differences in p. Only when these conditions are met should one begin to investigate the effects of differences in pattern on ecological processes.

Program Listing and Executables

The source code listing for RULE is contained in the distribution package as distrib.zip. This package is available from http://www.al.umces.edu//wwwz/download. The files may be extracted using PKZIP/PKUNZIP (http://www.pkware.com/l) utilities by typing:

pkunzip distrib.zip.

The files that will be extracted by PKZIP include 10 FORTRAN source files for rule (genmap.f90, hfract.f90, input.f90, lacun.f90, main.f90, mapanl.f90, module.f90, ranl.f90, stats.f90, and visual.f90), 3 script files for creating fractal, curdled and random maps, the associated log files produced by running the script files, and a single batch file, RUNALL.BAT, which will automatically run all three script files.

The executable. RULE.EXE, can generate and analyze maps as large as 512 rows and columns. A change in the size of the map to be analyzed, and consequently of the amount of memory required to run RULE, can be adjusted by editing the value of **maxprm** in module.f90. The appropriate value for **maxprm** should be equal to the maximum number of rows or columns (whichever is larger) plus 2. Thus, to analyze a map as large as 512 by 512, **maxprm** is set equal to 514.

Two calculations are time particularly time consuming, and the user should expect long execution times when map sizes are large. The calculation of **rms** requires the location and calculation of all distances within a cluster. Time for this calculation increases as the square of the size of the largest cluster. The generation of random maps without replacement (versus the simple random map) is also time consuming. Time for a single map will increase as the square of the map size. When maps sizes are large, (i.e., >300 by 300) the improvement offered by this method is probably marginal.

Use of Script Files

It is possible to use script files to increase the efficiency or running RULE. Script files are ASCII files that contain all the "answers" required to run

RULE (see example script files in distrib.zip). To run RULE using a script simply type

rule < “script file name”

For instance, to run the script file to generate a random map type

rule < random.scr

MS-DOS also allows commands to be imbedded into batch files. To run the batch file that automatically runs the three included script files, maps, type

runall

Using the above scheme, the user can generate numerous script files and then automate the generation and analysis of maps by RULE by an appropriate batch file. The author frequently sets up such files when execution times are long (it is satisfying to know your computer is working hard while you are enjoying other activities).

Example Class Exercise Using RULE

The following exercise was developed for a graduate class in landscape ecology. The purpose was to help students understand what a neutral model is, how it is generated, inferences one might draw from these models, and how to identify pitfalls and problems of map analysis. If you are uncertain about any of these processes, then performing this exercise will be instructive. The necessary data files (leclass.xls and leclass.txt) can be found within distrib.zip. Problem 1 uses a random number table—something which has become rather rare. The reader may wish to substitute a hand calculator with a random number function if a random number table can not be located. Also included in distrib.zip is the program that does the renormalization for you (Problem 3).

Class Exercise #1: Neutral Models and Landscape Pattern

Components of handout:
1. This sheet describing objectives and procedures.
2. A page of random numbers.
3. Computer diskette containing:
   RULE executable (rule.exe).
   Documentation for RULE (rule.txt).
   Example script file for running rule (random.scr).
   Data file in Excel spreadsheet format (leclass.xls) or ASCII format (leclass.txt).

Objectives are to:
1. Gain practical experience using random processes to generate pattern.
2. Explore statistical relationships among pattern descriptors.
3. Determine the effect of changes in scale (i.e., grain and extent) on the analysis of landscape data.
4. To quantify the effect of neighborhood rules on pattern description.

This exercise can be performed individually or in groups of 2 to 3 (no larger). It is advisable that at least one member of the group have experience with IBMH/personal computers and some level of programming ability. The product of your efforts will be a written report concerning relationships between random processes and observed pattern. Report should emphasize how these result are useful in interpreting actual landscape patterns. Final report should be no longer then 10 pages (excluding references) typed double-spaced, standard margins, with a font no smaller then 11 point. Report must be on paper, no E-mail. Grade will be based on completeness and creativity of analysis.

Problem #1: Hand Generation of a Random Map

It is quite likely that no one in this class has ever used a random number table. It would be a shame if you were to miss out on this experience. Examine the attached random number table which contains 500 random numbers. This table can be used to generate many thousands of numbers by a very simple process. The trick is to: (a) pick a random starting point and (b) a random direction for reading the numbers (up, down, left, right—or even diagonal directions, but we'll leave that for the advanced lessons). One way to do this is to put your finger at random (eyes closed) on the paper and read the number. For example: 86490. Taking the first 3 digits (864), find the 8th column, 6th row, and read down to the 4th number. The number is 53205! The direction is determined by the 4th number of the original sequence (the 9 of 86490). We use modular arithmetic to find the direction with 0 = left to right (traditional); 1 = north to south; 2 = right to left; and 3 = south to north. The result of mod(9,4) is 1, so we will read north to south. Thus, the first number is 53205, the second is 68257, the third is 82295, and so on. Easy. Now let us put this to practical use.

To generate a simple random map, make a grid with 5 rows and 5 columns for a total of 25 sites. We will wish to fill this grid with 0s (no forest) or 1s (forested land) with a given probability, p. Let's set $p = 0.4$, then 40% of the map should be filled with 1's and 60% with 0's. Using the random number table we imagine that there is a decimal in front of all the numbers. Thus the first number (53205) is really 0.53205. Because this number is greater than p the first site in the map is set to zero. Reading numbers quickly we see that the next 7 sites are all 1's, and not until the 8th number (24595) do we find a value $< p$, allowing us to set this site to 1!

Procedure: Using the above procedure, create 4 25 $\times$ 25 maps with p = 0.2, 0.4, 0.6, and 0.8. For each map count the number of clusters, the frequency distribution of cluster sizes, the number of edges, and the total number of clusters that touch the border of the map. Report your results in graphical form, including in the plot the numbers for p = 0.0 and 1.0. Explain the results and be sure to include the definitions you used for an edge and a cluster.

Problem #2: Statistical Relationship Among Pattern Descriptors

There is data set on the disk in 2 forms: an ASCII file (leclass.txt) or an Excel file (leclass.xls). These numbers were generated using RULE for 3 map types and the output was placed in a spreadsheet. An explanation of the variables can be found in the documentation for RULE (rule.txt). Using graphical and statistical procedures explore the differences between the map types as well as relationships among the pattern descriptors. There is a lot going on here! This is your chance to be creative. Hints: (1) Don't miss the obvious; (2) Not all information is provided here, but additional information can be obtained by running RULE. (3) These maps are large, so analysis by RULE is time consuming—therefore, explore relationships with maps no larger than 256 by 256. (4) In writing up the results, be sure to cite the appropriate literature.

Problem #3: Changes in Scale

Before attempting this exercise become familiar with RULE, especially the generation of different landscape types and the import/export of files.

A. *Changing grain size.* Here is an exercise that is a bit more challenging. Using RULE generate a random map with:

a. Dimensions 512 by 512.
b. 4 habitat types with the *p*s = 0.1, 0.2, 0.3, and 0.4 ($p_{(0)}$ = 0.0).
c. Output 3 maps (N = 3) to text file.

Then (here is the tricky part) design a program to change the grain size of the map by a factor of 2 using either a majority rule (i.e., the aggregate site is set equal to most common cover type) or a random rule (i.e., the aggregate site is randomly chosen from 1 of the 4 unaggregated sites). For instance, imagine that the pattern of an unaggregated set of 4 sites is: [1,2/2,2], then the majority rule would result in an aggregated site of [2] while the random rule would result in [2] with a probability of 0.75 and a [1] with probability 0.25. The probability of obtaining a 3 or a 4 for the random rule would be 0.0.

By setting the number of iterations in RULE to 3, you will be generating 3 maps which can then be read by your renormalization program. If the renormalization program saves the new maps in a format which can be read by RULE, then the resulting patterns can be analyzed. Perform 3 iterative levels of the aggregation process to transform the original 512^2 map into a 256^2 map, then transform the 256^2 map into a 128^2 map and finally transform the 128^2 map into a 64^2 map.

Repeat the sequence with a multifractal map with $H = 0.5$. Explain the results by using graphical and statistical methods and appropriate references to the literature.

B. *Change map extent.* Use RULE to run maps of different sizes (64, 128, 256, and 512) with the same parameters for random and multifractal maps as described in part A. How do these maps differ from those involving changes in grain size? Hint: RULE measures maps in terms on the number of rows and columns. However, each site for a real map will have some dimensional unit associated with it (i.e., meters, kilometers, etc.). Does the A activity affect the dimensional units in the same manner as the B activity?

Problem #4: Quantify the Effect of Different Neighborhood Rules on the Analysis of Pattern

For this problem, we wish to observe the effect of p, landscape pattern, and the neighborhood rule used to identify clusters. This requires a factorial sampling design.

The unique feature of RULE is the specificaiton of the "rule" used to define clusters (see documentation). Three different neighborhood rules are "wired" in the program. Run RULE for 7 iterations ($N = 7$), rows and columns = 256 and analysis of pattern with each of the 3 neighborhood rules for:

a. Random maps with $p = 0.4, 0.6, 0.8$.
b. Multifractal map (levels = 8, H = 0.5) and $p = 0.4, 0.6, 0.8$.

The total number of maps will be (2 map types) × (3 neighborhood rules) × (3 probability levels) × (7 iterations) = 126 maps. However, the total number of runs of RULE will only be (2 map types) × (3 neighborhood rules) × (3 probability levels) = 18 separate runs of RULE. (Time saving hint: create 18 script files for RULE and run the whole sequence with a single batch file).

Compare and contrast the effect of changes in grain and extent (Problem 3) with the changes in neighborhood rules. Explain the similarity and differences between these two different concepts of scale and the practical implications for analysis of landscape pattern. That is, how should what was learned here be applied to the actual analysis of landscapes?

Acknowledgments. The ideas and methods used in RULE developed over several years of collaboration with many individuals. I would like to especially acknowledge Bob O'Neill, Roy Plotnick, Monica Turner, Sandra Lavorel, and Bruce Milne. In addition, the comments of Curt Flather, who helped debug an earlier version of visualization methods used by RULE, are greatly appreciated. Support for the preparation of this document was provided in part by funding from the U.S. Environmental Protection Agency under contract R819640 to the Center for Environmental and Estuarine Studies, University of Maryland System and by NSF EAR-9506606 to the University of Maryland Center for Environmental and Estuarine Studies.

References

Allain, C., and M. Cloitre. 1991. Characterizing the lacunarity of random and deterministic fractal sets. Physical Review A, 44:3552–3557.

Dale, M.R.T., and D.A. MacIsaac. 1989. New methods for the analysis of spatial pattern in vegetation. Journal of Ecology 77:78–91.

*Dale, V.H., H. Offerman, R. Frohn, and R.H. Gardner. 1995. Landscape characterization and biodiversity research. In: Measuring and Monitoring Biodiversity in Tropical and Temperate Forests, Proceedings of a IUFRO Symposium helt at Chiang Mai, Thailand, August 27–September 2, 1994. T.J.B. Boyle and B. Boontawee (eds.). Center for International Forestry Research, Bogor, Indonesia.

*Gardner, R.H., and R.V. O'Neill. 1991. Pattern, process and predictability: the use of neutral models for landscape analysis. In: Quantitative Methods in Landscape Ecology, pp. 289–307. M.G. Turner and R.H. Gardner (eds.). Springer-Verlag, New York.

*Gardner, R.H., B.T. Milne, M.G. Turner, and R.V. O'Neill. 1987. Neutral models for the analysis of broad-scale landscape pattern. Landscape Ecology 1:19–28.

*Gardner, R.H., R.V. O'Neill, M.G. Turner, and V.H. Dale. 1989. Quantifying scale-dependent effects of animal movement with simple percolation models. Landscape Ecology 3:217–227.

*Gardner, R.H., M.G. Turner, R.V. O'Neill, and S. Lavorel. 1992a. Simulation of the scale-dependent effects of landscape boundaries on species persistence and dispersal. In: The Role of Landscape Boundaries in the Management and Restoration of Changing Environments, pp. 76–80. M.M. Holland, P.G. Risser, and R.J. Naiman (eds.). Chapman & Hall, New York.

*Gardner, R.H., V.H. Dale, R.V. O'Neill, and M.G. Turner. 1992b. A percolation model of ecological flows. In: Landscape Boundaries: Consequences for Biotic Diversity and Ecological Flows, pp. 259–269. A.J. Hansen and F. di Castri (eds.). Springer-Verlag, New York.

*Gardner, R.H., R.V. O'Neill, and M.G. Turner. 1993a. Ecological implications of landscape fragmentation. In: Humans as Components of Ecosystems: Subtle Human Effects and the Ecology of Populated Areas, pp. 208–226. S.T.A. Pickett and M.J. McDonnell (eds.). Springer-Verlag, New York.

*Gardner, R.H., A.W. King, and V.H. Dale. 1993b. Interactions between forest harvesting, landscape heterogeneity, and species persistence. In: Modeling Sus-

---

The asterisk next to a citation indicates that paper either used RULE (or a precursor of RULE) for the analysis of spatial patterns.

tainable Forest Ecosystems. D.C. LeMaster and R.A. Sedjo (eds.). Proceedings of a 1992 workshop in Washington, DC. Published by American Forests, Washington, DC.

*Lavorel, S., R.H. Gardner, and R.V. O'Neill. 1993. Analysis of patterns in hierarchically structured landscapes. Oikos 67:521–528.

*Lavorel, S., R.H. Gardner, and R.V. O'Neill. 1995. Dispersal of annual plants in hierarchically structured landscapes. Landscape Ecology 10:277–289.

*Lavorel, S., R.V. O'Neill, R.H. Gardner, and J.B. Burch. 1996. Coexistence of annual plant species with different spatio-temporal dispersal strategies in a structured landscape. Oikos.

Ludwig, J.A., and J.F. Reynolds. 1988. Statistical Ecology. John Wiley & Sons, New York.

Mandelbrot, B.B. 1983. The Fractal Geometry of Nature. Freeman, New York.

*O'Neill, R.V., B.T. Milne, M.G. Turner, and R.H. Gardner. 1988a. Resource utilization scales and landscape pattern. Landscape Ecology 2:63–69.

*O'Neill, R.V., J. Krummel, R.H. Gardner, G. Sugihara, B. Jackson, D.L. DeAngelis, B. Milne, M.G. Turner, B. Zygmutt, S. Christensen, R. Graham, and V.H. Dale. 1988b. Indices of landscape pattern. Landscape Ecology 1:153–162.

*O'Neill, R.V., R.H. Gardner, B.T. Milne, M.G. Turner, and B. Jackson. 1991. Heterogeneity and spatial hierarchies. In: Ecological Heterogeneity, pp. 85–96. J. Kolasa and S.T.A. Pickett (eds.). Springer-Verlag, New York.

*O'Neill, R.V., R.H. Gardner, M.G. Turner, and W.H. Romme. 1992a. Epidemiology theory and disturbance spread on landscapes. Landscape Ecology 7:19–26.

*O'Neill, R.V., R.H. Gardner, and M.G. Turner. 1992b. A hierarchical neutral model for landscape analysis. Landscape Ecology 7:55–61.

*Pearson, S.M., M.G. Turner, R.H. Gardner, and R.V. O'Neill. 1994. An organism-based perspective of habitat fragmentation. In: Biodiversity in Managed Landscapes: Theory and Practice, R.C. Szaro (ed.). Oxford University Press, New York.

Press, W.H., S.A. Teukolsky, W.T. Vetterling, and B.P. Flannery. 1995. Numberical Recipies in Fortran. Cambridge University Press, Cambridge.

*Plotnick, R.E., and R.H. Gardner. 1993. Lattices and Landscapes. In: Lectures on Mathematics in the Life Sciences: Predicting Spatial Effects in Ecological Systems, Vol. 23, pp. 129–157. R.H. Gardner (ed.). American Mathematical Society, Providence, RI.

*Plotnick, R.E., R.H. Gardner, and R.V. O'Neill. 1993. Lacunarity indices as measures of landscape texture. Landscape Ecology 8:201–211.

*Plotnick, R.E., R.H. Gardner, W.W. Hargrove, K. Prestegaard, and M. Perlmutter. 1995. Lacunarity analysis: a general technique for the analysis of spatial patterns. Physical Review Letter.

Saupe, D. 1988. Algorithms for random fractals. The Science of Fractal Images, pp. 71–113, H.-O. Petigen and D. Saupe (eds.). Springer-Verlag, New York.

*Turner, M.G., R.H. Gardner, V.H. Dale, and R.V. O'Neill. 1988. Landscape pattern and the spread of disturbance. In: Proceedings of the VIIIth International Symposium on Problems of Landscape Ecology Research, Vol. 1, pp. 373–382. M. Ruzicka, T. Hrnciarova, and L. Miklos (eds.). Institute of Experimental Biology and Ecology, CBES SAS, Bratislava, CSSR.

*Turner, M.G., R.H. Gardner, V.H. Dale, and R.V. O'Neill. 1989a. Predicting the spread of disturbance across heterogeneous landscapes. Oikos 55:121–129.

*Turner, M.G., R.V. O'Neill, R.H. Gardner, and B.T. Milne. 1989b. Effects of changing spatial scale on the analysis of landscape pattern. Landscape Ecology 3:153–162.

*Turner, M.G., W.H. Romme, R.H. Gardner, R.V. O'Neill, and T.K. Kratz. 1993a. A revised concept of landscape equilibrium: disturbance and stability on scaled landscapes. Landscape Ecology 8:213–227.

*Turner, M.G., R.H. Gardner, R.V. O'Neill, and S.M. Pearson. 1993b. Multiscale organization of landscape heterogeneity. In: Eastside Forest Ecosystem Health Assessment, Vol. II, pp. 81–87. Ecosystem Management: Principles and Applications. M.E. Jensen and P.S. Bourgeron (eds.). U.S. Forest Service Research Report, United States Department of Agriculture.

*Turner, M.G., W.H. Romme, and R.H. Gardner. 1994. Landscape disturbance models and the long-term dynamics of natural areas. Natural Areas Journal 13:3–11.

With, K.A., and A.W. King. 1997. The use and misuse of neutral landscape models in ecology. Oikos 79:219–229.

*With, K.A., R.H. Gardner, and M.G. Turner. 1997. Landscape connectivity and population distributions in heterogeneous environments. Oikos 78:151–169.

14
ClaraT: Instructional Software for Fractal Pattern Generation and Analysis

BRUCE T. MILNE, ALAN R. JOHNSON, AND STEVEN MATYK

Few recently developed quantitative approaches have enlivened the sciences as much as fractal geometry (Mandelbrot 1982) and related topics of scaling and universality (Stanley et al. 1996). The classic example of a fractal is a coastline, which by virtue of curves and crenulations, is a very jagged line weaving over the planar surface of the globe, thereby creating an object that is too crooked to be a one-dimensional line, but too straight to completely fill the plane. Moreover, magnification of a small part of the coastline reveals yet greater detail because the part is essentially a shrunken version of the whole. Thus, a coastline is neither one nor two dimensional because ruggedness occurs at many scales. Rather, coastlines have dimensions between 1 and 2; they are fractal.

Despite myriad reviews and books about the wonderful mysteries of fractals (Mandelbrot 1982; Stanley 1986; Feder 1988; Peitgen and Saupe 1988; Milne 1991a, 1997), there is a dearth of software that both simulates and analyzes classical fractal patterns. Together, simulation and analysis enable interactive learning and the development of an intuitive understanding of fractals. Thus, the goals of this chapter are to: (1) introduce the fundamental concepts of self-similarity, dimension, random growth processes, neutral landscape models, iterated function systems, fractal analysis, stochastic renormalization, and applications to time series; and (2) introduce the ClaraT software (Table 14.1) developed in our laboratory and available over the Internet at http://sevilleta.unm.edu/ ~ bmilne/compute.services.html. In the end, the student will have mastered the central concepts of fractal geometry and thereby gained an appreciation for how to apply fractals to many practical scientific applications (e.g., Milne 1997). This chapter is a tutorial to help the student progress through the central concepts of fractal geometry and to use the ClaraT (pronounced "clara-tee") software. Given the constraints under which the software was developed, it is not as robust as some commercial software. In general, it is best to invoke the programs from scratch each time that a new analysis is performed.

TABLE 14.1. Programs, examples, and help files included in ClaraT.

A. Fractals in the plane	
Analysis	
FDIM.exe	Estimate fractal dimensions (includes box dimension, multifractal Mandelbrot measures, and multiscale visualization tools)
PERCA.exe	Analyze binary percolation maps (spatial cluster identification and manipulation)
Diffusion limited aggregation	
DLAcell.exe	Generate diffusion limited aggregations
DLAtree.exe	Generate multispecies DLA trees
Iterated function systems	
IFS1.exe	Generate fractals by iterated function systems; includes graphical explanation of algorithm
IFS1.hlp	Help for IFS
btm2f.ifs	Example parameter file for IFS
btmfern.ifs	Example parameter file for IFS
Percolation and renormalization	
PERC.exe	Generate percolation maps (uniform and gradient)
RENORM2.exe	Spatial renormalization methods (stochastic rule definition)
RENORM.hlp	Help for Spatial renormalization
B. Self-affine fractals (time series, surfaces, etc.)	
BROWNIAN.exe	Generate self-affine functions in 1 + 1 and 2 + 1 dimensions
HURSTW.exe	Analyze self-affine functions by several methods; graph results
HURSTW.hlp	Help file for HURSTW.exe
brown.dat	Example data where column 1 is time and column 2 is intensity
C. Windows dll file	
bwcc.dll	Required by ClaraT

Central Concepts

Self-Similarity: An Exploration of Broccoli

Examine a piece of broccoli. Break off a small piece (about one-third the whole). Close one eye and hold the bigger of the two pieces at arm's length. Hold the smaller piece near your eye. Do they look similar? Broccoli is fractal in the sense that a small piece is similar to the whole. If magnified properly, the small piece is indistinguishable from the larger. The painter Edgar Dégas recognized the self-similar nature of clouds when he used a wadded up handkerchief as a model of a cloud. Both broccoli and clouds can be viewed as lumps made of lumps, made of lumps, of ever decreasing size. Indeed, Grey and Kjems (1989) show that the self-similar spirals of the Minaret cauliflower continue down to the arrangement of cells! Cloud lumps continue down to the size of atoms, where energy is finally dissipated as heat. In biological systems, one expects some lower limit to the scales over which self-similarity occurs (West et al. 1997).

Dimension

Leonardo da Vinci observed, somewhat incorrectly, that at a node, the squared diameter of the tree trunk equals the sum of the squared diameters of the two ascendant branches. If da Vinci were correct, then there would indeed be a remarkable similarity in the architecture of trees from the base to the very tips of the branches. Benoit Mandelbrot (1982) generalized da Vinci's idea to state that:

$$W_0^{\Delta} = W_1^{\Delta} + W_2^{\Delta}$$

where the *W*s are the diameters of the lower internode (W_0) and the two internodal stems coming off of the first stem; Δ is an exponent related to the shape of a tree. Mandelbrot surmised that $2 < \Delta < 3$ for trees. His logic was that the branches of bronchi in lungs nearly fill the three dimensional space (i.e., $\Delta = 3$). Thus, the bronchi maximize surface area for gas exchange (see West et al. 1997). On the other hand, the geometry of trees is regulated by the display of two-dimensional leaves to the sun, and there is an energetic imperative for trees to maximize presentation of leaf area to the sun while minimizing the investment of energy in the supporting structure. Thus, Mandelbrot argues that trees exhibit Δ closer to 2. Empirical estimates of Δ estimated for ornamental deciduous crabapple trees are about 2.2 (Milne, unpublished data). The exponent Δ tells how to relate smaller parts to the whole.

In Euclidean geometry, points, lines, planes, and solids have dimensions of 0, 1, 2, and 3, respectively. The dimension is useful for relating the length of the set to some property of the whole, as when the length of a rectangle is squared to estimate its area:

$$A = L^2$$

where L = the length. More generally, we can write:

$$A(L) = kL^D$$

where $A(L)$ indicates area measured at scale L, $D = 2$ is the dimension of a square, and k is a constant, equal to 1 in this case.

Fractal sets, such as a crooked line drawn on a two-dimensional piece of paper, have dimensions less than or equal to the space in which they are embedded. Examples of fractal dimensions are coastlines $D = 1.2$ to 1.5, areas of archipelagos $D = 1.4$, and disease occurrences through time $D = 0.2$ (see Burrough 1981). Thus, the concepts of self-similarity and dimension can be considered simultaneously. Specifically, a set, such as a coastline, is said to be fractal if a constant, noninteger dimension applies over 1 to 2 orders of magnitude in length scale, or more. The precise meaning of length scale is discussed below.

Creation of Fractals

The origin of self-similarity in nature is a topic of great interest (Creswick et al. 1992; Binney et al. 1993). A basic understanding is needed of simple processes that create complex fractal patterns exhibited by clouds, mountains, rivers, fluctuations in river flows, variation in tree density, changes in the stock market, distributions of word frequencies, errors in telephone transmissions, and music (Mandelbrot 1982; Voss 1988). Here, the authors describe the creation of several fractals and emphasize how the effects of a given process that operates on the smallest elements of a system are magnified to create the whole. In all cases, early events affect the ultimate pattern via feedbacks on the process. In this way, small, physiologically controlled interactions among cells can build trees.

Branching Clusters

This first kind of fractal, sometimes called a diffusion limited aggregation (DLA), is generated by a process analogous to the growth of root systems, snowflakes, and eroded land (Meakin 1986). You can use the ClaraT software to grow a fractal and learn how the process works.

Instructions

Assuming that the ClaraT software is installed on a personal computer,

1. Select the "DLAcell" icon. A window called "DLA Cluster" will appear.
2. Select "Options" in the DLA Cluster window, then "parameters."
3. Change the "Limit Radius" to 1 for faster execution.
4. Change the "Color cycle" to 100 for better visualization. This will change the color of the pattern after 100 steps.
5. You may modify the "seed" that controls the random number generator.
6. Click "OK."

What You Should See

A dendritic cluster should grow on your display. It will branch in a natural looking fashion. Several long branches will appear, along with many more shorter branches. Thus branches are made of branches, are made of branches . . . in a selfsimilar fashion, much like broccoli. Notice that the most recently added cells are on the extremities.

How DLAcell Works

Here is the basic method used to make the cluster (Meakin 1986), described in "pseudocode." The cluster has two basic parts, a kernal and N particles that are added to it.

1. Place a "kernel" at the center of the map. The kernel has four sides.
2. For *particles* = 1 to *N*, do the following:
   a. Release a randomly walking *particle* some distance from the kernel (actually, away from all existing particles on the DLA structure).
   b. Let the particle walk until it bumps into the kernel or the structure.
   c. Let the particle stick where it hits.

Release the next particle, and let it walk until it sticks to the evergrowing cluster. As the branches grow they "shield" the innermost particles such that new particles tend to be added near the tips of the branches.

Basically, a fractal is produced by interactions between random and deterministic processes. In the case of DLA, the randomly walking particle imparts randomness to the cluster, while the deterministic components include the four-sided kernel and the requirement that the particle hit the DLA cluster only once in order to stick. Once the cluster begins to grow, the process imparts a certain deterministic flavor to the branches but a random roughness to the perimeter.

Diffusion limited aggregation is a simple model of root growth. Plant roots grow toward higher concentrations of water or nutrients in the soil. At the molecular level, these resources occur with some randomness, so it is not always clear exactly which way a root will turn next as it grows toward the resource. Moreover, new roots are always added to existing roots. Thus, the development of roots depends on the pattern of earlier growth, which was very much influenced by the earlier distributions of resources. The rooting pattern reflects the historical development of the root system. Similar arguments can be made for erosion. If one begins with a flat surface, like polished marble, and water rains down upon it, there will be some single crystal that gives way first and is lost. The resulting pit exposes greater surface area to weathering, making it likely that future losses will come from this small beginning. Continued erosion magnifies this inauspicious beginning and creates macroscopic features that are recognized as gullies. Because a given process is involved regardless of the size of the valley, one expects to find the classic relations between drainage network size and various measures of stream geometry (Johnson et al. 1995).

The program *DLAtree* (Table 14.1) simulates growth of dendritic features by releasing the random walkers only from above. The resulting trees grow toward the "light source." Competition can be simulated by requiring some species of trees to receive two random walkers before growing. These slow growers are eventually outcompeted by other species that rise quickly to greater heights and intercept most of the incoming walkers.

Spatially Neutral Landscapes (Percolation Maps)

Scientists are generally trained to think of random processes as "simple" and without predictable consequences. For example, the classic experiment in statistics is to ask "what is the chance of drawing five black marbles in a row out of a jar containing 50% black and 50% red marbles?" If one assumes a well mixed jar of an infinite number of marbles, the probability is $(0.5)^5 = 0.03125$. This formula represents the idea that the chance of obtaining one black marble is not affected by the color of marble obtained in the previous try. This idea of independence is central to statistical theory. So far, so good.

Early in the studies of landscapes, Gardner et al. (1987) asked, "What would a landscape look like if no processes affected the distribution of forest on the land?" They hypothesized that a random map was a spatially neutral model of a landscape. Differences between real and random landscapes indicate that various processes regulate landscape pattern.

Here are the steps for making a random map as a neutral model of some particular real landscape:

1. Choose a cover type of interest from the real landscape (e.g., forest, agriculture, or water).
2. Measure the percent of the landscape covered by the chosen cover type.
3. Express the percent as a portion p ranging from 0 to 1. For example, 70% cover is expressed as $p = 0.7$.
4. In a computer, set up a *lattice*, or grid, composed of a large number of *cells*. Here, we use a rectangular lattice like a checkerboard. Let there be N cells on the lattice, say $N = 1{,}000{,}000$.
5. To make the random map, turn exactly pN of the cells on and $(1 - p)N$ cells off.

In our example that would be $0.7 \times 1{,}000{,}000 = 700{,}000$ cells on and 300,000 off. This is done such that the state of each cell is independent of the states of other cells. It turns out that real landscapes differ markedly from random maps, largely because many processes cause virtually all land cover types to be more clumped together than on a random map of the same p value (Gardner et al. 1987).

Unlike marbles in a jar, random maps do not have "simple" behavior by any means, and there is an entire field in physics, called percolation theory (Stauffer 1985; Orbach 1986) dedicated to the study of random spatial patterns. One remarkable property of random maps is the occurrence of a "spatial phase transition" (Zallen 1983) at a critical density. Spatial phase transitions may form a basis for ecological theories of ecotones and landscape responses to environmental changes (Milne et al. 1996).

Investigation of the Percolation Transition

Here, you will make a random percolation map and then analyze it to find evidence of a spatial phase transition.

1. Select the *PERC* program to generate a random map.
2. Select "Type" and then "Random percolation." The "Percentage" default is 59.28 which equals $p = 0.5928$. Set the "random seed" to 2.

3. Select "OK." This will produce a map. Red cells are "occupied," black are "empty."
4. Select the *PERCA* program (percolation analysis) to study the map you just made. *PERCA* will pop up an empty black window that just says "Edit" and a percolation analysis control window. Move them to separate locations so that you can see both windows clearly.
5. Now, copy the percolation map into the black window that says "Edit" so that the map can be analyzed. To do so,
   a. Go to the "Percolation generator" window and select "Edit."
   b. Then "Copy."
   c. Next, go to the black window, select "Edit."
   d. Then "Paste."

What You Should See
The random map should now be in the black window of "Percolation Analysis."

6. In the Percolation Analysis window, select "Identify clusters." The map will update with a set of new colors. Each color indicates occupied cells that are connected *vertically* or *horizontally* to other occupied cells. Groups of connected cells are called *clusters*. The software identifies the clusters by the von Neumann rule (i.e., occupied cells form clusters with occupied neighbors only in the north, south, east, or west directions, not diagonally) which has profound implications for the quantitative, but not qualitative, results (Zallen 1983). There are other rules which, in principle, could be applied for various ecological purposes. For example, the Moore rule allows both diagonal connections and those of the von Neumann rule.
7. A spatial phase transition is evidenced by an "infinite" cluster, sometimes called the "spanning" or "percolation cluster." By definition, a percolation cluster spans from one side of the map to the other. We want to test to see if the map you just made contains a percolation cluster. If the map were infinitely long the cluster would be too. Some maps have them, others do not, depending on the probability of having occupied cells, the rule that is used to connect occupied cells, and the lattice geometry (e.g., square, triangular, hexagonal) (Zallen 1983). To find out if your map has an infinite cluster, look at the "# of Inf. Clusters" in the lower left of the Percolation Analysis Window. If you have an infinite cluster (rarely will there be more than one) you can view it by selecting "Infinite clusters" in the "percolation analysis" window (this button is grayed-out if there is no infinite cluster). This will pop up a small window called "Cluster Display Options." Choose "Remove All Finite Clusters." Whatever is left will be an infinite cluster.

Experiment: Detection of p_{crit}

The infinite cluster of occupied cells appears at a critical density called p_{crit}. This is analogous to a critical temperature at which water freezes. Imagine that the occupied cells represent some land cover of interest, such as the forest habitat of migratory birds; the unoccupied cells are nonhabitat. If the density of habitat is less than the critical density, the habitat is "fragmented" while above the critical density it is "connected." In Forman and Godron's (1986) terminology, the occupied cells are the matrix of the landscape when they appear above the critical density. We might expect parasites like mistletoe to spread easily above the percolation threshold, although we must realize that species and their movements may be governed by effectively different lattice geometries (e.g., continuous coordinate systems) and different neighborhood rules such that the critical densities of real systems may not correspond to $p_{crit} = 0.59$, as in this example.

Here, the percolation threshold can be studied systematically to find p_{crit} and to see how the occurrence of the infinite cluster behaves as a function of p.

1. In the classroom, students at each workstation should be assigned a p value for study. Select from the following list: $p = 0.3, 0.5, 0.55, 0.59, 0.65, 0.7$, and 0.8.
2. For each probability, generate 6 maps (of size 32×32 for speed, larger for precision). For each map, repeat steps 1–7 above and record whether the map has an infinite cluster or not.
3. Have someone summarize the results on the board as a graph showing p on the x-axis and "Number of maps with infinite cluster" on the y-axis. Sketch the graph. Label the axes.
4. The class should see that the curve starts low, with zero maps exhibiting a percolation cluster at low probabilities. As $p = 0.59$ is approached, the curve turns sharply upwards until all six attempts exhibit percolation clusters. The critical value can be estimated as the probability at which 3/6 or 50% of the maps contain a percolation cluster.

This exercise shows how a random process cast into the 2-dimensional space of a map produces nonintuitive results. It also suggests the appearance of "spatial phase transitions" in landscapes. The class may discuss some cases where phenomena like this might occur in nature and what environmental conditions would be associated with the critical percolation density (Milne et al. 1996).

Iterated Function Systems

Barnsley's (1988) Iterated Function Systems (IFS) provide one of the most flexible ways of making fractals. IFS teach us that a huge world of fractals can be represented with a handful of simple equations.

To generate a fractal fern by IFS:

1. Select the "IFS" icon.
2. Select "File," then "Open." Select "btmfern.ifs."

What You Should See

You will see a series of dots drawn on the screen. These will eventually combine to make a fern-like picture. How is it made?

How *IFS* Works

1. Select "File," then "Open." Select "btm2f.ifs." This is a simple fractal generated entirely from two equations!
2. Under "IFS control," click on "Trajectory."

To see how the IFS works, look at the lower righthand corner. You will see a string of red dots forming a connected line toward the top right. Notice the other lines of red dots that converge on one point. This is the "fixed point." Start with a point exactly on the lower right corner with coordinates $x = 100$, $y = 100$, and then transform it according to:

$$\begin{aligned} x' &= 0.85x + 0.04y + 8 \\ &= 85 + 4 + 8 \\ &= 97 \qquad \text{which is the new value of } x. \end{aligned}$$

Similarly,

$$\begin{aligned} y' &= -0.0401x + 0.85y + 16 \\ &= 96.9 \qquad \text{which is the new value of } y. \end{aligned}$$

These are the coordinates of the first red dot next to the corner. In matrix algebra, these two equations can be rewritten as a single function called $w(1)$:

$$w(1) = \begin{bmatrix} x' \\ y' \end{bmatrix} = \begin{bmatrix} 0.85 & 0.04 \\ -0.0401 & 0.85 \end{bmatrix} \begin{bmatrix} x \\ y \end{bmatrix} + \begin{bmatrix} 8 \\ 16 \end{bmatrix}$$

If the function is applied to the coordinates a second time, the point would move over to the second red point in the line, then the third, all the way to the fixed point. Then it would be fixed and unchanging even under repeated application of function $w(1)$. Similarly,

$$w(2) = \begin{bmatrix} x' \\ y' \end{bmatrix} = \begin{bmatrix} 0.4 & -0.359 \\ 0.53 & 0.22 \end{bmatrix} \begin{bmatrix} x \\ y \end{bmatrix} + \begin{bmatrix} 46 \\ 5 \end{bmatrix}$$

In the computer, we can start with any point on the map, select function $w(1)$ or $w(2)$ *randomly*, subject the x,y coordinates to the function, and redraw the point at its new location. Amazingly, the points only occur at certain locations seen as a fern, or some other pattern!

You can manipulate the numbers in the matrices. In ClaraT these are called "coefficients." Simply select an equation to manipulate, then use the mouse to increase or decrease the coefficients. Notice how the trajectories and the fractal change with each change of the coefficients. The geometry is continuously dependent on the coefficients.

Milne (1991b) explains how to apply IFS in landscape design and provides a graphical explanation for how the IFS works; much of that explanation is built into the display capabilities of ClaraT. With Windows, it is possible to use the clipboard to paste an image or map created previously with another program into the IFS window. Then, the coefficients can be adjusted interactively to create fractals that blend with the background image. This capability may enable fractal harvesting or planting patterns to be customized for a given landscape. Saving the fractal as an ASCII file of x,y coordinates specifies the locations of points on the ground where plants can be added or removed from the existing landscape.

An ecological model can be made using an IFS assuming that we have a discrete time system (e.g., Tilman and Wedin 1991). For example, rangelands in the southwest are mixtures of grass, shrubs, and bare soil. Winter precipitation is affected by the El Niño Southern Oscillation (ENSO) which produces wet winter and spring conditions during the El Niño phase, average moisture during medial years, and dry winters during La Niña years. We expect a continuous string of El Niño years to promote grass by increasing available soil moisture. In contrast, continuous La Niña conditions would promote shrubs and ultimately bare soil. In reality, the climate fluctuates among these three possibilities. Thus, climatic oscillations could force the rangeland to approach the associated fixed points (Fig. 14.1) while tracing out a finite, fractal set of state vectors (i.e., the amount of grass, shrubs, and bare soil).

With year-to-year variation in climate and land use, the trajectory of the system switches towards one equilibrium or another depending on the prevailing conditions. For example, a La Niña moves the system from point $t - 1$ to t (Fig. 14.1). Then a medial year moves the system to point $t + 1$ along a trajectory (dashed) toward the medial fixed point. Trajectories are implicit from any point to each of the fixed points.

It is apparent from construction of the IFS that a particular set of coefficients represents the effects of relevant ecological factors (e.g., soil moisture holding capacity, available light, temperature, seed supply, etc.) and will vary throughout the landscape. Thus, if the landscape is represented in raster form (i.e., as a lattice of cells), then each cell would contain a vector of IFS functions and the IFS associated with each cell would describe the dynamics at that location.

Self-Affine Fractals

Fractals come in two distinct types. The first type, introduced above, are called statistically self-similar. They reside in a real coordinate system $\mathcal{R}^d$ where d is the number of coordinate axes in the space. Each coordinate has the same units of measure, such as degrees latitude and longitude. For

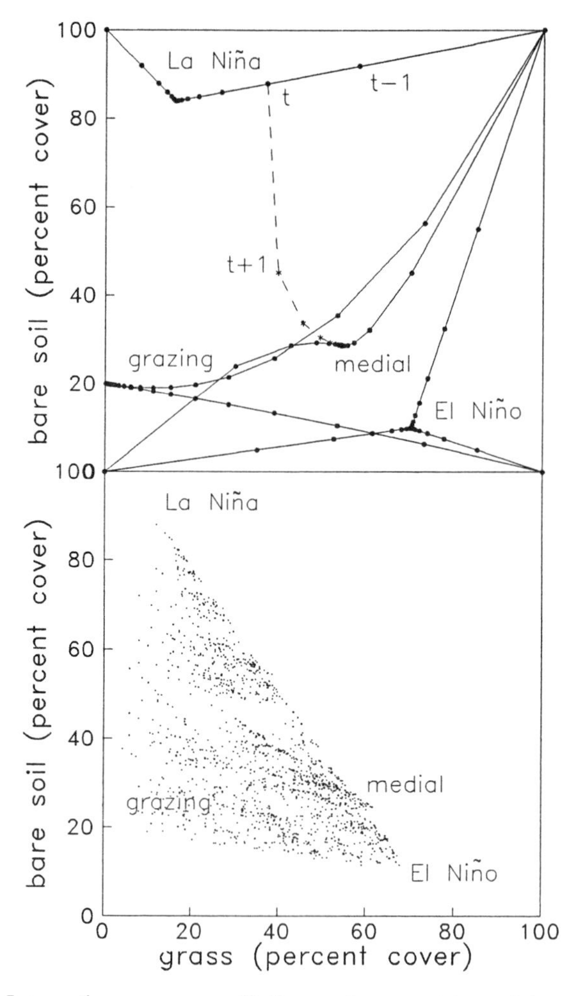

FIGURE 14.1. Interactions among oscillating environmental conditions produce a set of allowable system states defined by the percentage cover of bare soil, grass, and shrubs, (Top) Under constant, ungrazed La Niña conditions, any initial values of grass and bare soil converge to a unique equilibrium (La Niña). Other equilibria would obtain under constant ungrazed E1 Niño conditions, ungrazed medial conditions (i.e., years that are neither El Niño nor La Niña), and constant grazing. (Bottom) Interactions among the functions constrain the system to visit a subset of "allowable" states (dots) and preclude visitation to some "unreachable" states (gaps).

example, a coastline is a set of points in a latitude-longitude coordinate system $\mathcal{R}^2$ in which x and y coordinates are in degrees. The fractals generated by *PERC* and *IFS* are of this type.

The second type of fractal is called "self-affine," in which a variable, say the volume of water in a reservoir, varies through time. Here, flow and time are measured in different units; a unit of volume is not the same as a unit of time. Affine refers to a transformation such as $y = mx^{\alpha} + b$ in which x is rescaled by some constant m and exponent α. The familiar case with $\alpha = 1$ is the equation of a line, which can be thought of as a stretched and displaced x axis. In this context, one coordinate of a self-affine fractal must be rescaled in a particular way to relate it to the other. Thus, different conceptualizations underlie the generation and analysis of self-affine fractals.

Generation of Self-Affine Fractals

Feller (1951) studied affine functions by conducting a coin flipping experiment that anyone can try. Heads (H) and tails (T) are assigned values of 1 and −1, respectively. A series of coin tosses produces a string of outcomes such as H, H, T, H, T, T. Translated into 1 and −1, the string becomes 1, 1, −1, 1, −1, −1. The running sum of the string is $X(t) = 1, 2, 1, 2, 1, 0$ for $t = 1, 2, \ldots 6$ (Fig. 14.2B). Continuing the experiment for thousands of flips produces a sum that can take on surprisingly large positive or negative values (Fig. 14.2D) because the variance of the series grows as time goes on. For this reason, an arbitrary tenth of Fig. 14.2B (magnified in Fig. 14.2A) does not have as much variation as the whole. Rather, an affine transformation has to be applied to Figure 14.2A to make it resemble the whole (i.e., by multiplying $X(t)$ by $10^{1/2}$).

Thus, like diffusion limited aggregation and iterated function systems, Feller's experiment contains a random component, namely the coin flips, and a rule that is applied consistently to the random series (i.e., the serial summation). In that sense, the sum at any moment is determined by the cumulative history of the coin flips, much as the addition of particles to a growing DLA is affected by the history of its development.

Self-affine functions can be created by transforming empirical data. For example, DNA is composed of purine (adenine and guanine) and pyrimidine (cytosine and thymine) bases. Reading along a DNA sequence, the pyrimidines can be represented as +1 while purines are −1. A summation of the series creates a function analogous to Feller's summation. Stanley et al. (1996) hypothesized that DNA sequences are fractal with dimensions corresponding to nonrandom selection processes that induce correlations in the bases over vast distances along the gene.

ClaraT includes a program called *BROWNIAN* (Table 14.1) for generating self-affine fractals. Self-affine fractals involve one dependent variable which can be thought of as a measure of "intensity" and one or more independent variables that are measures of location or time. For example, the number of sunspots varies through time, and are said to reside in a space of 1 + 1 dimensions. Similarly, an affine surface, such as an elevation map, resides in a 2 + 1 space, in which there are two spatial dimensions and one dependent variable (i.e., elevation). *BROWNIAN* simulates self-affine functions in both embedding spaces.

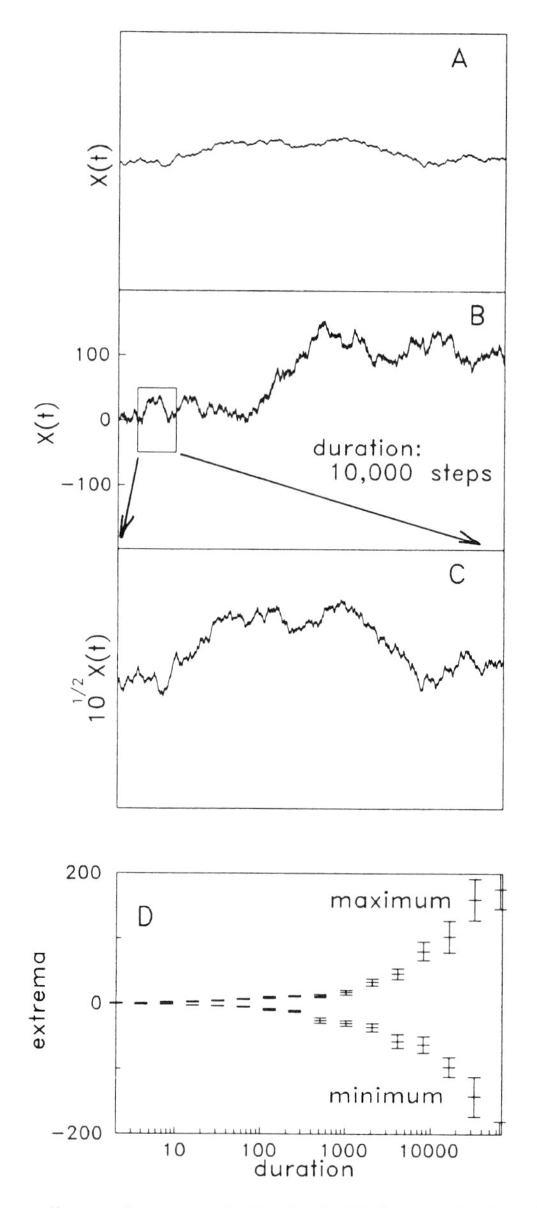

FIGURE 14.2. Rescaling and extreme behavior in Feller's coin flipping experiment. (A) Improperly magnified 1000 observations out of the 10,000 in (B)which is the running sum of random digits with values of +1 or −1. (C) Shows the rescaled $X(t)$ of panel (A), illustrating how proper rescaling recovers the jaggedness of (B). (D) Maximum and minimum values of $X(t)$ grow as the duration of the simulation increases. Vertical bars represent 1 standard error of the mean from a series of replicate simulations.

Steps for Generating Self-Affine Fractals

1. Open "Brownian.exe."
2. Select "Methods." There are several methods to choose from, but for now,
   a. Select 1-D Midpoint displacement.
      i. Notice that the Hurst exponent, which is a measure of the jaggedness of the pattern, is set to 0.5.
      ii. Adjust the number of levels to 8. The number of levels n produces 2^n data points. In this case, there will be $2^8 = 256$ points.
      iii. Click "OK."

What You Should See

A graph will appear showing the dependent variable (called "Intensity") versus the independent variable, called "Time." The ordinate and abscissa could as well have been called "Amount" and "Location," respectively. The curve represents measurements of something as a function of time or location. The fractal structure is apparent in that there are fluctuations up and down at all scales such that the largest dips in the curve contain excursions upward as well. You can experiment with the number of levels while leaving the Hurst exponent constant, or change the exponent.

This section provided four fundamental ways to make fractals. The diversity of approaches demonstrates several things. First, DLA growth processes involve both random and deterministic aspects such that the ontogeny of the object affects its future growth. This feedback between pattern and process is a quintessential aspect of fractals. Second, analyses of random maps demonstrate many surprising and nontrivial properties of spatial systems, such as critical behavior. Thus, percolation theory opens new and productive theoretical avenues concerning the origin, nature, and dynamics of ecotones or the sustainability of rare species on fragmented landscapes (Bascompte and Solé 1996). Third, iterated function systems are versatile and powerful means by which to represent a wide array of systems ranging from landscape patterns to regular grammars (Barnsley 1988) that can be used in studies of development and growth.

Analysis

The general basis of fractal analysis entails study of patterns at multiple scales. Statistical determinations are made of exactly how the measurements at different scales relate to one another. Fractals exhibit power law dependence between the measurements and the scale at which they are made.

Scale was discussed thoroughly in Chapter 11. For our purposes scale involves two components, namely extent and grain. *Extent* is the breadth of a map or the entire region represented. *Grain* is the smallest resolvable unit within a map. For example, the 30 × 30 m cells of Thematic Mapper satellite image are the grains.

The statistical characterization of fractals involves manipulation of either the length of windows (i.e., quadrats, plots) or the distance between grains from which measurements are made. Here, *window length* is the length of a square window within which grains are counted. *Lag*, a word from geostatistics (Cressie 1991), is the distance between two grains that are compared, as by subtracting a variate at one location from that at another. Typically, the observer manipulates lag or window length and quantifies how measurements vary with scale. The methods explored below rely on purposeful changes in length or lag, rather than use of a single length scale for measurement.

The Box Dimension

The box counting method is perhaps the simplest of techniques for measuring a fractal dimension. For reference, the box dimension equals 0 if the map contains but a single point (as expected from Euclidean geometry). The box dimension equals 2 when 100% of the map contains the set of interest (e.g., forest) or if the set is randomly distributed. Fragmented forests have box dimensions less than 2 (Milne 1997).

Estimation of the box dimension is straightforward. First, a rigid grid of boxes larger than the grain size is superimposed on a map, just as one might overlay a piece of transparent graph paper on a map. Boxes are counted that contain the cover type of interest, in any small amount. The process is repeated with different box sizes until 1 to 2 orders of magnitude in box size have been explored (say boxes of size 1 to 100). Then, the logarithm of the number of occupied boxes of each length is regressed against the logarithm of box length. The slope of the regression is the exponent in the power law:

$$N(L) = kL^{-D_b}$$

The exponent D_b is negative because the number of boxes decreases as the box length increases, regardless of the pattern. The simplest case is when just one grain is occupied. Then, $N(L) = 1$ for all box lengths and the slope $D_b = 0$, the dimension of a point.

Conversion of the box counts to areas $A(L)$ is simple: multiply each side of the equation by L^2 which is the area of a box of length L. This gives a new equation:

$$A(L) = kL^{2-D_b}$$

This transformation is especially useful for revealing ranges of L over which the power law is a poor fit (in which case a fractal model is not appropriate). More interestingly, graphing the areas at each L may reveal a scale at which the slope of the power law changes. Changes in slope are good indicators that the landscape pattern is governed over short scales by one process and over broader scales by another process (Krummel et al. 1987; Johnson et al. 1992). If the assumption of self-similarity can be made at scales below the grain size (a tenuous, but testable assertion), the scaling relation for area could be used to estimate the available forest cover at subgrain scales; this approach would be useful for estimating population densities or habitat availability for species whose home range sizes are less than the grain size (Milne 1997).

Mandelbrot (1982) emphasizes that the dimension is not the only parameter of interest. Rather, the coefficient k in the power law is crucial for helping to estimate the expected number of boxes $N(L)$ of length L. The coefficient has been referred to as a measure of lacunarity (Mandelbrot 1982), which characterizes the gaps or lacunae in the set (Plotnick et al. 1993).

The box counting technique is sensitive to the orientation of the grid of boxes. A small jostling of the grid could change the occupancy of some grid cells and thereby induce variation in the counts. Thus, the software uses a small random displacement to orient the grid and slightly different estimates of the dimension are obtained from repeated trials with a given map. In theory, it is best to orient the grid to obtain the minimum number of boxes necessary to cover the map. Fractal dimensions estimated by means of Mandelbrot measures (below) always achieve the minimum covering.

Tutorial: The Box Dimension: Exercise 1

Goals

1. Learn to use *FDIM* for the analysis of the box dimension.
2. Learn how the box dimension is obtained by regression.

Steps

1. Use *PERC* to make a gradient percolation map
   a. Choose “Type,” then “Gradient Percolation,” then “cubic function.”
   b. Set coefficient b to 1.0, and all others to zero. This creates a map on which the probability of an occupied (red) cell is zero on the left and then changes smoothly to a value of 1 on the right. The map is a simple model for an elevation gradient along which tree density increases from left to right (low to high elevation).
   c. Set the map size to 128×128.
   d. Select “OK.” (You should now have a gradient percolation map that is mostly black on the left and red on the right.)
2. Invoke *FDIM* for “fractal dimension analysis.”
3. Click on “Edit” in *PERC* and “copy” the map to the clipboard.
4. Click on “Edit” in *FDIM* and “paste” the map into *FDIM*. (You should now have the gradient map in the *FDIM* window.)
5. In *FDIM*

   a. Select "box counting." Accept the default window size and pixel value parameters by clicking "OK." (A number should now appear beside Box dim; this is the box dimension of the map.)
   b. Select "options"
      i. Then select "Box counting regression plot." This shows a regression of the number of occupied boxes versus box length. A straight line here is diagnostic of a fractal (note the logarithmically scaled axes). The absolute value of the slope of the line equals the box dimension.

What You Should Learn from This Exercise
1. That the box dimension equals the absolute value of the slope of the regression between the number of occupied boxes and the box length.
2. How to use *FDIM* to measure the box dimension.

Tutorial: The Box Dimension: Exercise 2

This exercise should be tried after Exercise 1 for the Box Dimension.

Goals
1. Show that the box dimension of a spanning cluster from a percolation map equals the predicted value of 1.89 (Feder 1988).
2. Use three tools in combination.

Steps
1. Use *PERC* to make a random map of size 256 × 256 with $p = 0.6$.
2. Invoke *PERCA* (for percolation analysis).
3. Use copy and paste to transfer the map to *PERCA*.
   a. Use "IDENTIFY CLUSTERS" and then
   b. Remove all the finite clusters, leaving only one large cluster that spans from top to bottom.
4. Invoke FDIM (fractal dimension tool).
5. Use copy and paste to transfer the infinite cluster to the fractal dimension tool.
6. Select "Box Counting." Note the box counting dimension of approximately 1.89, which is the theoretical dimension for a spanning cluster (Feder 1988) and is diagnostic of a percolation process.

Mandelbrot Measures

The fractal dimensions of natural objects such as coastlines, mountains, and stream networks vary somewhat from place to place. For example , if cities A, B, and C occur in order along a coastline, the dimension of the coast between point A and B will differ from that between B and C, and both of these will differ somewhat from the dimension estimated for the coast between A and C. The variation results from the stochastic processes that create coastlines, leaving behind a coastline that exhibits "statistical self-similarity." Statistical self-similarity differs from the simpler case of "exact self-similarity" seen in deterministic fractals, such as the Sierpinski triangle which appears as the default in the *IFS* (Iterated Function System) program in ClaraT. Small pieces of exactly self-similar fractals are clearly miniature copies of the larger fractal. However, in natural fractals such exact miniaturization is not possible because of random fluctuations. Thus, the notion of statistical self-similarity accommodates the demonstrable variation in fractal dimensions from place to place (e.g., the variation in coastline scaling observed among continents).

The notion of "multifractals" (Feder 1988) accommodates the multitude of fractal dimensions that occur within a broad extent (i.e., along the coastline between cities A and C). Multifractal analyses enable the investigator to partition the map into locally dense and locally sparse subregions, each of which have unique scaling behavior (i.e., each has a scaling exponent that characterizes its scale dependence) (Milne 1991a). Mandelbrot measures (Voss 1988) are the basis of a rich set of scaling relations with interesting applications. Mandelbrot measures are constructed by scanning a map of a cover type, say forest, with a series of sliding windows. Beginning with a small window, say $3 \times 3 = 9 = L^2$ cells, a window is passed over the map of forest. Each window that contains a forest cell at the center is inspected, and the forest cells within the window are counted to provide a value m which must be $\leq L^2$. Dividing each m by the sum of all m values produces a Mandelbrot measure.

Tutorial: Visualization of Mandelbrot Measures

Maps can be made of scale-dependent density. In true fractal spirit, the locations of "dense" subregions depend on the scale at which density is measured. This has implications for the coexistence of species, given allometric dependence between body mass and home range size. The latter can be considered a scale at which species perceive habitat density (Milne 1992, 1997).

Steps

1. Use *PERC* to create a gradient percolation map.
2. Use *PERCA* to extract the spanning cluster.
3. Copy and paste the spanning cluster map into *FDIM*. Select the PML button.
4. Enable the "Mandelbrot measures" capability within *PML*. Then, choose three window lengths and assign them to the red (R), green (G), and blue (B) colors of the display. A simple mnemonic to aid interpretation is the following. Assign the longest window to red, which is a *long* wavelength, and assign the shortest window to blue, which is a *short* wavelength. This is a computationally intensive procedure, so you may have to wait a while.

The program will scan the image with sliding windows of each size and tally the number of occupied cells within the window if it contains an occupied cell at its center. The tally will be stored at the center location of the window. Counts from the short window will be displayed in the blue color of the display while counts obtained from long windows will appear in red. Numbers are scaled by dividing the tally within a window by the area of the window.

Interpretation
An RGB display (red, green blue) produces white wherever high intensities are displayed in all three colors. Locations that have relatively high densities within small windows appear blue while locations with relatively high densities within large windows appear red. It is easy to imagine that animals that use habitat in small home ranges will congregate where the image is blue, while large species will center their activities in the red regions.

Statistical Treatment of Mandelbrot Measures

Frequency distributions of the m values are made after scanning the image with several window sizes. Then, for each L, the frequencies are divided by the integral of the frequency distribution to obtain the probability density function $P(m,L)$ which is the probability of obtaining m cells in a window of length L. The $P(m,L)$ distribution has moments $M^q(L)$ analogous to the mean, variance, and so on, indexed by an integer $-\infty < q < \infty$. Moments indexed by $q < 0$ characterize sparsely populated windows and moments indexed by $q > 0$ characterize locally dense windows (Voss 1988; Milne 1991a; Milne et al. 1996). The moments for $q \neq 0$ scale according to:

$$<M^q(L)>^{1/q} = kL^{Dq}$$

Thus, a scaling exponent D_2 for $q = 2$ characterizes the scale dependence of fairly dense subregions that were measured at several scales. This is because the moment for $q = 2$ is formed by multiplying $P(m,L)$ for each value of m by m^2 and summing the products; thus large values of m act as large weights. Negative values of q create weights that vary inversely as $1/m^q$; thus the scaling of moments for q much less than 0 represent the fractal geometry of sparse partitions of the original map.

Tutorial: Moments and Scaling of Moments

Goals
1. Use *PML* to compute various moments and scaling relations.
2. Plot the moments versus window length.
3. Compare the various moments.
4. Examine the probability density functions at each scale.
5. Plot the scaling exponents D_q versus q to see the multifractal spectrum.

Steps
1. Use *PERC* to create a gradient percolation map.
2. Use *PERCA* to extract a spanning cluster.
3. Copy and paste the spanning cluster into *FDIM* and select *PML*.
4. Change the default maximum window size to 21.
   a. Hit "OK" to execute.
5. Select the first moment (the mean, $q = 1$) by selecting PML(1) from the drop down menu, rather than the default PML(−1).
6. Under "Options" at the top of the window:
   a. Choose PML Regression Plot, then "plot a single moment" to see a plot of the first moment. This is the expected number of occupied cells as a function of window width.
   b. Close the graph and choose "all moments" from PML Regression Plot to see regressions for all moments within the range of selected q values.
   c. From "Options" choose "D_q plot" to see all the scaling exponents versus q.
      i. An exact fractal, such as a Sierpinski triangle, has equal values of D_q for all q; even the sparsest subregions have geometries that are identical to the most dense subregions (e.g., the Sierpinski triangle that appears as the default attrator in the *IFS* program).
      ii. A statistical fractal, such as a fragmented forest, has unequal values of D_q for all q. Moreover, a fragmented map that is too sparse to produce a spanning cluster has a monotonically ascending D_q function. A map above the critical density contains many subregions which are dense, so that even the *relatively* sparse subregions are prevalent, yielding high scaling exponents. Most of the information about map geometry appears for $q < 0$ because the exponents for positive q generally reach an asymptotic value.
   d. From "Options" choose "*PML*(m,l) plot" to see all the statistical distributions.
      i. Select "rough axes" to expand the distributions along the y-axis.

Spatial Renormalization

One of the great challenges in ecology is to relate fine-scale processes to broad scale patterns and to translate information from one scale to another. For example, methods are needed to relate measurements of vegetation on the ground to the coarse resolution grains (30–1000 m) used in satellite imagery. Spatial renormalization (Gould and Tobochnik 1988; Creswick et al. 1992; Milne and Johnson 1993; Milne et al. 1996) provides several approaches for meaningful scale transformation. The key concept in renormalization is that a high resolution map is aggregated to coarser scales while preserving a particular geometrical aspect of the map. In the process, inconsequential detail is eliminated, leaving a coarser map that retains some essential characteristic.

For example, assume that the investigator is interested in aggregating a map of forest cover. This might enable a high resolution vegetation map (30 m resolution) to be compared directly to a coarse scale satellite image (1 km resolution). The investigator could inspect each 2×2 block of cells or grains on the 30-m map and substitute the block with a single, larger cell if the block satisfied some criterion such as "the block contains at least one occupied cell." Repeating the process several times would change the cell resolution from 30 to 960 m through a series of resolutions: 30, 60, 120, 240, 480, 960 m. Perhaps the most important question at this point is, "What geometrical aspect should be preserved?"

By the "presence/absence renormalization rule" new cells are occupied if they contain one or more cells of forest cover. Other rules exist that enable preservation of: (1) horizontal connections between neighboring occupied cells within a block, (2) the majority class, (3) percolation clusters, if they exist, and (4) the similarity fractal dimension which is simply $D_s = \ln N(1)/\ln E$, the ratio of (1) the logarithm of the number of occupied cells at $L = 1$ and (2) the logarithm of the extent of the map (Mandelbrot 1982). Upon renormalization to cells of length L', there are $N(L') = (E/L')^{D_s}$ occupied cells, thereby ensuring that the dimension is preserved (Milne and Johnson 1993).

Goals

1. Introduce the routine *RENORM2* that performs renormalizations in ClaraT.
2. Learn about renormalization trajectories and nontrivial fixed points.
3. Interrelate renormalization and the box fractal dimension.
4. Implement a strategy to preserve map density while reducing resolution.
5. Use thresholding to recode grayscale images as binary maps for renormalization.

Tutorial: How to Renormalize a Map

Steps

1. Use *PERC* to make a random percolation map with $p = 0.55$. Choose a large map size (e.g., 128×128 or 256×256).
2. Copy and paste the map into *RENORM2.*
3. Open the "Rule Box." You will see a display of all the 2×2 configurations of occupied (white) and unoccupied (black) cells that can appear on a map.
4. Click on "Go" to apply the rule. You can click on "Go" again to reapply the rule to the new map.

While renormalizing the map, the program will turn the selected (checked) configurations into "occupied" cells at the next coarser resolution.

What You Should See
Because the map started with $p < p_{crit}$ and the default rule preserves connections between occupied cells across each block, the density of occupied cells should decrease because there is no percolation cluster at this density. Had a percolation cluster existed, the rule would preserve it, leading to ever higher densities with each renormalization. Repeat the analysis with a random map at say $p = 0.65$. It should have a percolation cluster and therefore renormalization will drive the map to higher density.

Fixed Points and Nontrivial Fixed Points

As a map is renormalized it undergoes changes in the overall density of points. Some renormalizations, such as the presence/absence rule, drive any map with an initial probability >0 towards a final probability of 1 (i.e., 100% cover). The final density is reached asymptotically as the number of renormalization cycles approaches infinity (but usually requires only a few cycles). The final value is called a fixed point. Other methods, such as the majority rule, drive maps that start above $p = 0.5$ towards 1 but drive maps below 0.5 towards 0; a map that starts at 0.5 remains there. This special initial density (0.5) is a nontrivial fixed point. To implement the majority rule, we need a way of resolving ties (i.e., when a block contains two occupied and two unoccupied cells). In ClaraT, this is accomplished in the custom rule editor by setting the chance for all 2-cell blocks to 50%. In half such cases the block will be turned on, as though a coin were flipped to determine its state. Because of the stochastic nature of this procedure, slightly different maps may appear from one trial to another.

Another important example of a nontrivial fixed point comes from the "percolation renormalization rule" which accepts configurations that include a set of orthogonally connected occupied cells. The percolation rule is the default in *RENORM2*. In cases where two cells in the block are occupied, the default rule includes only those configurations with horizontal connections. This particular rule has a nontrivial fixed point of 0.61, which is close to the theoretical critical percolation density of 0.59 (under the von Neumann neighborhood rule). The small deviation from theory is the result of artifacts produced by renormalizing with 2×2 blocks; some of the actual connections between cells in adjacent blocks are ignored by the procedure (Gould and Tobochnik 1988). Alternatively, the "symmetric" rule of Nakanishi and Reynolds (Creswick et al. 1992) can be used in which a block remains occupied if it contains a cluster that spans vertically or horizontally. In that case, the nontrivial fixed point is 0.879. Thus, the exact value of the critical point has little direct physical significance. Rather, the geometrical properies and diagnostic scaling exponents of the system can be modelled relative to a given critical value (Binney et al. 1993; Milne et al. 1996).

Typically, the fate of a map after repeated applications of a given rule indicates a particular condition present on the map. In the case of the percolation rule, a map that contains a spanning cluster will renormalize to $p = 1$; a map without the cluster will go to $p = 0$. Renormalization eliminates extraneous degrees of freedom related to irrelevant or random control variables, thereby leaving the tell-tale structure because of the relevant deterministic process (e.g., in a model of forest fires) (Loreto et al. 1995).

Renormalization can be used to discover the nontrivial fixed point. The general strategy is to start with a series of maps that range in p from 0 to 1. Then, renormalize each just once using a particular rule, such as the default percolation rule in *RENORM2*. For each map, measure the resulting p value and subtract the original value from the new. These deviations will equal 0 at the trivial fixed points (i.e., $p = 0$ and 1) and at the nontrivial fixed point, p_{crit}.

In the method used to obtain the box counting dimension, grids of various sizes are placed over the map and the number of occupied cells contained at each scale is counted. If the grid cell sizes change as powers of 2, then the procedure is identical to using the presence/absence renormalization rule. Because the number of occupied boxes varies as $N(L) = kL^{-D_b}$ the portion of the map occupied $p(L) = N(L)/(E/L)^2$, where E is the extent of the map. Thus, $p(L)$ varies as $kL^{-D_b}/(E/L)^2 = k/E^2L^{2-D_b}$ and the box dimension governs the probability trajectory.

Because the software enables the user to change the renormalization rule after each cycle, it is possible to use a combination of rules to alter the map density. For example, a map that starts at $p = 0.2$ is below the nontrivial fixed point (0.5) of the majority rule and would normally go to $p = 0$ after repeated applications of the rule. However, use of the presence/absence rule would rapidly move the density above 0.5 and subsequent use of the majority rule would create a final map with p closer to 1. Graphical monitoring of map density (using the plot option) enables the user to "drive" the map towards or away from p values by evaluating the effect of a given rule.

Some rules automatically produce density oscillations. In general, a binary map is created by thresholding a continuous variable. For example, a continuous variable z_i that is less than some threshold may be mapped as $x_i = 0$ while $z_i \geq$ threshold is mapped as a 1. Consequently, the binary map sacrifices information about the magnitude of the difference between neighboring z values, leading to uncertainty about the nature of patch edges and structures on the landscape. Nonetheless, the geostatistical properties (Cressie 1991; Deutsch and Journel 1992) of a binary map (in which the cells have values of $x_i = 0$ or 1) are represented by the indicator semivariance:

$$\gamma(h) = 1/(2\mathrm{N}(h)) \sum_{i=1}^{N(h)} (x_{\mathrm{i}} - x_{\mathrm{i+h}})^2$$

where $N(h)$ is the number of pairs of points separated by the lag distance h. In adopting a renormalization approach to this statistic, one can ask how the semivariance is affected by pattern at different scales and how uncertainties in the thresholding process contribute to the statistical properties of the map.

When lag $h = 1$, one can examine the sum of all squared differences $(x_i - x_{i+h})^2$ to find the contribution that a block makes to the semivariance. Empty and full configurations contribute nothing to the sum (because there is no variation) while configurations with 1, 2, or 3 occupied cells do contribute. The latter occur on a random map at rates equal to $4p(1 - p)^3$, $6p^2(1 - p)^2$, and $4p^3(1 - p)$, respectively. Upon renormalization, maps beginning with p occupied cells emerge with p' occupied cells according to:

$$\begin{aligned} p' &= 4p(1-p)^3 + 6p^2(1-p)^2 + 4p^3(1-p) \\ &= 2p(2 - 3p + 2p^2 - p^3) \end{aligned}$$

(see Gould and Tobochnik 1988; Creswick et al. 1992; or Milne and Johnson 1993 for construction of such functions) with attendant changes in the sum of the squared deviations. In *RENORM2*, the rule can be implemented by using the rule box to select all configurations with 1, 2, or 3 occupied cells, that is, click on each check box next to the configurations such that all are checked, except for the completely filled and completely empty blocks. Applying the rule to random maps will show jagged behavior at first, a tendency to hover around 0.722, and ultimately, oscillation between just two values. Amazingly, study of many maps with various initial probabilities reveals transient behavior which can be quite varied for the first few iterations, but ultimately the function oscillates between two values, $p = 0.441$ and 0.864 (Fig. 14.3). Standard analysis of discrete nonlinear systems can help to elucidate the dynamics of this renormalization function. Specifically, we subscript p according to the iteration number n plus an increment k. Comparison of p_n to p_{n+1} reveals an unstable equilibrium, i.e., $p_n = p_{n+1} = 0.722$. It is unstable because the absolute value of the derivative $|dp_{n+1}/dp_n| = 1.465$ at $p_n = p_{n+1}$ is greater that 1 (Fig. 14.3). Comparison of p_n to p_{n+2} (Fig. 14.3) reveals equilibria at 0.44 and 0.86 with derivatives equal to -1, within experimental error, indicating a stable period-2 oscillation.

The parabolic curve (Fig. 14.3) constructed from the first few interations suggests that the classic logistic model $p_{n+1} = \lambda p_n(1 - p_n)$ with growth parameter λ which controls the type of dynamical behavior expected. Using $p_n = 0.5$ in a polynomial model of the parabola $p_{n+1} = 0.01116 + 3.4345p_n - 3.3975p_n^{\ 2}$ gives $p_{n+1} = 0.879035$. This same mapping of 0.5 to 0.879035 is accomplished with a logistic model and $\lambda = 3.516$. A parameter value of $\lambda = 3.516$ in the logistic model leads to a period-4 oscillation (Turcotte 1992:119). The fact that a period-2 oscillation is observed for the renormalization function must be due to its cubic and quartic terms which cause a slight deviation from a true parabolic shape.

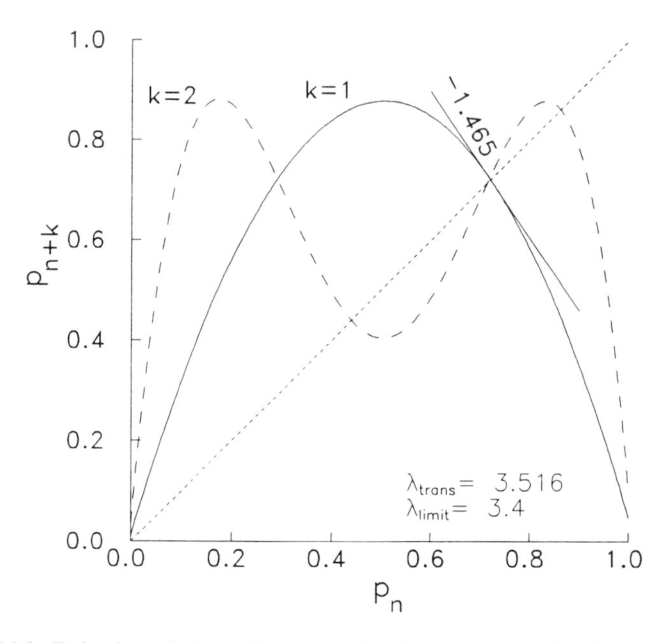

FIGURE 14.3. Behavior of the indicator semivariance renormalization for random maps prior to reaching a stable oscillation. Dependence between the proportion of map cells occupied at one scale (k = 1) versus another (k = 2) reveal equilibria wherever the curves intersect the dashed line. The tangent with slope = −1.465 indicates an unstable equilibrium that prevents the renormalized maps from obtaining an asymptotic proportion of occupied cells.

Thus, renormalization enables the geostatistics of the map to be examined at different scales. The transient behavior stemming from a renormalization rule based on the semivariance criterion tells us that random maps of various p values may have semivariances that show very different dependencies over small lag distances. For a given map, behavior over the first few cycles depends crucially on the initial p value, p_0. Once these initial conditions are renormalized out, maps with p_0 other than 1 or 0 oscillate between two values as an indication of the uncertainty associated with its "true" p value. This is indeed an unusual but fascinating application of renormalization.

The *RENORM2* program includes a tool that enables an image, such as a digital photograph, to be thresholded; values below a selected value are turned to 0. Aggregation of the remaining nonzero values is done by averaging.

Analysis of Self-Affine Functions

The fractal dimension of self-affine functions, such as the sum from Feller's coin flipping experiment (Fig. 14.2), is related to the changes in the function (i.e., the differences between successive times or locations). If we take the sum in Feller's experiment to be our function, then the changes from one flip to the next are the +1 and −1 values of the coin flips. We can also examine changes that occur over 2, 3. or more flips where the number of flips is a lag h. The average change, measured at lag h, will vary as h^H. The function has a fractal dimension $D = 2 - H$ (Voss 1988). As H approaches 0, D approaches 2, and the function becomes more jagged.

A more general explanation of a self-affine function and its scaling behavior is to first consider a string of numbers chosen randomly from a distribution. The sum of the numbers is analogous to Feller's sum. Assume the randomly chosen numbers come from a Gaussian distribution with a mean of zero and a standard deviation of 1 (i.e., a population of random deviates from 0). The probability of obtaining value z is described by the normal probability distribution $\text{prob}(z) = 1/(2\pi)^{1/2}\ \text{EXP}(-z^2/2)$. If 1000 such numbers are chosen randomly, 68.3% of them will fall within one standard deviation of the mean and only 0.03% will be more than 3 standard deviations from the mean. Summing the numbers for $t = 1, 2, 3, \ldots 1000$ produces $X(\mathrm{t})$ which is a self-affine function and t is the time step. An example is given in brown. dat (Table 14.1). To view the file:

1. Invoke the "HURSTW.exe"program.
2. Select "Open Time Series." Use the file browser to select the file called brown.dat. The series will be displayed automatically.

Analysis of brown.dat and other self-affine functions involves comparing the mean change as a function of lag. The "displacement method" in pseudocode is:

1. For each lag from $h = 1$ to $E/3$, where E is the total length of the series, do the following.
2. Compute the average change (i.e., change $= x_{\mathrm{i}} - x_{\mathrm{i+h}}$) by analyzing all pairs of points. Continue for the next lag.
3. Graph the average change for each lag versus the lag. The slope of the doubly logarithmically transformed plot is the scaling exponent H.

To implement this procedure in *HURSTW*, select "Compute Hurst Exponent" from the "Hurst Rescaled Range Analysis" window. Select "Displacements" which are the changes in the function. A new window entitled "Hurst Regression Options" will appear. Use the defaults.

A window entitled "Displacement Plot" will appear showing each change versus the lag. The scaling exponent $H = 0.5068$ will be reported in the widow entitled "Hurst Rescaled Range Analysis."

Notes on Input/Output

The Windows environment enables one to use other software to manipulate images. The images can be cut to the clipboard from other programs and then pasted into the ClaraT software. Printing and file saving are supported through the Windows environment.

Input files for *HURSTW* include a short header followed by the data. To create a new file, one could copy brown.dat to a new file, remove the data portion (which are the lines containing pairs of numbers), insert one's own data, and edit the header to reflect the title and names of the variables. The new file can then be read by *HURSTW*, displayed, and analyzed.

Summary

ClaraT was designed to provide beginning students an opportunity to interact with fractals, both during their creation and analysis. The diversity of generating processes and analysis tools make it clear that there is a wealth of ways in which fractals can be applied to everyday problems in ecology (Milne 1997). Rather than relying on a single dimension, such as the perimeter-area dimension that is used frequently in landscape ecology, ClaraT illustrates methods that are appropriate for habitat estimation, statistical partitioning of patterns by means of Mandelbrot measures, and time series analysis. Moreover, ClaraT introduces a general, stochastic approach to real-space renormalization, itself a doorway into the realm of scaling and universality (Stanley et al. 1996) that has had profound impacts on physical disciplines that routinely address complex systems rivaling those found in ecology.

Acknowledgments. The National Science Foundation (BSR-9058136, BSR-8811906, and BSR-9411976 to BTM and BSR-9107339 to BTM and ARJ) and the Electric Power Research Institute provided generous assistance for the development of ClaraT. Sherri Holmes, John Bissonette, and Bob Gardner made helpful comments on early drafts. Students in Biology 200 and 310 at the University of New Mexico provided helpful feedback regarding the design of ClaraT. Sevilleta LTER publication no. 102.

References

Barnsley, M. 1988. Fractals Everywhere. Academic Press, London.

Bascompte, J., and R.V. Solé. 1996. Habitat fragmentation and extinction thresholds in spatially explicit models. Journal of Animal Ecology 65:465–473.

Binney, J.J., N.J. Dowrick, A.J. Fisher, and M.E.J. Newman, 1993. The Theory of Critical Phenomena: An Introduction to the Renormalization Group. Oxford Science Publications, Oxford.

Burrough, P.A. 1981. Fractal dimensions of landscapes and other environmental data. Nature 294:241–243.

Cressie, N.A.C. 1991. Statistics for Spatial Data. John. Wiley & Sons, New York.

Creswick, R.J., H.A. Farach, and C.P. Poole Jr. 1992. Introduction to Renormalization Group Methods in Physics. John Wiley & Sons, New York.

Deutsch, C.V., and A.G. Journel. 1992. GSLIB: Geostatistical Software Library and User's Guide. Oxford University Press, New York.

Feder, J. 1988. Fractals. Plenum Press, New York.

Feller, W. 1951. The asymptotic distribution of the range of sums of independent random variables. Annals of Mathematical Statistics 22:427.

Forman, R.T.T., and M. Godron. 1986. Landscape Ecology. John Wiley & Sons, New York.

Gardner, R.H., B.T. Milne, M.G. Turner, and R.V. O'Neill 1987. Neutral models for the analysis of broad-scale landscape pattern. Landscape Ecology 1:19–28.

Gould, H., and J. Tobochnik. 1988. An Introduction to Computer Simulation Methods: Applications to Physical Systems, Part 2. Addison-Wesley, Reading. MA.

Grey, F., and J.K. Kjems. 1989. Aggregates, broccoli and cauliflower. Physical D 38:154–159.

Johnson, A.R., B.T. Milne, and J.A. Wiens. 1992. Diffusion in fractal landscapes: simulations and experimental studies of Tenebrionid beetle movements. Ecology 73:1968–1983.

Johnson, A.R., C.A. Hatfield, and B.T. Milne. 1995. Simulated diffusion dynamics in river networks. Ecological Modelling 83:311–325.

Krummel, J.R., R.H. Gardner, G. Sugihara, R.V. O'Neill, and P.R. Coleman. 1987. Landscape patterns in a disturbed environment. Oikos 48:321–324.

Loreto, V., L. Pietronero, A. Vespignani, and S. Zapperi. 1995. Renormalization group approach to the critical behavior of the forest-fire model. Physical Review Letters 75:465–468.

Mandelbrot, B. 1982. The Fractal Geometry of Nature. W.H. Freeman, New York.

Meakin, P. 1986. A new model for biological pattern formation. Journal of Theoretical Biology 118:101–113.

Milne, B.T. 1991a. Lessons from applying fractal models to landscape patterns. In: Quantitative Methods in Landscape Ecology, pp. 199–235. M.G. Turner and R.H. Gardner (eds.). Springer-Verlag, New York.

Milne, B.T. 1991b. The utility of fractal geometry in landscape design. Landscape and Urban Planning 21:81–90.

Milne, B.T. 1992. Spatial aggregation and neutral models in fractal landscapes. American Naturalist 139:32–57.

Milne, B.T. 1997. Applications of fractal geometry in wildlife bilolgy. In: Wildlife and Landscape Ecology, pp. 32–69. J.A. Bissonette (ed.). Springer-Verlag, New York.

Milne, B.T., and A.R. Johnson. 1993. Renormalization relations for scale transformation in ecology. In: Some Mathematical Questions in Biology: Predicting Spatial Effects in Ecological Systems, pp. 109–128. R.H. Gardner (ed.). American Mathematical Society, Providence, RI.

Milne, B.T., A.R. Johnson, T.H. Keitt, C.A. Hatfield, J. David, and P. Hraber. 1996. Detection of critical densities associated with piñon-juniper woodland ecotones. Ecology 77:805–821.

Orbach, R. 1986. Dynamics of fractal networks. Science 231:814–819.

Peitgen, H.-O., and D. Saupe (eds.). 1988. The Science of Fractal Images. Springer-Verlag, New York.

Plotnick, R.E., R.H. Gardner, and R.V. O'Neill. 1993. Lacunarity indices as measures of landscape texture. Landscape Ecology 8:201–211.

Stanley, H.E. 1986. Form: an introduction to self-similarity and fractal behavior. In: On Growth and Form: Fractal and Non-Fractal Patterns in Physics, pp. 21–53. H.E. Stanley and N. Ostrowsky (eds.). Martinus Nijhoff Publishers. Dordrect.

Stanley, H.E., L.A.N. Amaral, S.V. Buldyrev, A.L. Goldberger, S. Havlin, H. Leschhorn, P. Maass, H.A. Makse, C.-K. Peng, M.A. Salinger, M.H.R. Stanley, and G.M. Viswanathan. 1996. Scaling and universality in animate and inanimate systems. Physica A 231:20–48.

Stauffer, D. 1985. Introduction to Percolation Theory. Taylor & Francis, London.

Tilman, D., and D. Wedin. 1991. Oscillations and chaos in the dynamics of a perennial grass. Nature 353:653–655.

Turcotte, D.L. 1992. Fractals and chaos in geology and geophysics. Cambrideg University Press. Cambridge.

Voss, R.F. 1988. Fractals in nature: from characterization to simulation. In: The Science of Fractal Images, pp. 21–70. H.-O. Peitgen and D. Saupe (eds.). Springer-Verlag, New York.

West, G.B., J.H. Brown, and B.J. Enquist. 1997. A general model for the origin of biometric scaling laws in bilogy. Science 276:122–124.

Zallen, R. 1983. The Physics of Amorphous Solids. John Wiley & Sons, New York.

Part V
The Teaching of Landscape Ecology

15 Effective Exercises in Teaching Landscape Ecology

Scott M. Pearson, Monica G. Turner, and Dean L. Urban

The development of landscape ecology and its many applications to land management created a need for courses that address both the conceptual and practical sides of the discipline. Graduate seminars and full-fledged courses in landscape ecology are now featured at many colleges and universities; undergraduate ecology courses may include an introduction to principles of landscape ecology. Because landscape ecology involves the study of spatially explicit ecological patterns and processes along with much larger regions than ecologists have typically studied, landscape ecologists often employ a variety of new quantitative analysis techniques in their work. In particular, metrics are used to quantify spatial patterns, and the importance of spatial heterogeneity for ecological processes is evaluated. Modeling also plays an important role in landscape ecology because it is logistically impossible to conduct truly replicated experiments across entire landscapes. Students of landscape ecology, even at the undergraduate level, need some familiarity with the tools of the discipline to gain confidence in the practice of landscape ecology and to develop a critical understanding of the strengths and weaknesses of these techniques.

This chapter contains six exercises created to teach concepts in landscape ecology. All three authors currently teach ecology at the undergraduate and/or graduate levels and incorporate landscape ecology principles in their specialized and general courses. The text of each exercise is written for general use in a class; notes specifically to the instructor and recommended readings are included in the appendices.

This collection of exercises stresses three main aspects of landscape ecology. Exercises I and II emphasize the quantification of landscape pattern. The first exercise is designed to familiarize students with straightforward techniques for quantifying the similarities and differences between landscapes. The second demonstrates the important influence of spatial scale (both grain and extent) and classification scheme on landscape metrics. Exercises III and IV address the interpretation of landscape patterns. The third exercise allows students to quantify changes through time in a landscape, challenging them to consider where and why these changes occur.

The fourth exercise has students interpret a landscape from the perspective of four nonhuman species which vary in their vulnerability to human influences. This exercise demonstrates how the same landscape can functionally be quite different for various species. Exercises V and VI foster understanding dynamic landscapes and lead students through the process of generating working hypotheses about drivers and mechanisms of landscape change (i.e., landscape models). These last two exercises help students bridge the intellectual gap between quantifying pattern and understanding the processes underlying landscape pattern and change.

Exercise I: Neutral Models and Landscape Connectivity

Background

This exercise is about modeling and spatial heterogeneity, with particular reference to landscape ecology. Landscape ecology is defined by two characteristics: (1) landscape ecology often studies ecological processes over very large areas (such as the upper midwest, or all of Yellowstone National Park, or the southern Appalachian Mountains) that include a variety of different ecosystems or habitats, rather than focusing only one type of ecosystem; and (2) landscape ecology explicitly studies the effects of spatial patterning—heterogeneity—on ecological processes (such as the movement or dispersal of organisms or the spread of natural disturbances). Therefore, landscape ecological studies may involve studying the amount and spatial distribution of a particular habitat type over a large geographic area and understanding the effects of different habitat arrangements on particular species or ecological processes. For example, the study of how the amount and spatial arrangement of old-growth habitat affects population dynamics of the northern spotted owl in the Pacific Northwest is an example of a landscape study.

To understand the relationship between spatial pattern and an ecological process, ecologists need to know how to quantify spatial patterns and also have some "yardstick" against which the effects of particular spatial patterns can be measured. Considerable effort has gone into developing pattern metrics that can be compared across different landscapes or monitored through time. These include intuitive attributes like number of patches of a habitat type, the size distribution and mean size of the patches, and the ratio of edge to area for the habitats. It is important to be able to tease apart the effects of the total amount of the habitat from the effect of its spatial arrangement. Students will examine these effects in this exercise using a neutral model. The neutral model serves as the yardstick for comparison with actual landscapes.

Purpose

This lab will introduce you to a neutral landscape model based on percolation theory (Gardner et al. 1987; Gardner and O'Neill 1991). Percolation theory was developed in the physical sciences to explain and predict the processes that lead to connectivity across a two-dimensional space (Stauffer 1985). Its development was motivated by questions such as, How much metal must be plated across a surface so that electricity can flow across it? A physicist would want to have just enough gold to maintain conductivity, but perhaps not extra because of the cost. Percolation theory studies the properties of clusters, or patches as ecologists would say, across a two-dimensional space. Ecologists also are interested in questions that deal with connectivity or conductivity across two-dimensional space. For example, How much habitat must be present for a red-backed vole to move across a given landscape? How much flammable forest must there be for a fire to spread (or stop spreading) across a landscape? Because of the similarity in the questions and the well-developed theory in the physical sciences, percolation theory has been applied in ecology to develop neutral models for landscape patterns (e.g., Gardner et al. 1987; Turner et al. 1989; Andren 1994; With et al. 1997).

Why develop neutral models? One approach to modeling is to develop very simple models to compare with empirical data to see how well they fit. If the predictions of a very simple neutral model fit satisfactorily with the data, it may not be necessary to develop more complex approaches. However, it is often more informative and interesting if there is a relevant difference between the model predictions and the empirical data. Then it is possible to expand the neutral model and learn what additional features must be included to achieve agreement with the data—that is, What other parameters are important?

The exercise contains two parts. In the first part, students will develop percolation maps (that is, the neutral model of a landscape) and observe how the spatial characteristics on these maps change with the abundance of a particular habitat type. The set of characteristics describing the pattern will be plotted (on the *Y* axis) against the proportion of the map occupied by the habitat (on the *X* axis), and the shapes of the curves will be examined. In the second part, students will quantify the spatial patterns of land and water from different portions of the Wisconsin landscape by using topographic maps provided. To illustrate an important concept—that the scales at which we conduct our studies influences our answers, something true for science as a whole—these patterns will be quantified at two spatial resolutions on each topographic map. Results from the whole class will be synthesized to make two comparisons: (1) How different or similar are random (i.e., the neutral model) maps and actual landscapes, and (2) How different or similar are the spatial patterns of land and water in Wisconsin when quantified at different scales from the same maps?

Procedure

Read the instructions for the exercise in advance. The Introduction and Discussion sections from Gardner et al. (1987) should be included with the handout as background material.

Random Maps

Percolation theory provides a framework for examining landscapes as two-dimensional grids, usually square grids of size $m \times m$ containing m^2 unique sites or cells. Gardner in chapter 13 uses computers to create random two-dimensional maps of various sizes ranging from 50×50 to 500×500 cells. Here, you will use pencil and papers to generate smaller 10×10 grids containing 100 unique sites or cells.

Consider the probability, $\boldsymbol{p}$, that any of the 100 cells is occupied by a particular habitat type (e.g., forest or grassland). In an empty (homogeneous) 10×10 grid, $\boldsymbol{p} = 0$. If you choose two random coordinates (x,y) and fill in that cell on the grid, then $\boldsymbol{p} = 0.01$. If $\boldsymbol{p} = 0.10$, then 10 cells are occupied; when $\boldsymbol{p} = 1.0$, all 100 cells are occupied. When the grid contains some cells of the habitat of interest, several properties about its spatial arrangement can be measured. For example, we can measure the number of habitat patches or clusters, their sizes, and the amount of edge surrounding the habitat patches. On our hypothetical 10×10 landscape where $\boldsymbol{p} = 0.01$, we observe:

C = number of clusters = 1
L = size of largest cluster = 1 cell
O = number of outer edges = 4
I = number of inner edges = 0

In this exercise, the following definitions and rules for describing the spatial patterns will be followed. These calculations are illustrated for the simple maps shown in Figure 15.1.

1. An edge is a surface of an occupied cluster adjacent to an unoccupied area.
2. Outer edges lie along the outside of a cluster.
3. In contrast, inner edges are adjacent to unoccupied areas completely enclosed by a cluster, like the holes in Swiss cheese.
4. Two clusters only merge into one when they share an horizontal or vertical edge; a diagonal does not connect clusters.
5. When it falls on the edges of the grid, the outside-most edge of the patch should always be included in your count of outside edges, but not for your count of inside edges.

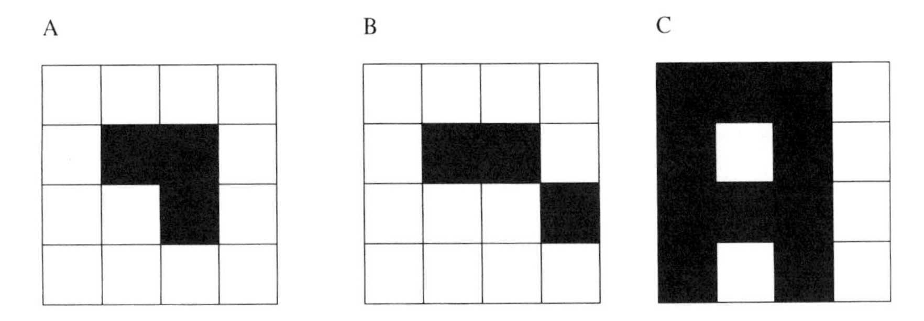

FIGURE 15.1. Simple 4 × 4 grids illustrating the rules for identifying habitat clusters and quantifying their inner and outer edges. (A) one cluster, with 8 outer edges and 0 inner edges; (B) two clusters with a total of 10 outer edges and 0 inner edges; and (C) one cluster with 4 inner edges (not 7) and 16 outside edges (not 7).

To be sure that you understand how to count clusters and outer and inner edges, consider the following example (Fig. 15.2), where $\boldsymbol{p} = 0.45$. (Answers are at the end of this portion of the exercise.)

What is C? ——— L? ——— O? ——— I? ——— $O + I$ ———

To further check your understanding, consider a completely filled grid ($\boldsymbol{p} = 1.0$):

What is C? ——— L? ——— O? ——— I? ——— $O + I$ ———

Look back at Figure 15.2, where $\boldsymbol{p} = 0.45$, and note that although the grid is almost half full, it is not possible to "travel" from one edge of the grid to the other edge on occupied (filled) clusters. (Remember that a diagonal does not connect clusters; travel across a diagonal is disallowed). When the habitat is not connected, we say that the grid does not percolate. From percolation theory, we know that on a random map with the rules defined above, habitat will suddenly become connected at $\boldsymbol{p} = 0.5928$ (Stauffer 1985). This value is called the critical threshold for percolation, or $\boldsymbol{p}_c$. In the exercise explained below, students should begin to watch for percolation at approximately $\boldsymbol{p} = 0.50$.

Work in groups of five or six to generate random maps with $\boldsymbol{p}$ ranging from 0.0 to 1.0 and tracking the habitat patterns as follows. One or two students will draw random x, y coordinates and fill in the habitat on the grid using an erasable marker. As the grid is filled in, three students should track the pattern, quantifying C, L, O, I and $O + I$. The first can count the number of clusters and the largest cluster size and watch for percolation when $\boldsymbol{p} > 0.50$. The value of $\boldsymbol{p}$ at which percolation is observed should be recorded. The second student should count the total number of outer edges, and the third student can count the total number of inner edges. Be particularly attentive to these edge calculations, since filling in "holes" or creating

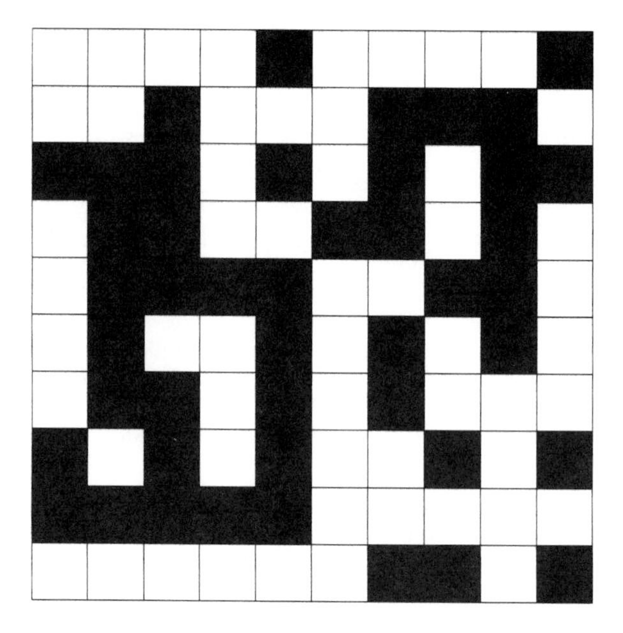

FIGURE 15.2. A simple random 10 × 10 map for which $P = 0.45$.

clusters can dramatically change the number of edges on the map. Another student should record the data for all 100 "moves" from $p = 0.0$ to $p = 1.0$. (Data sheets and write-on, wipe-off 10 × 10 grids should be provided by the instructor.) You should always work all the way through the exercise (not just stopping when "percolation" is reached). Ideally, your class should have several groups of students do this exercise twice, then compile all the results. Results should be plotted as a function of the p value of the map.

"Real" Maps at Different Scales

This part of the exercise asks whether the characteristics of "real" landscapes are similar to the characteristics of landscapes produced by the random neutral model. In addition, the exercise examines how the quantification of the landscape pattern may differ with spatial resolution. A series of topographic maps is provided along with transparent 10 × 10 transparent grids of two different sizes that will be superimposed on the topographic maps. Grids will be overlaid on the maps beginning in the top left corner and working along to the bottom right on each map. Twelve large grids are indicated by pins on the maps; four small grids fit within each of the large grids (48 small grids per map).

On each grid, estimate five characteristics: (1) p, the proportion of land covered by lakes; (2) the size of the largest lake in the grid; (3) the number of lakes in the grid; (4) whether it is possible to traverse the grid horizontally; and (5) whether it is possible to traverse the grid vertically. Data should be recorded (see Fig. 15.3) for at least one fourth of the map with the smaller grid, or the entire map should be used with the larger grid. (In a larger class, each laboratory section can analyze a different map.) The size of the largest lake, number of lakes detected, and the presence or absence of percolation (both horizontal and vertical) should be plotted against ***p*** for each topographic map and compared with the neutral models. In your plots,

DATA SHEET - FINER SCALE

MAP LAYOUT:
Note that 12 larger grids fit across the 7.5′ topo map, and 4 smaller grids fit within each larger grid.

MAP NAME:

A1a	A1b	A2a	A2b	A3a	A3b
A1c	A1c	A2c	A2d	A3c	A3d
B1a	B1b	B2a	B2b	B3a	B3b
B1c	A1c	B2c	B2d	B3c	B3d
C1a	C1b	C2a	C2b	C3a	C3b
C1c	C1c	C2c	C2d	C3c	C3d
D1a	D1b	D2a	D2b	D3a	D3b
D1c	D1c	D2c	D2d	D3c	D3d

GRID	Lakes (P)	# of Lakes	Size of Largest	Trav. Vert?	Trav. Hor?	GRID	Lakes (P)	# of Lakes	Size of Largest	Trav. Vert?	Trav. Hor?
A1a						C1a					
A1b						C1b					
A1c						C1c					
A1d						C1d					
A2a						C2a					
A2b						C2b					
A2c						C2c					
A2d						C2d					
A3a						C3a					
A3b						C3b					
A3c						C3c					
A3d						C3d					
B1a						D1a					
B1b						D1b					
B1c						D1c					
B1d						D1d					
B2a						D2a					
B2b						D2b					
B2c						D2c					
B2d						D2d					
B3a						D3a					
B3b						D3b					
B3c						D3c					
B3d						D3d					

FIGURE 15.3. Example of data sheet indicating the position of both large and small grids to be positioned across a 7.5′ topographic map and the format for recording data on the spatial pattern of land and water on for each of the small grids.

use an open circle if the data came from the large grids and a solid dot if the data came from the small grids. Should plots from the random and "real" map data look the same? Does the scale of the sampling affect the results for each map?

Summary and Discussion

Answers to Sample Exercises

$p = 0.45$, C = 10, L = 23, I = 10; $p = 1.00$, C = 1, L = 100, O = 40, I = 0.

Questions for Discussion

1. Why might percolation be observed at values other than the critical threshold, $p_{crit} = 0.5928$?
2. Why should real landscapes differ (or not differ) from random maps? How might these differences relate to the forces, both natural and anthropogenic, that create the pattern?
3. What kinds of ecological processes might be affected by thresholds of connectivity, and how might you detect their responses?
4. Why should the manager of a wilderness preserve or a regional planner be concerned about critical thresholds of habitat connectivity?
5. Can ecologists compare data collected at different scales? Why or why not, and under what conditions?

Exercise II: Constraints on Landscape Pattern Analysis

Purpose

The objectives of this exercise are (1) to gain hands-on experience with the analysis of landscape structure on digitized maps by using some standard (representative) landscape metrics; (2) to explore the implications of changes in grain and extent of the landscape data on the results of the analyses; and (3) to explore the effects of altering the classification scheme on the results of the analyses.

Procedure

Work in groups of four. The analyses can be conducted on raster data that you already have, such as from individual research projects, or 100 × 100 cell subsets of larger GIS data bases provided in class. Landscape metrics can be computed by using (1) stand-alone code provided by the instructors, such as SPAN (Turner 1990); (2) FRAGSTATS (McGarigal and Marks 1995); (3) r.le (Baker and Cai 1992), if you have access to this interface with the GRASS geographic information system; or (4) other code to which the students have access. The instructor should provide detailed instructions on accessing the data set and for running the analysis program to be used. For illustration, the following text assumes the use of a 100 × 100 landscape to be analyzed with SPAN.

Effects of Changing Grain and Extent on Landscape Metrics

Two sets of analyses are to be completed here. Copy the initial data file to a new file name, then edit the new file to change its grain size. (If you are good at programming and can write a quick code to do this, it can be done on the computer; however, editing the file manually is fine, and actually makes the point well).

First, the map will be reduced from 100 × 100 to a 50 × 50 by taking each 2 × 2 "window" and replacing the four grid cells in the window with a single value. The replacement will be by majority rule, that is, the dominant cover type "wins"; if there is no dominant, roll a die or do some other random assignment. For example, the following 2 × 6 array would be reduced to a 1 × 3 with the following composition:

```
223456    236
233343
```

where the 2 and 3 are obtained from the majority rule, and the 6 is a random assignment. This can be done manually in a word processor (make sure you save the file as text only!). Note that the number of rows and columns must be adjusted in the spatial analysis program. students follow the same procedure for a 4 × 4 window and a 5 × 5 window (which give you matrices of 25 × 25 and 20 × 20, respectively. The original and each of the new maps should be analyzed with SPAN, and selected metrics (students' choice) plotted as a function of grain size to show how they change with this component of scale. NOTE: For the interested, you can also experiment with alternative assignment rules to see how the mode of aggregation influences results (for ideas, see Gardner and O'Neill 1991).

Second, leave the original grain size alone but successively reduce the size of the landscape array by units of 10 rows and 10 columns. Run SPAN on each new map from the 100 × 100, 90 × 90, 80 × 80, ..., 10 × 10. Again, plot the metrics as a function of extent of the map to determine how the results are influenced by spatial extent.

Effects of Classification Scheme on Landscape Metrics

In this part of the exercise, the grain and extent will be left alone (e.g., the matrix will remain 100 × 100 in size), but the categories of land cover used for the analyses will be reclassified. You should explore the effects of at least two alternative ways of aggregating the data; for our purposes, students will always be reducing rather than increasing the number of categories. The aggregations can be done by lumping like categories into a single category. For example, with data on forest composition and age, one might aggregate by species (i.e., lumping age classes) or by age classes (i.e., lumping species). Landscape metrics should then be presented in a table by the classification scheme employed; results for the original landscape map should be included for comparison.

Summary and Discussion

Products

Results should be submitted as group reports. Reports should contain three parts: (1) a description of what was actually done for each problem, (2) graphs depicting the results, and (3) a thoughtful interpretation/discussion of the implications of changes in grain and extent and of sensitivity to the classification scheme for landscape analyses. Pay particular attention to part (3). Your interpretation is one of the most important efforts for this exercise. Be sure to cite the appropriate literature.

Questions for Discussion

1. How important is the selection of categories used in an analysis of landscape pattern? What are the implications of different classification schemes for the comparison of different landscapes or changes through time in a given landscape?
2. What are the advantages and limitations of various metrics of landscape pattern with regard to their sensitivity (or lack thereof) to changes in grain, extent and classification?
3. One metric alone is not sufficient to describe a landscape adequately, but how many are needed and why?
4. Can the results of a landscape pattern analysis be extrapolated to other scales? How?

Exercise III: Quantifying Land-Cover Change

Background

Landscapes change through time because of natural processes (e.g., disturbance, succession) and human use (e.g., urban growth). The type and rates of these changes can be quantified from remotely sensed data taken at different times. This lab exercise is designed to familiarize the student with a technique for quantifying land-cover change from a time series of land-cover maps. The land-cover maps used in this exercise show change around Franklin, North Carolina. The maps were developed from Multispectral Scanner (MSS) images taken in 1975 and 1986 and show the distribution of forest, grassy/brushy, and unvegetated/urban land covers. The dimensions of both maps are 210 rows × 180 columns; pixel size is 90 × 90 m.

Purpose

This exercise will address the following research questions:

1. For a given land-cover type, what is the probability of change during the period 1975 to 1986?
2. Which land-cover type is the most stable through time? Which one is the most unstable?
3. Given its 1975 land cover, what is the projected land-cover for the same location in 1986?

Procedure

These questions will be answered by constructing a transition probability matrix by sampling random locations on the two land-cover maps.

Sampling the Maps

1. Use a random number table to select 50 points from the sampling-grid transparency provided with this exercise. ⟨For example, number the rows and columns of the grid. Then, draw a series of two-digit, random numbers. Use each number to designate the row or column address of a sampling point (if the random number exceeds the number of rows or columns, just take the next number).⟩
2. Place the sampling-grid transparency over the 1975 map (Fig. 15.4). Use a pen to mark the corners of the map on the transparency so it can be placed on top of the 1986 map (Fig. 15.5) in the same manner. Use paper clips to secure the transparency to the map. Working from the top left toward the bottom right, record the land-cover class for each point in Table 15.1. Number the points on the transparency with your pen as you record each one.
3. Place the transparency on the 1986 map (Fig. 15.5). Line up the corners of the map with the marks on the transparency. Record the land-cover class for each point, working through the same sequence of points used for the 1975 map.

Calculating the Transition Probabilities

4. Using the data in Table 15.1, tally the number of occurrences for each of 1975 to 1986 land-cover combinations.
5. Total each row and column of Table 15.2. Divide the row and column totals by the number of sampling points to estimate the frequencies of

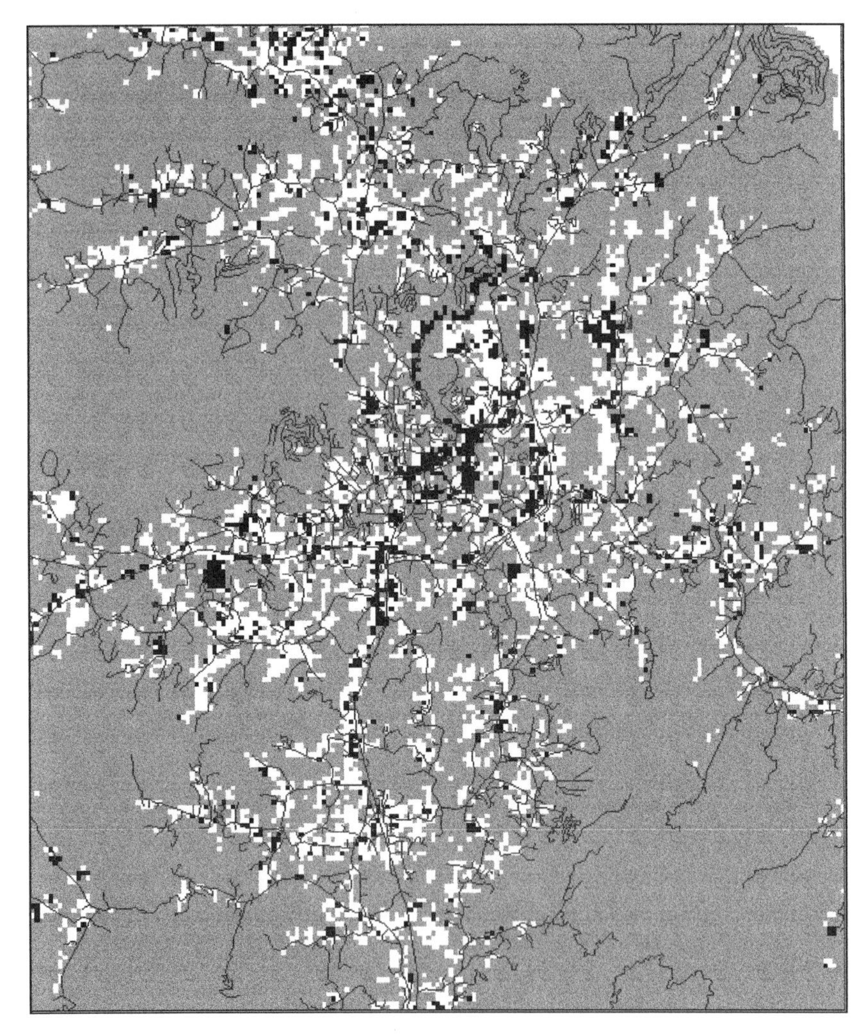

FIGURE 15.4. Land cover around Franklin, NC, in 1975. This land-cover map was developed from a MSS image. The land-cover classes are as follows: forested (gray), grassy/brushy (white), and unvegetated/urban (black). Roads are shown as black lines.

each land-cover type for each year. Did the frequencies change across time?

6. Calculate the transition probabilities for each 1975 land-cover (row in Table 15.2) by dividing the number in each cell by the row total. Record the result in Table 15.3. This conditional probability estimates the likelihood of the 1986 land cover, given a particular 1975 land cover at a location.

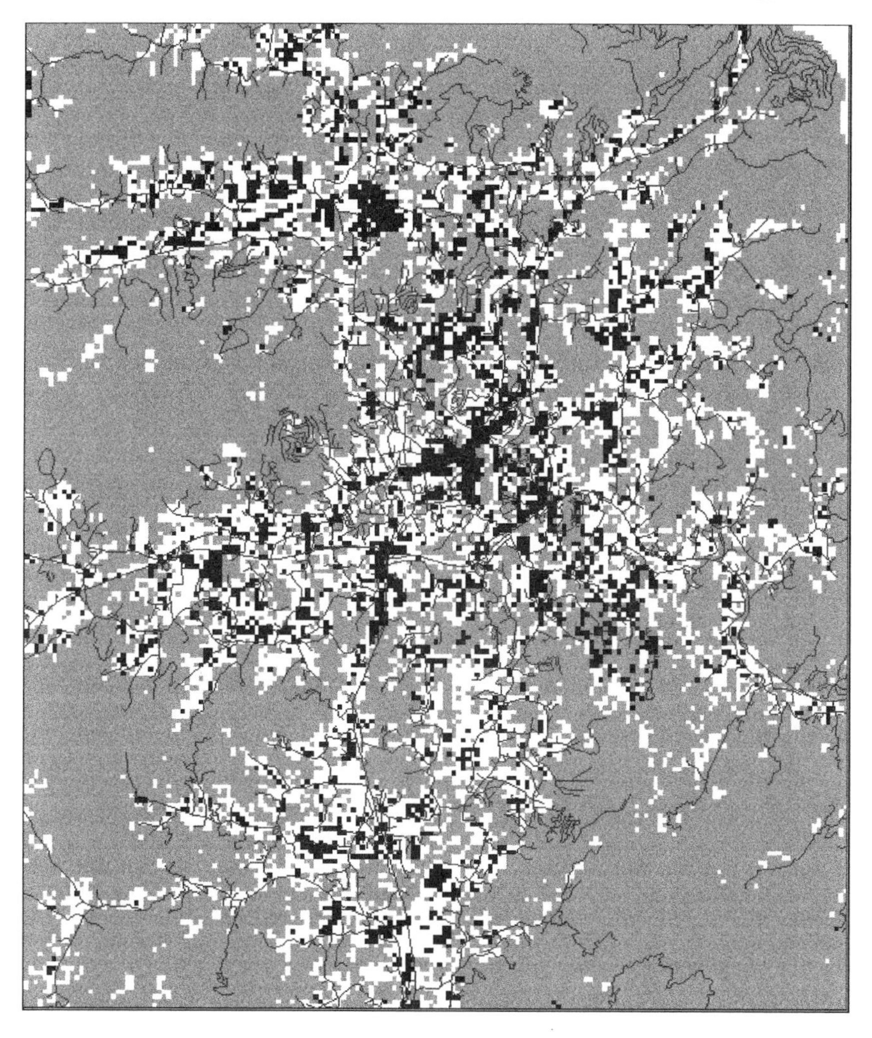

FIGURE 15.5. Land cover around Franklin, NC, in 1986. This land-cover map was developed from a MSS image. The land-cover classes are as follows: forested (gray), grassy/brushy (white), and unvegetated/urban (black). Roads are shown as black lines.

Summary and Discussion

1. Did the frequency of land-cover types change across time? Which cover types showed increases/decreases?
2. Given a site was in forest during 1975, what is the probability it remains in forest in 1986? What is the probability it becomes grassy or unvegetated?

TABLE 15.1. Land-cover at each sample point for the 1975 and 1986 maps.

Point	1975	1986	Point	1975	1986
1			26		
2			27		
3			28		
4			29		
5			30		
6			31		
7			32		
8			33		
9			34		
10			35		
11			36		
12			37		
13			38		
14			39		
15			40		
16			41		
17			42		
18			43		
19			44		
20			45		
21			46		
22			47		
23			48		
24			49		
25			50		

TABLE 15.2. Land-cover change frequencies.

1975 Land cover	1986 Land cover			1975 Totals
	Forest	Grassy	Unvegetated	
Forest				
Grassy				
Unvegetated				
1986 Totals				

TABLE 15.3. Transition probability matrix.

1975 Land cover	1986 Land cover		
	Forest	Grassy	Unvegetated
Forest			
Grassy			
Unvegetated			

3. Which land cover class was most stable (cover type is likely to remain unchanged) through time? Which one was most unstable? Speculate about the reasons some cover types are more stable than others.
4. This analysis assumes that the processes affecting land-cover change in this map are homogeneous across space. Is this assumption valid?

Exercise IV: Organism-Based Views of the Landscape

Background

One of the challenges of ecosystem management is understanding the effects of landscape-level changes on biological diversity. Depending on their habitat requirements and life-history attributes, species may respond quite differently to landscape changes. Changes that favor one species may reduce the habitat for others. The abundance and spatial pattern of habitat in a landscape can vary between species because species have different habitat requirements (e.g., preferences for late versus early successional stages). Moreover, life-history attributes, such as area requirements and vagility, can interact with the spatial pattern of habitat (i.e., fragmented vs. connected) to affect population dynamics on a landscape. Therefore, an organism-based perspective (e.g., Wiens 1989; Pearson et al. 1996) is needed to estimate the effects of landscape pattern on nonhuman species.

Purpose

The goal of this laboratory exercise is to illustrate how landscape patterns, recorded on land-cover maps, can be interpreted from the perspective of different species. Habitat maps will be produced for four species: mountain dusky salamander (*Desmognathus ochrophaeus*, a native amphibian), princess tree (*Paulownia tomentosa*, an exotic tree), showy orchis (*Orchis spectabilis*, a native herb), and wood thrush (*Hylocichla mustelina*, a forest-interior breeding bird). The following research questions will be addressed:

1. Is the abundance and spatial pattern of habitat similar for both native and exotic species?
2. Does the area requirement of native species affect the suitability of landscapes?

Procedure

The land-cover map used for this exercise was produced from a 1986 Multispectral Scanner (MSS) image of a region northeast of Franklin, North Carolina (Fig. 15.6). The land-cover types include: mixed forest, mesic forest, unvegetated, and grassy/brushy (see map legend). Landscape metrics for these land covers are listed in Table 15.4. The forests of this area are mostly deciduous interspersed with occasional pines. Mesic forests (cove forests) are found on slopes and ravines with north-facing aspects.

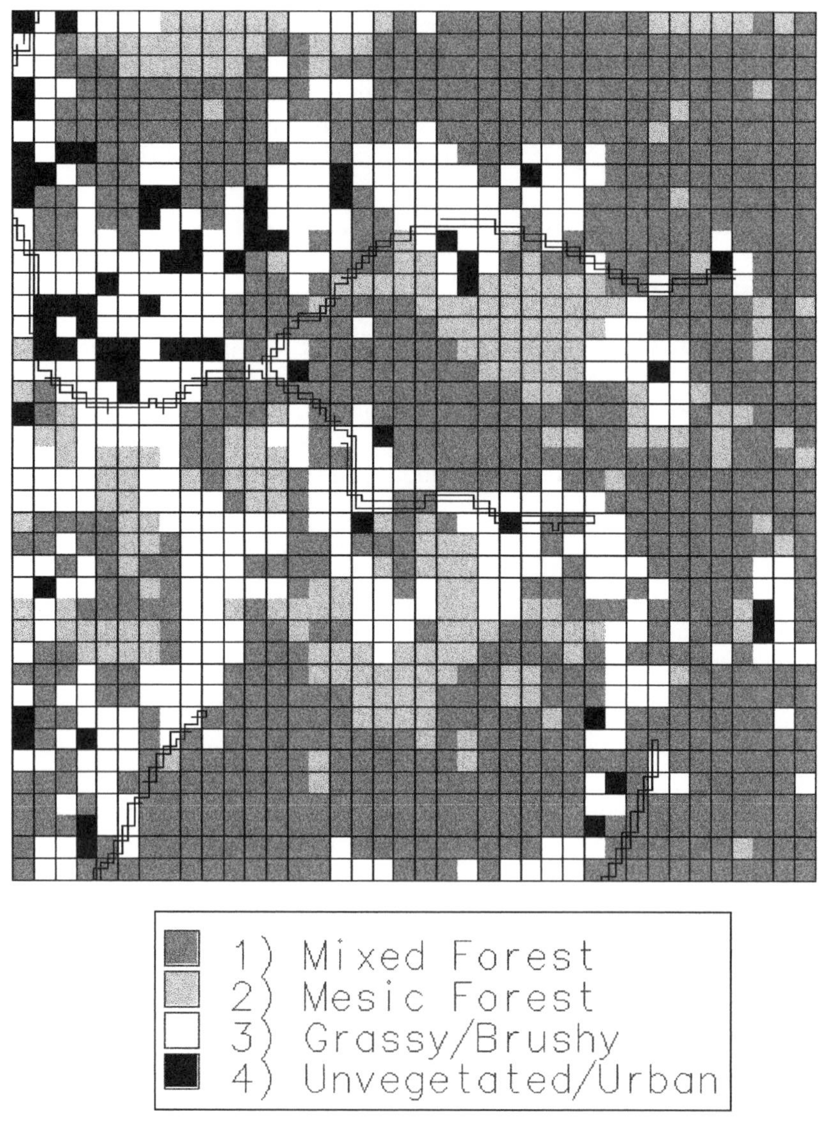

Figure 15.6. Land-cover map of region east of Franklin, NC. Pixel size is 90 × 90 m.

The elevation ranges from 638 to 900 m above sea level. This map can be used to produce habitat maps for each species by applying a habitat recipe based on requirements (Table 15.5).

Making Habitat Maps

Obtain four copies of the land-cover map; use one copy for each species. Secure a piece of mylar over the map using paper clips. Use a marking pen

TABLE 15.4. Landscape metrics for land-cover map. Units are in cells. Each cell is 0.81 ha.

Land cover	Total area in cells	Number of patches	Mean patch size	Area of largest patch
Mixed forest	839	5	167.8	792
Mesic forest	219	40	5.5	48
Crassy/brushy	400	27	14.8	169
Unvegetated/urban	62	28	2.2	13

to color in all map cells that are suitable for the species of interest. Make a map for each species; be sure to label the mylar sheets with the species names.

Quantify Habitat Abundance and Pattern

A *patch* of suitable habitat is defined as a group of contiguous cells. For each patch, record its size by counting the number of cells. Record the patch number and sizes in Table 15.6. Calculate the total area of habitat (in cells) and mean patch size, and note the size of the largest patch for each species.

Summary and Discussion

1. Compare the abundance of habitat among the species. Which species has the most habitat in this landscape? Which one has the least?
2. The fragmentation of species' habitats can be compared by examining the mean patch size, number of patches, and size of largest patch. For which species is its habitat most connected?—most fragmented?
3. The wood thrush can use both types of forest in this landscape; however, it is restricted to forest-interior cells. Compare the total number of cells of thrush habitat to the total number of forested cells (Table 15.4). What percentage of the forest cells are unsuitable for the thrush because of edge effects?

TABLE 15.5. Habitat requirements and mapping recipes for species.

Species	Habitat required	Mapping recipe
Mountain dusky salamander	Forests with streams	Forest cells crossed by or adjacent to streams.
Princess tree	Open habitats, disturbed sites	Unvegetated and grassy cells
Showy orchis	Rich woods and stream banks	Mesic forest cells and mixed-forest cells adjacent to streams
Wood thrush	Forest-interior sites	Forest cells at least two cells away from unvegetated and grassy cells

Requirements taken from Wofford (1989), Hamel (1992), and Robinson et al. (1995).

TABLE 15.6. Patch-based statistics for each species' habitat map. Record patch sizes in number of cells.

Species	List of patch sizes	Total cells	Number of patches	Mean patch size	Area of Largest patch

4. Suppose that we evaluate the landscape from a perspective of another species, such as a broad-winged hawk (*Buteo platypterus*), that requires the same habitat as the wood thrush but has a minimum area requirement (e.g., territory size) of 50 cells (40.5 ha). What proportion of the patches would be too small? What proportion of the forested cells would therefore be unsuitable? What effect would an expansion of nonforest land covers have on this species?
5. Limitations in dispersal ability may prevent some species from recolonizing patches that have experienced local extinctions. Lungless salamanders are such species because they can seldom cross dry, open land covers. If we assume that mountain dusky salamanders cannot cross more than two cells of unsuitable habitat, how many of the existing patches of salamander habitat are isolated with respect to potential colonists from other patches?
6. If urban expansion in this landscapes increases the extent of grassy and unvegetated land-covers, how will each of these species be affected? Will these effects depend on the spatial pattern (where and how much) of urban expansion?
7. Given a scenario of future urban growth and the potential to regulate the location of that growth, what portions of the landscape would you protect? Which species would influence your strategy?

Exercise V: Agents of Landscape Pattern

Background

The agents of pattern formation on landscapes include the physical template (abiotic gradients such as temperature and precipitation as influenced by elevation; edaphic heterogeneity), biological processes (demographic processes such as: establishment, growth, and mortality; competition; dispersal), and disturbance (natural as well as anthropogenic regimes). Inferences about the relative importance of these agents in shaping any particular landscape are confounded by interrelationships among the agents (e.g., fire regimes that are conditioned by forest pattern and by topography), and also by the sheer logistical difficulties of collecting data at landscape scales. The central problem in this issue is to devise analysis strategies that can partition the relative importance of pattern-generating agents most efficiently, that is, to provide the most information for the least amount of hard-bought data.

Purpose

The objective of the exercise is to develop a logical framework for quantitative analysis of landscape pattern, partitioning the relative importance of the physical template, biotic processes, and disturbances in governing the distribution of vegetation types or focal species.

Procedure

Approach

The strategy for landscape analysis focuses on an additive regression model, such as forward-selection stepwise regression or regression trees (see below). The approach is to add explanatory variables into the analysis sequentially, choosing the variables and the sequence according to a priori hypotheses (choosing the most likely predictors first) and also according to logistical considerations (specifying the necessary data strategically). In general, this approach amounts to choosing a likely predictor variable, specifying how and where it should "fail" (misclassify) under given circumstances, and then adding predictions about these residuals as the next stage of the analysis. This process is iterative, with additional layers added until no further improvements can be anticipated. This approach is also consistent with a "levels of activity" program funded at varying levels and thus with varying capacity for fieldwork and analysis. For example, one might propose to perform only a few iterations of this process under a low level of funding (i.e., few personnel and little time), but pursue the analysis to additional levels if more funding (personnel, time) was available.

Preparation

The key concepts related to this exercise are concerned with methods for characterizing the physical template (e.g., terrain analysis, geometric models of solar radiation, methods for interpolating climate over complex landscapes); the action of demographic processes, competition, and dispersal in generating or amplifying pattern; and the role of disturbance acting alone and disturbance as it interacts with other agents.

The multiple regression methods tend to be most helpful in this area (e.g., a forward-selection, stepwise model). Classification and regression trees (CART; e.g., Michaelsen et al. 1987; Venables and Ripley 1994; MacNally 1996) are especially appealing because the "flowchart" or tree structure of these methods are a natural fit for this approach. Consequently, CART will seem natural even if you have no prior experience with this analytical technique. Some familiarity with GIS (overlays, buffering) will also be helpful.

Protocol

For this exercise, you will read a paper describing the distribution of some species or land cover type, and then outline an analysis to explain the observed distribution. You should work in a small group of students—three to six participants, with one student acting as moderator—to develop these analyses. Specifically, your group should:

1. Outline the sequence of steps in the analysis in terms of which variable would be entered, how it would be quantified from field or map data (i.e., what data would be required), and the form and direction of the expected relationship.
2. Detail the field or map data needed to verify the predictive model (this data collection effort could include a combination of pilot studies, the main field campaign, and any follow-up studies implied by the analysis). Emphasize *where* these data would be collected.
3. Explain how the results of the analysis would be interpreted, with particular attention to model failures (predictive residuals or misclassifications). It is the residuals or misclassifications that serve as the point of departure for the next stage of the analysis.
4. Summarize the analysis in terms of a flowchart that illustrates the logical flow of the analysis, with key decision points (branches of the tree) explained.

Example 1: Live Oaks in California Foothills

One example of this approach can be reconstructed by embellishing an analysis conducted by Davis and Goetz (1990) (with sincere apologies to the authors for willful recasting of their study to meet this need). The problem is concerned with predicting the distribution of live oaks in the foothills of California. The facts relevant to this contrived example are these. The oaks tend to be found on more mesic sites, which are defined by topographic moisture as driven by solar radiation (a function of slope and aspect), drainage (a function of slope and upslope contributing area), and soil water-holding capacity (estimated from parent material). Thus, the physical template is derived via terrain analysis and a geology map. But oaks occur frequently on sites not predicted to be oak habitat, and also fail to occur on sites predicted to support oaks. The second step of the analysis is to add variables to explain these misclassifications, and so on. The analysis might produce a regression tree and flowchart that looks like Figure 15.7.

In this example, the logic is that some oaks might occur on "non-oak" sites if there was a sufficient dispersal rain to support them in habitats that are demographic sinks (Fig. 15.7). On the other hand, oaks might fail to occur on mesic sites if there was some natural (fire) or anthropogenic disturbance (development, firewood harvesting) operating on those other-

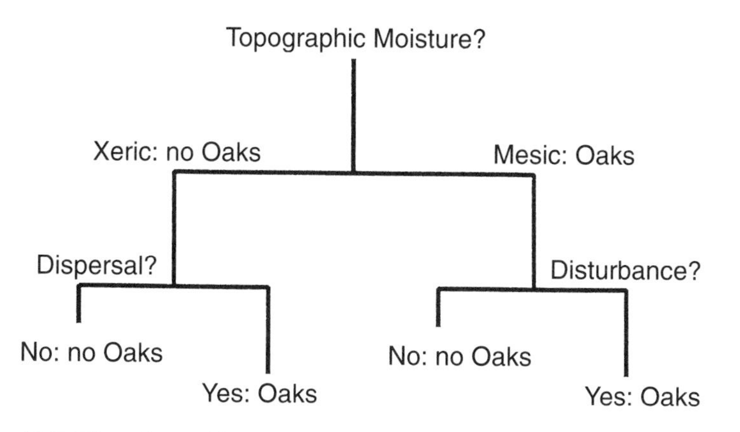

FIGURE 15.7. Flow chart of logic used to relate moisture, dispersal, and disturbance conditions for oaks.

wise appropriate sites. In each of the second-tier stages, the approach might involve buffering the maps to focus on particularly informative locations (zones within the presumed dispersal range of sites supporting dense oaks; zones within a specified distance of roads or urban areas).

Your summary of the analysis would include the flowchart as well as a more detailed explanation of the logic and interpretation of the analysis.

Example 2: Relic Populations of the Rare *Fusilli puttanesca*

Fusilli puttanesca is a rare herb found on limestone bluffs and outcrops in the Shawnee Hills of southern Illinois. Its current distribution is very patchy. Human disturbance does not seem to be an issue, as these sites are too rugged for agriculture or development. Conservationists would like to be able to predict its occurrence so that they can locate potential sites for reserves. A key concern is to maintain connectedness among relic populations, which are presumed to operate as a classical metapopulation.

Assume that you have or can obtain reasonable data (a DEM, an accurate map of the plant's current distribution, etc.). Devise an approach to explain (predict) the distribution of the species across these landscapes. Outline the approach as a sequence of steps, being specific about your hypotheses and how you would test and interpret them. Summarize the analysis as a flowchart.

Exercise VI: Modeling Landscape Dynamics

Background

Much of landscape ecology is concerned with predicting how landscape pattern might change under various future scenarios including natural succession, alternative management, or anthropogenic climatic change. As many of these future scenarios are without historical precedent, this goal implies an emphasis on models that incorporate at least some level of landscape-scale processes and forcings. Even for ecologists with no plans to actually build and use models, an appreciation of landscape models is crucial because of the increasingly widespread use of models in the discipline.

Purpose

The objective of this exercise is to acquaint you with the basic stages of model building, and also to introduce you to the variety of modeling approaches currently being used in landscape ecology. The objective is not so much to convert you to modelers, but rather to give you a more sophisticated appreciation for how models are developed and applied in landscape ecology.

Procedure

Model Building Basics

This exercise follows an overview of the model-building process, which itself recognizes discrete stages of model development: *conceptualization* (a narrative model), *formulation* (choosing state variables, key processes, and the equations that describe these), *parameterization* (assigning empirical estimates to the state equations and auxiliary functions), and *verification* (initial tests to ensure that the model can adequately reproduce the data used to build it). Subsequent stages of model analysis (sensitivity, uncertainty) and validation (tests against independent data) are discussed in lecture but not addressed in this exercise.

In preparation for this lab, review your lecture notes on the types of models commonly applied in landscape ecology: Markov models, cellular automata, and patch transition simulators. Look in recent journal articles for examples of studies using these models. Also, review your notes on the use of Forrester diagrams or similar notations. This diagrams are used to provide "box and arrow" representations of models.

Protocol

Select one of the papers provided that describe factors affecting change in a particular landscape. These papers were selected to illustrate key issues in landscape dynamics. You should work in a small group of students. The group should follow the steps below. Your group should evaluate alternative conceptual models or opinions about what needs to be included in the model. However, in the end the group should reach consensus on formulation to be used. The steps in the model-building exercise are:

1. State the general goal of building the model, and a small number of specific objectives for initial applications (these may be dictated by the instructor, simply to provide a common focus for the class). Objectives should be few and specific, and should define the spatial scale, resolution, and information content required of predictions, as well as the time scale over which these predictions will be made.
2. Write a concise narrative description of the conceptual model—one paragraph at most.
3. Outline the conceptual model schematically, using Forrester or similar conventions (see below). In this diagram, include the state variables, the key interactions (fluxes, transitions), and auxiliary variables that influence these states or processes.
4. In a companion table, itemize the parameters of the model, specifying their units and their nominal values (if known), or identify the data needed to estimate the parameter. In most cases, the values will not be known and a short explanation of how the data could be collected to parameterize that part of the model will be required. This step is one of the more sobering stages of model building, as landscape-scale models are often more data intensive than is logistically practical.
5. Specify how the model could be verified, by itemizing the comparisons between model output and empirical measurements that would corroborate its behavior, and also specify the criteria by which you would accept or reject the model's predictions. If data are already available, describe the test; if test data are not already available, describe data that could be collected to verify the model.

Example

Figure 15.8 shows an example of conventions for diagramming models. Here, cover type *X1* (a state variable) undergoes a transition to cover type *X2*, as modified by the auxiliary parameter *b1* (e.g., elevation or soil type). The influence of *b1* might be specified as a scaling function (e.g., linear or

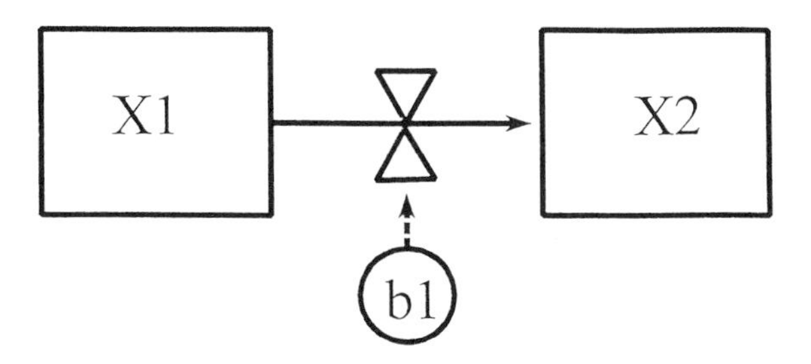

FIGURE 15.8. Conceptual model for relating the transition of a parcel of land from one cover type (*X*1) to another (*X*2).

some other form) or as a conditional probability, depending on the model. In a spatial model, *b1* might be the proportion of the neighborhood already occupied by cover type *X2*. It is at this level of implementation that most of the crucial decision in model building take place, and this stage is the focus of this exercise.

Appendix I. Origin and Acknowledgments for Exercises

Exercise I

This exercise is currently used in the undergraduate General Ecology course at the University of Wisconsin-Madison. The exercise was initially developed by Dr. Timothy F.H. Allen and graduate teaching assistant Hillary Callahan. Dr. Monica G. Turner subsequently modified the lab, and laboratory coordinator Dr. Susan Will-Wolf has supervised its implementation.

Exercise II

This exercise is currently used in the graduate Landscape Ecology course at the University of Wisconsin-Madison which is jointly taught by Monica G. Turner and David J. Mladenoff.

Exercise III

This exercise is being used in introductory and advanced courses in ecology for undergraduates at Mars Hill College, Mars Hill, North Carolina. It was prepared by Dr. Scott Pearson. The exercise is designed to demonstrate a straightforward technique for quantifying the frequency of land-cover types in complex landscapes. After the students complete this exercise, they are introduced to geographic information systems (GIS) explaining that computers provide means to conduct the same types of measurements with greater speed and accuracy. See Brewer and McCann (1982) for another simple exercise that uses aerial photographs.

Exercise IV

This exercise is being used in an introductory course in ecology for undergraduates at Mars Hill College, Mars Hill, North Carolina. It was created by Dr. Scott Pearson. The exercise is designed to demonstrate that species respond to landscape-level changes in different ways. Ideas for this exercise came from collaborations with R.H. Gardner, R.V. O'Neill, and V.H. Dale at Oak Ridge National Laboratory. The data for the maps has been provided and research related to Exercises III and IV has been supported by the Temperate Ecosystems Program of the U.S. Man-and-the-Biosphere Program, U.S. Department of State, and by a grant from the National Science Foundation DEB 9416803.

Exercise V

This exercise was produced by Dean Urban for his Landscape Ecology course. This survey course is intended for beginning graduate students at the Nicholas School of the Environment at Duke University. The School confers a professional degree, a Master's in Environmental Management (MEM), and these students comprise the bulk of the class roster (the remainder being Ph.D. students and an occasional advanced undergraduate). The MEM program emphasizes environmental problem solving and tries to instill in students a proficiency in the logic and tools of environmental analysis.

The Landscape Ecology course typically fills up with about 35 students. The format is a combination of lectures and student-moderated small-group discussions. In lieu of a formal laboratory session in a computer lab, the strategies and technical methods for problem solving are developed in "dry lab" exercises in which students work on the initial set-up and design of landscape analyses—that is, they outline the approach, specify how the analysis would proceed, and how the results would be interpreted. A combination of real examples from published analyses and hypothetical examples are contrived to illustrate specific points.

The example exercises outlined here (Exercises V and VI) are the capstone exercises for two units of the course and are concerned with (1) inferring the relative importance of various agents of pattern formation on landscapes, and (2) building models of landscape dynamics. The full course syllabus and a guided survey of key concepts and literature in landscape ecology are currently being made available over the Internet via http://www.env.duke.edu/lel.

Exercise VI

This exercise was prepared by Dean Urban for his landscape ecology course at the School of the Environment at Duke University.

Appendix II. Recommended Readings and Notes to Instructors

Exercise I

Recommended Reading

Andren, H. 1994. Effects of habitat fragmentation on birds and mammals in landscapes with different proportions of suitable habitat: a review. Oikos 71:355–366. (The author reviews the main results from percolation theory and asks whether empirical studies of birds and mammals are in agreement with the results.)

Turner, M.G., R.H. Gardner, V.H. Dale, and R.V. O'Neill. 1989. Predicting the spread of disturbance across heterogeneous landscapes. Oikos 55:121–129. (This paper links a neutral model of spatial pattern with the spread of disturbance and identifies different disturbance dynamics related to the threshold of connectivity.)

Notes to Instructors

This exercise assumes at least one prior lecture on elementary concepts and approaches to landscape ecology. Students should be familiar with what constitutes a landscape; why we study the effects of landscape heterogeneity; the use of models as a component of scientific inquiry; and notions of habitat connectivity and why it would be important for processes like the movement of organisms or spread of a disturbance.

An advantage of the lab is that it is clearly "low tech." That is, even though much of the landscape literature is replete with elegant computer-based explorations of various types of real and artificial maps, this exercise is pencil-and-paper based, requiring no computer resources, and the results are readily interpretable and intuitive. Also, the students work in groups of approximately five providing an excellent opportunity for interaction.

The instructor should assemble the following materials in advance: (1) A handout describing the lab and including a practice sheet on which students make sure they understand what is meant by defining patches, counting edges, and so on. The text provided in this chapter can serve as a foundation for an exercise based on local landscapes. (2) A random number table or generator from which to draw (x, y) coordinates ranging from 1 to 10. (3) Either many copies of 10×10 blank grids or erasable 10×10 grids for generating the random maps. (4) A set of topographic maps (USGS 7.5′ quads work just fine) or other mapped source of data from real landscapes. For Wisconsin, we use topographic maps and have students look at the spatial distribution of land and water in different regions of the state. For other regions, however, one might choose other categories, such as forest versus nonforest, or developed versus undeveloped land. (5) A set of acetate 10×10 grids at two spatial scales that will be overlain on the real landscape maps.

Exercise II

Recommended Reading

Gardner, R.H., and R.V. O'Neill. 1991. Pattern, process, and predictability: the use of neutral models for landscape analysis. In: Quantitative Methods in Landscape Ecology, pp. 289–307. M.G. Turner and R.H. Gardner (eds.). Springer-Verlag, New York.

Moody, A., and C.E. Woodcock. 1995. The influence of scale and the spatial characteristics of landscapes on land-cover mapping using remote sensing. Landscape Ecology 10:363–379.

Turner, M.G., R.V. O'Neill, R.H. Gardner, and B.T. Milne. 1989. Effects of changing spatial scale on the analysis of landscape pattern. Landscape Ecology 3:153–162.
Wickham, J.D., and D.J. Norton. 1994. Mapping and analyzing landscape patterns. Landscape Ecology 9:7–23.
Woodcock, C.E., and A.H. Strahler. 1987. The factor of scale in remote sensing. Remote Sensing Environment 21:311–332.

Notes to Instructors

Prior to implementing this exercise, students should have had an in-depth introduction to the quantification of spatial pattern and a basic introduction to scale issues. The following topics would be appropriate to cover in advance: definition of grain and extent; why scale is important; why quantify pattern; data used in landscape analyses; metrics of landscape pattern; temporal change in landscape patterns; and neutral models of landscape patterns.

Prior to the exercise, the instructor should assemble the following materials: (1) A data set or sets for the class to analyze. These should not be too large (100×100 is plenty) and should be in a format that is ready to go. (2) A source and executable code for conducting spatial pattern analyses OR a set of very simple but sensitive metrics that can be applied by pencil and paper. Ideally, a set of computers available for the class would be loaded with the data and programs. (3) Visualizations of the original data file (hard copy, overhead, or slide). (4) Detailed handout of instructions, and a readiness to deal with computer problems! (5) Group assignments. Students enjoy doing this lab collaboratively. However, the instructor should form the groups, recognizing that the computer/GIS expertise within a class of graduate students is extremely variable! Make sure that a computer-experienced student is in each group. Four students is an optimal group size.

As presented here, completing this exercise requires between 50 and 60 person hours, or about 15 hours per student. Rescaling the data set—either by writing an algorithm or by doing it manually in a word processor—was very time consuming. To reduce the amount of time required by the students, a program to do this could be supplied or the data could be distributed initially at the various scales.

Some students prefer to receive more explicit instructions on what metrics to use and compare, and how to go about this. Leaving the exercise open-ended may be unsettling, yet in the "real world" one must make choices about what to consider and learn about how sensitive the metrics may be to various manipulations of the data. However, the instructor should decide what will be most effective for his or her students.

Exercise III

Recommended Reading

Baskent, E.Z., and G.A. Jordan. 1995. Characterizing spatial structure of forest landscapes. Canadian Journal of Forest. Research 25:1830–1849.

Gustafson, E.J., and G.R. Parker. 1992. Relationship between landcover proportion and indices of spatial pattern. Landscape Ecology 7:101–110.
Jelinski, D.E., and J. Wu. 1996. The modifiable aerial unit problem and implications for landscape ecology. Landscape Ecology 11:129–140.
Kienast, F. 1993. Analysis of historic landscape patterns with a Geographical Information System—a methodological outline. Landscape Ecology 8:103 118.
O'Neill, R.V., J.R. Krummel, R.H. Gardner, G. Sugihara, B. Jackson, M.G. Turner, B. Zygmunt, S.W. Christensen, V.H. Dale, and R.L. Graham. 1988. Indices of landscape pattern. Landscape Ecology 1:153–162.
Pastor, J., and M. Broschart. 1990. The spatial pattern of a northern conifer-hardwood landscape. Landscape Ecology 4:55–68.
Turner, M.G., and R.H. Gardner. (eds.), 1991. Quantitative Methods in Landscape Ecology: The Analysis and Interpretation of Landscape Heterogeneity. Springer-Verlag, New York.

Notes to Instructors

Sampling grid: The sampling-grid transparency is a piece of mylar or transparency film with a grid of points. The rows and columns of this grid may be numbered ahead of time, or the students can do the numbering as part of the exercise.

Question (1): Students can perform a goodness-of-fit test to test the statistical significance of the change in land-cover frequencies recorded in Table 15.2. Given a null hypothesis of no change, we can expect the 1986 row totals (observed) to closely match the 1975 column totals (expected). Plug the totals for each land-cover type into the following equation:

$$X^2 = \sum\left[(\text{total}_{1986} - \text{total}_{1975})^2 / \text{total}_{1975}\right] \quad \text{d.f.} = 2$$

Reject the null hypothesis of no significant change if $X^2 > 5.991$ ($p \leq 0.05$). See a statistics text such as Bailey (1995) for more information.

Question (4): The mechanisms of land-cover change for this area are not homogeneous. Wear and Flamm (1993), Turner et al. (1996), and Wear et al. (1996) demonstrate that a number of site characteristics, including sociological and economic qualities, influence the frequency and trajectory of land-cover changes in this study area. Students may notice that most of the conversion of forest to non-forest covers occurs along the existing road network. Therefore, the rate and pattern of change along roadsides was different than the rate and pattern of changes away from roads. Students could test this hypothesis by repeating the analysis to compare the results from a set of random points near roads to a set of points some maximum distance away from roads.

Exercise IV

Recommended Reading

Andren, H. 1992. Corvid density and nest predation in relation to forest fragmentation: a landscape perspective. Ecology 73:794–804.

Blake, J.G., and J.R. Karr. 1987. Breeding birds of isolated woodlots: area and habitat relationships. Ecology 68:1724–1734.
Flather, C.H., S.J. Brady, and D.B. Inkley. 1992. Regional habitat appraisals of wildlife communities: a landscape-level evaluation of a resource planning model using avian distribution data. Landscape Ecology 7:137–147.
Hansen, A.J., and D.L. Urban. 1992. Avian response to landscape pattern: the role of species' life histories. Landscape Ecology 7:163–180.
Hansson, L., and P. Angelstam. 1991. Landscape ecology as a theoretical basis for nature conservation. Landscape Ecology 5:191–201.
Kadmon, R. 1993. Population dynamic consequences of habitat heterogeneity: an experimental study. Ecology 74:816–825.
Levin, S.A. 1992. The problem of pattern and scale in ecology. Ecology 73:1943–1967.
Pearson, S.M., J.M. Walsh, and J. Pickering. 1992. Wood stork use of wetland habitats around Cumberland Island, Georgia. Colonial Waterbirds 15:33–42.
Price, M.V, P.A. Kelly, and R.L. Goldingay. 1994. Distance moved by Stephen's kangaroo rate (*Dipodomys stephensi* Merriam) and implications for conservation. Journal of Mammalogy 75:929–939.
Robinson, S.K., F.R. Thompson III, T.M. Donovan, D.R. Whitehead, and J. Faaborg. 1995. Regional forest fragmentation and the nesting success of migratory birds. Science 267:1987–1990.

Notes to Instructors

Rather than using the mylar template for making the habitat maps, you can provide students with extra photocopies of the land-cover map. They can use ink markers or grease pencils (red or orange) to color in the cells that meet the habitat criteria for a given species. Students need one additional map for each species.

Having each student make a map for each species is time consuming. You can divide the students into small groups (two–four students each) and assign one or two species to each student. When they finish making the maps, have them compare maps within and between groups.

The questions listed above can be used for group discussions or to from the basis of a lab report to be prepared for each group or individual student. Instructors are encouraged to use alternative land-cover maps and/or develop mapping recipes for species native to their geographic region.

Exercise V

Recommended Reading

Davis, F.W., and S. Goetz. 1990. Modeling vegetation pattern using digital terrain data. Landscape Ecology 4:69–80.
Gardner, R.H., B.T. Milne, M.G. Turner, and R.V. O'Neill. 1987. Neutral models for the analysis of broad-scale landscape pattern. Landscape Ecology 1:19–28.

Notes to Instructors

The key issue to underscore in this exercise is that data at the landscape scale are logistically expensive and by focusing the analysis as much as possible we can derive the most information from minimal data, carefully selected. If nothing else, the students should appreciate that all data are not created equal, that some data are more informative and hence more valuable than others.

The example concerned with oaks in California foothills also illustrates the utility of CART analysis in analyses like this. CART is a recursive procedure which, for categorical response variables, executes a logistic regression at each "branch" of the regression tree, yielding a split between, say, mesic "oak sites" and more xeric "non-oak sites" (Fig. 15.7). Importantly, the analysis also provides a summary of how many sites classified as "oak sites" were not observed to support oaks, and reciprocally, how many "non-oak sites" actually had oaks on them. The next step in the analysis would be to refine these branches, that is, to distinguish the misclassified sites on either branch, improving the model's classification accuracy recursively. As a tree diagram, this procedure highlights the take-home message that information about dispersal limitations is best expressed on sites that are potential habitat but are not occupied by oaks. Reciprocally, it is impossible to gain any information about dispersal limitations from sites that do not qualify as potentially usable habitat in the first place. Thus, the regression tree can graphically enforce the notion that landscape analysis often requires highly selective subsets of site conditions to provide useful answers to questions about agents of landscape pattern. (The instructor should note that there are analysis scenarios that can be sufficiently complicated that CART still works as an analysis but may fail miserably as a heuristic device!)

Given real data and adequate computing facilities, this exercise could be expanded into a "live" analysis. In this, students actually would analyze data using either partial regression or regression trees. (This is how it's done in the more advanced, second-year classes in Duke's MEM curriculum.)

Exercise VI

Recommended Reading

Baker, W.L. 1989. A review of models of landscape change. Landscape Ecology 2:111–133.

Sklar, F.H., and R. Costanza. 1991. The development of dynamic spatial models for landscape ecology: a review and prognosis. In: Quantitative Methods in Landscape Ecology, pp. 239–288. M.G. Turner and R.H. Gardner (eds.). Springer-Verlag, New York.

Usher, M.B. 1992. Statistical models of succession. In: Plant Succession: Theory and Prediction, pp. 215–248. D.C. Glenn-Lewin, R.K. Peet, and T.T. Veblen (eds.). Chapman & Hall, London.

Weinstein, D.A., and H.H. Shugart. 1983. Ecological modeling of landscape dynamics. In: Disturbance and Ecosystems, pp. 29–45. H.A. Mooney and M. Godron (eds.). Springer-Verlag, New York.

Notes to Instructors

The challenge in teaching landscape modeling, of course, is that few students will have the technical skills needed to actually build a model (for example, programming language, algorithms), and the empirical effort in parameterizing and testing a model are even more intimidating. Commercial software packages that make simple models easy (for example, STELLA)©, can be quite useful for labs such as this, but the initial investment in getting students acquainted with the package might require more time than is available for a single lab exercise (Duke's Masters in Environmental Management program defers STELLA© to a separate course, Principles of Ecological Modeling).

In this exercise, students build models by concentrating on the conceptual stages of model development, but stopping short of actual coding. This approach argues that the conceptual stages are the most crucial steps in model building, and also presumes that an appreciation of models at this level might be adequate for many students' needs. Models are developed from purely empirical, descriptive papers that document particular landscapes.

Because this exercise requires a prior familiarity with some landscape, it is difficult to provide a facile example of this exercise that can be explored in just a short time. Some example landscapes that might provide useful tutorials: First, Foster (1992) provides a nice reconstruction of the history of landscape change in New England. This paper underscores an important point, that the rules that drive landscape change vary over time. New England has undergone a shift from deforestation to reforestation during the past century. Implemented as a simple Markov model, this would imply nonstationary transition probabilities; to circumvent this problem, a model must either become more than first-order (i.e., transitions depend on past states as well as current states), or multiple transition matrices could be used (one for each time period of interest). Second, gradient studies (there are countless examples) provide easy empirical patterns for use in building models in which transitions among cover types or vegetation zones are conditioned by environmental variables such as those derived from digital elevation models. Finally, in more complicated scenarios, transitions might include disturbances (pest outbreaks, fires) that include feedbacks to vegetation status or environmental variables. (This level of complexity matches many current landscape models.)

The exercise of building a model prototype in a small-group setting nicely illustrates the trade-offs between realism and simplicity in model construction. A further benefit of doing this exercise in multiple small groups is that the groups can compare models in a follow-up discussion session. Different groups invariably will devise different models, and it is especially fruitful to force groups to justify the approach they adopted over other alternatives.

The emphasis on the initial, largely conceptual aspects of modeling allows students with limited math and computer skills to participate equally with their more technically advanced peers.

The next level of activity beyond this exercise would be to actually use models. There are two approaches to this. The easier would be to provide students with simple models that they could use to perform various demonstration runs or model experiments. This approach would require a well-documented model and adequate computer facilities, and would also require a minimal level of computer familiarity of the students. A more in-depth approach would be to have students build and encode a model themselves. This is clearly beyond the scope of most introductory courses.

As an example of the former approach, that of using an existing model, we have had quite good experiences by providing the students with a simple Markov model of succession in a forested landscape. Usher (1992) provides an excellent overview of the construction and analysis of Markov models such as this. The example is drawn from a Pacific Northwestern landscape that has been classified from Landsat Thematic Mapper imagery into discrete age classes (see http://www.env.duke.edu/lel for similar lab exercises on landscape change). Students are given an array of cell values that indicates the age class of the cell in each of the three time periods; these data are provided for 200 cells randomly sampled from the images. Students then build the transition tally matrix from these data, summarizing the number (ultimately, the proportion) of cells that changed from type (age) i to type j during each time interval. Students then normalize these transitions to an annual timestep, and construct the transition matrix $\boldsymbol{P}$, which gives the probability of a cell (equivalently, the proportion of cells) that change from type i to j in each timestep.

$$P = \begin{bmatrix} p_{11} & p_{12} & p_{13} & \cdots \\ p_{21} & p_{22} & p_{23} & \cdots \\ p_{31} & p_{32} & p_{33} & \cdots \\ \cdots & \cdots & \cdots & \cdots \end{bmatrix}$$

The students also tally the initial state vector x, which is the proportion of cells in each type (age class) for the first time period. For model testing, they also tally the state vectors for the second and third time steps.

The solution of a Markov model is given by:

$$x_{(t+1)} = x_{(t)}P$$

where $x(t)$ is the initial state vector. Similarly,

$$x_{(t+2)} = x_{(t+1)}P = x_{(t+1)}P^2$$

and, in general,

$$x_{(t+\mathrm{k})} = x_{(t)}P^{\mathrm{k}}$$

for k timesteps after the initial condition. The steady-state solution can be solved by eigenanalysis, that is, by finding the vector x^* such that $x^* = x^*\boldsymbol{P}$.

For our purposes, students are provided with a simple Fortran program that iterates the model and provides output in a format suitable for graphics packages. They initialize the model with data from the first time period, verify it against the second time period (which works nicely), and then validate the model using data from the third time period (it does not validate because the timber harvest rates have increased). They are then asked to find the steady state, and to speculate on how the model might be extended to address landscape-scale issues such as stationarity (they see the lack of this when they attempt to validate the model with data from the third time period) and spatial contingencies in forest harvest or other land use change.

This exercise is especially effective because it allows students to parameterize a model, test it to discover its weaknesses, and then to speculate on how they would improve the model. Still, the exercise does not require any special skills such as programming. It should be noted that commercial packages such as STELLA© (High Performance Systems Inc., Hanover, NH) could also be used in this exercise; STELLA© would solve the model as a system of differential equations as compared to a Markov model, but the parameters and the solution are equivalent.

References

Andren, H. 1994. Effects of habitat fragmentation on birds and mammals in landscapes with different proportions of suitable habitat: a review. Oikos 71:355–366.

Bailey, N.T.J. 1995. Statistical methods in biology, 3rd ed. Cambridge University Press, New York.

Baker, W.L., and Y. Cai. 1992. The r.le programs for multiscale analysis of landscape structure using the GRASS geographic information system. Landscape Ecology 7:291–302.

Brewer, R., and M.T. McCann. 1982. Ecological use of remote sensing. In: Laboratory and Field Manual of Ecology, pp. 159–169. Saunders College Publishing, Fort Worth, TX.

Davis, F.W., and S. Goetz. 1990. Modeling vegetation pattern using digital terrain data. Landscape Ecology 4:69–80.

Foster, D.R. 1992. Land-use history (1730–1990) and vegetation dynamics in central New England USA. Journal of Ecology 80:753–772.

Gardner, R.H., and R.V. O'Neill. 1991. Pattern, process, and predictability: the use of neutral models for landscape analysis. In: Quantitative Methods in Landscape Ecology, pp. 289–307. M.G. Turner and R.H. Gardner (eds.). Springer-Verlag, Ner York.

Gardner, R.H., R.T. Milne, M.G. Turner, and R.V. O'Neill. 1987. Neutral models for the analysis of broad-scale landscape pattern. Landscape Ecology 1:5–18.

Hamel, P.B. 1992. The Land Manager's Guide to Birds of the South. The Nature Conservancy; Chapel Hill, NC.
MacNally, R. 1996. Hierarchical partitioning as an interpretative tool in multivariate inference. Australian Journal of Ecology 21:224–228.
McGarigal, K., and B.J. Marks. 1995. FRAGSTATS: spatial pattern analysis program for quantifying landscape structure. USDA Forest Service, General Technical Report PNW-GTR-351.
Michaelsen, J.F., F.W. Davis, and M. Borchert. 1987. A nonparametric method for analysing hierarchical relationships in ecological data. Coenoses 2:39–48.
Pearson, S.M., M.G. Turner, R.H. Gardner, and R.V. O'Neill. 1996. An organismbased perspective of habitat fragmentation In: Biodiversity in Managed Landscapes: Theory and Practice, pp. 77–95. R.C. Szaro (ed.). Oxford University Press, New York.
Stauffer, D. 1985. Introduction to Percolation Theory. Taylor & Francis, London.
Turner, M.G. 1990. Spatial and temporal analysis of landscape patterns. Landscape Ecology 4:21–30.
Turner, M.G., D.N. Wear, and R.O. Flamm. 1996. Land ownership and land-cover change in the Southern Appalachian Higlands and the Olympic Peninsula. Ecological Applications 6:1150–1172.
Usher, M.B. 1992. Statistical models of succession. In: Plant succession: Theory and Prediction, pp. 215–248. D.C. Glenn-Lein, R.K. Peet, and T.T. Veblen (eds.) Chapman & Hall. London.
Venables, W.N., and B.D. Ripley. 1994. Modern Applied Statistics with S-Plus. Springer-Verlag, New York.
Wear, D.N., and R.O. Flamm. 1993. Public and private disturbance regimes in the Sourthern Appalachians. Natural Resource Modeling 7:379–397.
Wear, D.N., M.G. Turner, and R.O. Flamm. 1996. Ecosystem management multiple owners: landscape dynamics in a southern Appalacian watershed. Ecological Applications 6:1173–1188.
Wiens, J.A. 1989. Spatial scaling in ecology. Functional Ecology 3:385–397.
With, K.A., R.H. Gardner, and M.G. Turner. 1997. Landscape connectivity and population distributions in heterogeneous environments. Oikos 78:151–169.

Part VI
Synthesis

16
The Science and Practice of Landscape Ecology

John A. Wiens

In his essay, "Second thoughts on paradigms," Thomas Kuhn (1974) illustrated the power of paradigm shifts in the sciences with an analogy to the child's puzzle in which one is asked to find the animal shapes or faces hidden in a drawing of shrubbery or clouds. The child, Kuhn observed, "seeks forms that are like those of the animals or faces he knows. Once they are found, they do not again retreat into the background, for the child's way of seeing the picture has been changed." In the same way, a shift in paradigms changes forever the way scientists view the phenomena they study. Previous theories and methodologies are replaced by new ones. This change is more than a simple shift in emphasis—the scientific worldview and, ultimately, the way scientists conduct their investigations have been altered.

The emergence of landscape ecology as a discipline has catalyzed a shift in paradigms among ecologists and (to the extent that they care about such things as paradigms) resource managers and land-use planners. Having now seen the faces of spatial patterning and scale—the features of landscapes—lurking in our pictures of natural or human-dominated systems, we can never go back to the old ways of viewing things. We can no longer ignore spatial variation and pattern, nor can we continue to cling to the belief that the scale on which we view systems does not affect what we see. We may consider whether the effects of spatial variation and scale are likely to be important in a particular situation (e.g., Kareiva and Wennergren 1995), but we must begin with the presumption that they *do* matter. This is quite a different way of viewing the world than that which was in vogue a decade ago, and it is by no means yet widely embraced by everyone (see, e.g. the statistics on scale awareness compiled by Wiens 1992; Schneider 1994; Chapter 11). But the paradigm shift is inevitable, if we are to further our understanding of ecological systems and their management and sustainable use.

It is remarkable that landscape ecology has played such a role, in view of its ongoing identity crisis and its immaturity as a discipline. This chapter provides some perspective for the other contributions to this volume, first by considering what I mean by an "identity crisis" and "immaturity," and then by exploring some aspects of the relationship between landscape ecology as a science and as a practice. I will conclude by noting some themes of the emerging paradigm of landscape ecology, which, if Kuhn's (1970) thesis applies, will guide our thinking and research into the next millennium.

The Identity of Landscape Ecology

As Klopatek and Gardner have noted, "landscape ecology" means different things to different people. To some, it is characterized primarily by the tools that are used, especially geographical information systems (GIS; Johnston 1990; Haines-Young et al. 1993). To others, particularly many Europeans, it represents a holistic integration of human geography and ecology with land systems; humans are considered as an integral part of functioning landscape ecosystems (Naveh and Lieberman 1994; Zonneveld 1995). Even within a more strictly biological frame of reference, there are differing views. Thus, the "landscape" may be a level of organization falling somewhere between the community or ecosystem and the biome (O'Neill et al. 1986; Gosz 1993; Lidicker 1995), or a scale somewhere between "local" and "regional" (i.e., tens to thousands of ha; Forman and Godron 1986; Hobbs 1994), or the templet on which spatial patterns influence ecological processes, regardless of scale (Turner 1989; Pickett and Cadenasso 1995; Wiens 1995).

In some ways, this diversity of views about what landscape ecology is represents a strength of the field. It enhances a cross-fertilization of ideas and approaches that contributes to the vitality of the discipline, and it serves to integrate basic science with applications and with humanistic perspectives on landscapes (e.g., Bissonette 1997; Nadenicek 1997; Nassauer 1997). This diversity is much in evidence in the contributions to this volume. There is little doubt that, at a time when "interdisciplinary" science has become the fashion. landscape ecology is among the most interdisciplinary of the sciences. Yet landscape ecology continues to suffer from something of an identity crisis. With so many faces, which one is the "real" landscape ecology? Some European landscape ecologists disparage what they perceive as the "narrow bioecological conceptual and methodological framework" of North American landscape ecology (Naveh and Lieberman 1994), while some North Americans complain (at least in private) about the absence of "science" in the more humanistic approaches the Europeans bring to landscape ecology. Landscape ecology remains, in Hobbs' (1994) words, "a science in search of itself." Why is this?

The Conceptual Immaturity of Landscape Ecology

Scientific disciplines are distinguished by their concepts and theories rather than their "facts." In a mature discipline, these concepts and theories are unified into a framework that provides a foundation for both research in and applications of the science. Fields such as genetics, biochemistry, molecular biology, and geology have such conceptual unity. Ecology, on the other hand, does not, and it has therefore been accused of being intellectually immature (Strong et al. 1984; Peters 1991). Hagan (1989) has attributed this apparent immaturity to the historical development of ecology about differing intellectual perspectives, one focused on population processes, the other on flows of materials and energy through ecosystems. Much in the manner of paradigms, these perspectives influence the types of questions ecologists ask and the types of explanations they find acceptable. They are now well established and institutionalized perspectives, and, recent attempts (e.g., Allen and Hoekstra 1992; Jones and Lawton 1995; Ulanowicz 1997) notwithstanding, they continue to defy conceptual integration.

Landscape ecology is a much younger discipline than ecology. Although nascent threads (as well as the name) existed much earlier, the real emergence of landscape ecology as a recognizable field of study dates only from the 1970s (Schreiber 1990; Zonneveld 1995). Nonetheless, Zonneveld (1995), Forman (Chapter 4) and Hobbs (Chapter 2) have suggested that landscape ecology has already developed a solid conceptual and theoretical foundation. The recent development of new technologies and tools for describing, analyzing, and modeling the spatial patterns and dynamics of landscapes (some of which are illustrated in the previous chapters) has also provided a degree of methodological unity to the discipline. Several authors (e.g., Forman 1995; Lidicker 1995; Turner et al. 1995; Golley 1996) have even identified general principles of landscape ecology.

Does landscape ecology therefore have the conceptual unity we expect of a mature science? I think not. In my view, the diversity of approaches and viewpoints in landscape ecology has yet to coalesce about a clear central theme or unifying conceptual structure, much less a body of predictive theory. Thus, although the concept that "scale matters" is a central concern of landscape ecology, we have only fragments of a theory of scaling (Meetenmeyer and Box 1987; Wiens 1989, 1995), and few studies have gathered empirical information that directly addresses scaling effects. Neither are principles such as "all ecosystems in a landscape are interrelated" or "different species may respond differently to a given patch array" really theories. Rather, they are provisional statements that may or may not have an empirical foundation. While they may call attention to particular aspects of landscapes that merit investigation, they do not generate predictions or testable hypotheses. To do that, we need coherent concepts and formalized theories, at least according to conventional views of scientific progress.

Landscape Ecology as a Science

One can argue, however, with the premise that conceptual unification and the development of predictive theories are really valid measures of the intellectual maturity of a science (e.g., Laudan 1977; Hagan 1989), especially one that deals with complex relationships such as those that exist in landscapes (Pickett et al. 1994; Hobbs 1997). Landscape ecology is rapidly developing a methodological foundation for dealing with this complexity. As the chapters in Part IV of this volume demonstrate, we now have powerful ways to describe the spatial patterns of landscapes over multiple scales, to conduct rigorous statistical analyses of these patterns, and to incorporate landscape complexity into spatially explicit computer simulation models. But being a science involves more than methods. Landscape ecology also requires a consistent philosophy that stipulates how we should go about gaining knowledge of landscape systems while maintaining rigor.

Following the example of physics, most contemporary scientific disciplines are firmly rooted in causality—the belief that the phenomena and patterns that we see are founded upon simple cause-effect relations and that the power of science lies in deriving the rules or theories ("principles") that capture the essence of this causation. This philosophy is the basis for Urban's (1993) observation that "ecological processes generate patterns, and by studying these patterns we [landscape ecologists] can make useful inferences about the underlying processes" (see also Wiens 1997). This philosophy is the basis for the central role of hypothetico-deductive approaches and experimentation in science. A well-designed experiment can factor out potentially complicating sources of variation and permit the investigator to isolate key elements of a proposed process-pattern (i.e., causal) linkage. In this way, incorrect explanations can be falsified and new theories and hypotheses proposed.

The systems studied by landscape ecologists, however, are characterized by complexity and scale dependencies. The complexity means that the potential process-pattern linkages are likely to be sensitive to the details of the spatial configuration of a mosaic and that simple experiments are likely to omit key components among a web of interacting elements. Scale dependency means that it is logistically difficult to conduct experiments at the broad scale of many landscapes. More importantly, it means that what happens at a particular scale may be strongly influenced by the character of the broader-scale landscape as well as by events occurring at finer spatial scales. Hierarchy theory (O'Neill et al. 1986; Ahl and Allen 1996) has called attention to these complicating effects of scale, but it has not developed an effective recipe for dealing with them. Although our ability to discern cause-effect relationship in such complex systems will probably be enhanced if we consider patterns and processes that vary on similar spatial and temporal scales (Meentemeyer and Box 1987; Wiens 1989; Withers and Meentemeyer, Chapter 11), this approach does not ensure that the complicating effects of cross-scale influences will be considered.

The difficulty of establishing simple cause-effect relationships in landscape systems does not mean that we should abandon a philosophy of causality, but it does imply that a strict falsificationist and experimental model of "proper" scientific investigation may be inappropriate for landscape ecology. Following the arguments of ecologists such as Macfadyen (1975), McIntosh (1987), and Pickett et al. (1994), Hobbs (1997, Chapter 2) has argued for a pluralism of approaches to studying landscapes that embraces their complexity rather than imposes simplicity upon them. Thus, while studies on experimental model systems occupying fine-scale "microlandscapes" (e.g., Wiens and Milne 1989; Barret et al. 1995; Wiens et al. 1995; McIntyre 1997; With 1997) may yield useful insights, they are inevitably limited by their scale and by the particular features of the systems that lend themselves to experimentation. Broad-scale experiments or anthropogenic landscape modifications (e.g., Saunders et al. 1987; Settele et al. 1996; Laurance and Bierregaard 1997) may have greater realism but generally suffer from both a lack of replication and from the confounding effects of covarying factors (Nicholls and Margules 1991; Hargrove and Pickering 1992). Other existing approaches, such as spatial simulation modeling or GIS modeling, or some of the novel techniques described in this volume, need to be combined with new and innovative ways of assessing landscape interrelationships that do not attempt to simplify the complexity that is the essence of landscapes. And, because we still know relatively little about the variety of patterns that exist in landscapes, careful description still has an important role to play (Burke 1997). A relaxation of adherence to a philosophy of strict and simple causality, coupled with a diversity of scientific approaches, probably means that a "conceptual unification" of landscape ecology is still some time away, if indeed it is even a desirable goal.

The problem with such a pluralism of approaches, of course, is that it does not conform to the image of science-as-experiments held by the public as well as by many other scientists. The work of landscape ecologists may therefore be perceived as lacking in rigor or "soft," being "just descriptive," or as offering only uncertainty instead of useful predictions. But science does not need to be experimental to be good. What it *does* need is clear logic, sound design, careful measurement, quantitatively rigorous and objective analysis, and thoughtful interpretation. These features are all the more critical if landscape ecology as a science is to mold landscape ecology as a practice.

Landscape Ecology as a Practice

In the Conclusion to her book, *Placing Nature: Culture and Landscape Ecology*, Joan Nassauer (1997) observed that "landscape ecology should be as much about doing as it is about thinking." Landscape ecology, from its very beginnings, has been a discipline that transcends the boundary between science and application. Its focus, to varying degrees, is on human land use and its effects, and on how scientific findings and principles can be applied to real-world problems in natural resource management and land-use planning. Notwithstanding the view that the ecosystem level is an inappropriate level for landscape planning (Zonneveld 1995; Ahern, Chapter 10), landscape ecology has become an important component of ecosystem management (Bartuska, Chapter 3; Coulson et al., Chapter 5; see Christensen et al. 1996). Perhaps because of its relative youth as a science, however, the "principles" that landscape ecology offers are general statements that provide only vague predictions. They do not specify the ways in which spatial phenomena are contingent not only upon landscape structure, but on other features of the environment, the organisms or systems involved, the management objectives, and (of course) the scales of interest. It is fine to argue for the benefits of large natural patches in a developed landscape, as Forman (Chapter 4) has done, but it is important to realize also that these benefits vary as a function of the above features and that bigger is not invariably better. Depending on the characteristics of the organisms and systems considered and one's objectives (e.g., conservation of an endangered species, enhancement of biodiversity, reduction of soil or nutrient loss), the scales of land management may or may not coincide with the scales on which landscape patterns and processes are actually important.

This is not to say that landscape ecology has not yet had important applications. The very recognition that land management or conservation efforts might be directed toward entire landscape mosaics rather than isolated parcels of "habitat" (e.g., McKelvey et al. 1993; Bissonette 1997) is a major advance. Forest ecologists and managers are now considering the spatial consequences of various management activities (Franklin and Forman 1987; Gustafson, Chapter 7) or are incorporating information on natural patterns of landscapes or disturbances into their practices (Lämås and Fries 1995; Forman, Chapter 4). Perhaps the greatest success in putting landscape thinking into practice is in the area of designing nature reserves. Beginning with simple patch-matrix conceptualizations of landscapes derived from island biogeography theory, reserve design has progressed to incorporate the structure of entire landscape mosaics into specifying the properties, configuration, and placement of reserves, as well as recommending compatible uses of the surrounding landscapes (Margules et al. 1988; Pressey et al. 1993; Settele et al. 1996; Csuti et al. 1997; White et al., Chapter 8; Polasky and Solow, Chapter 9).

Landscape ecology, of course, is more than simply a branch of ecology and more than ecologically based land-use management. Its historical roots are, in fact, much broader. In Chapter 1 Paul Risser argues forcefully that landscape ecology must broaden its reach even further, to include economic analysis and social and political forces as explicit components of its structure. For landscape ecology to avoid becoming a science of marginal importance, it must undergo a fundamental reconstruction. A discipline that is arguably not yet even mature is already undergoing a midlife crisis!

As a counterpoint, I would argue that there are real dangers in further diffusing the focus of a discipline that already has some difficulty in establishing its identity. Yes, decisions affecting landscapes *are* made within a social, political, and economic setting, and yes, landscape ecology *must* contribute its findings and expertise and principles (such as they are) to such decisions. The question is, How should this be done? Two (of many) alternatives are depicted in Figure 16.1. Environmental problems, broadly considered, arise from the ways in which the natural patterns and processes of ecological systems interact with an amalgamation of human factors and forces that affect land (or water; Naiman 1996) use. In one view (Risser's; Fig. 16.1A), landscape ecology should function as an umbrella discipline, integrating the perspectives on environmental problems held by traditionally separate disciplines such as conservation biology, resource economics, environmental ethics, and the like to produce a cohesive picture of the forces that affect landscapes and, from that, possible solutions to the environmental problems. In another view (mine; Fig. 16.1B), landscape ecology represents one of several ways of addressing environmental problems. Each of these disciplines brings to the debate a particular perspective on the problems and a particular set of values, and developing potential solutions involves integrating and balancing information from each. The distinction between the two views represented in Figure 16.1 is important, because it affects how we identify landscape ecology as a discipline, how we do landscape ecology, and how we train students and practitioners.

The reality, of course, and the impetus behind Risser's appeal, is that traditionally separate disciplines usually approach environmental problems independently of one another. The power of Risser's thesis is that landscape ecology could act as an integrating discipline, bringing together disparate viewpoints to produce a broad-based and unified approach to environmental problems. I maintain, however, that this breadth will weaken landscape ecology, leading to a loss of rigor and credibility. The strength of landscape ecology lies in the distinctiveness of its approach—its emphasis on spatial patterns and relationships, scaling, heterogeneity, boundaries, and flows of energy and materials in space. This approach should not be diluted, but should be brought to bear on major issues in a coherent way, with a strong foundation in science. Landscape ecology can have this impact, however, only if there is vigorous communication among the various disciplines addressing environmental problems. This will not happen without an active effort to break down the boundaries of traditional disciplines and to make the ideas and findings of landscape ecology clear and comprehensible to workers in other arenas (and vice versa). As Joan Nassauer (1997) has observed, we must cross disciplinary boundaries to ask cultural questions informed by science and to ask scientific questions informed by culture.

Being "scientific," however, does not mean that landscape ecology should focus exclusively or even primarily on scientific questions. I concur

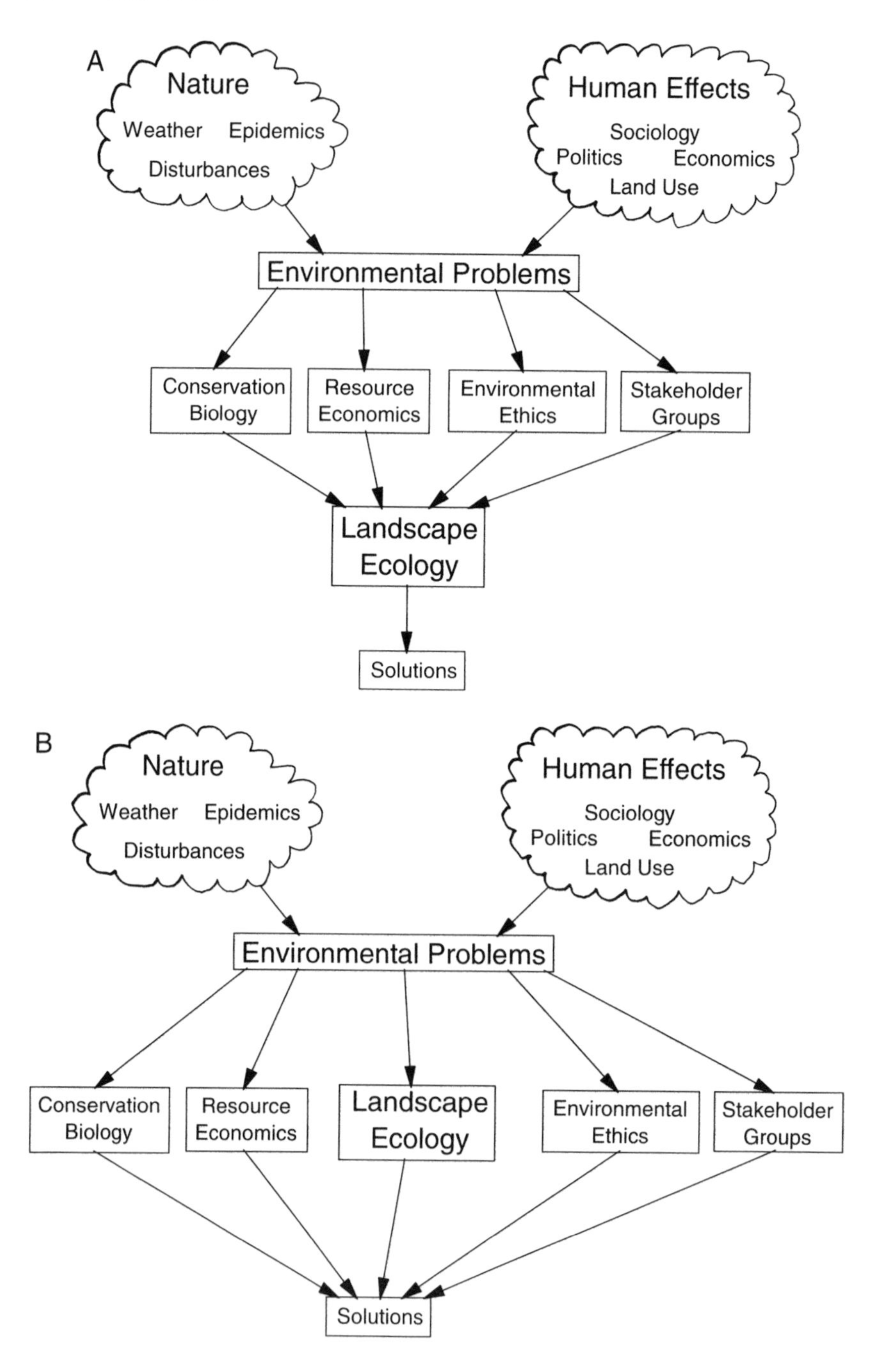
A
Nature
Weather
Epidemics
Disturbances
Human Effects
Sociology
Politics
Economics
Land Use
Environmental Problems
Conservation Biology
Resource Economics
Environmental Ethics
Stakeholder Groups
Landscape Ecology
Solutions
B
Nature
Weather
Epidemics
Disturbances
Human Effects
Sociology
Politics
Economics
Land Use
Environmental Problems
Conservation Biology
Resource Economics
Landscape Ecology
Environmental Ethics
Stakeholder Groups
Solutions

with Risser's concern that landscape ecology should not stand aside dealing with traditional science while landscapes are converted to urban sprawl and parking lots or are planned or managed in ways that clearly are not sustainable. Even at its present stage of development, landscape ecology has ample insights to offer to those charged with land or resource management or involved in land use or urban planning or concerned about conservation issues.

Conclusions: Elements of the Emerging Paradigm

"Embrace complexity" might well be the mantra of landscape ecologists. In fact, landscape ecologists have little choice but to deal with complexity. It is what comes of an explicit focus on spatial patterning and scale. If anything serves to unify the various approaches to landscape ecology, it is the shared concern about the causes and consequences of spatial heterogeneity and the effects of changes in scale on these relationships. Beyond this focus, it may not matter much whether or not landscape ecology develops "conceptual unity" (there are some advantages in never growing up!). What *is* important, however, is that the discipline develop formalized concepts and predictive theories of some sorts, and that answers to questions be founded upon a firm base of science. Clearly, there is no single "best" way of doing landscape ecology. The questions landscape ecologists ask require a diversity of tools and methods to answer, and they may be approached from a variety of directions. The key point is that the methods and approaches must be quantitatively rigorous and logically sound.

And what are the questions? In 1983, a workshop was held at Allerton Park, Illinois, to develop recommendations concerning the current and future status and the importance of landscape ecology (Risser et al. 1984; see Risser 1995). The workshop participants identified four central questions of landscape ecology:

- How are fluxes of organisms, or material, and of energy related to landscape heterogeneity?

FIGURE 16.1. The role of landscape ecology in contributing to solutions to environmental problems. Environmental problems arise from a conflict between nature and land use, which in turn is influenced by factors such as politics, sociology, and economics. In (A), different disciplines or interest groups respond to the environmental problems from particular perspectives. Landscape ecology then serves as an integrating discipline, pulling together the separate perspectives to produce comprehensive solutions. In (B), landscape ecology is viewed as making its own distinctive contribution to solutions, in concert with the perspectives of other approaches or interest groups.

- What formative processes, both historical and present, are responsible for the existing pattern in a landscape?
- How does landscape heterogeneity affect the spread of disturbances?
- How can natural resource management be enhanced by a landscape ecology approach?

To see how things have changed, consider the questions posed by Turner in a chapter on landscape ecology in a current undergraduate ecology text (1998):

- How does the spatial arrangement of habitat influence the presence and abundance of species?
- Does the surrounding landscape influence local populations?
- Do landscape patterns affect the transport of materials from land to water?
- How do ecosystem processes vary spatially?
- How are disturbances an integral part of landscapes?

There is a slight difference in emphasis, perhaps reflecting an increased influence of conservation issues in landscape ecology, but not much has really changed. Both sets of questions stress the ecological aspects of landscape ecology, and people interested in land use policy, landscape geography, or landscape architecture would frame the questions using different terms. But the thrust of the questions, I suspect, would be similar. They define the central themes of landscape ecology: spatial variation, scaling, boundaries, and flows. These are, indeed, the core elements of the emerging paradigm of landscape ecology—a science *and* a practice.

References

Ahl, V., and T.F.H. Allen. 1996. Hierarchy Theory: A Vision, Vocabulary, and Epistemology. Columbia University Press, New York.

Allen, T.F.H., and T.W. Hoekstra. 1992. Toward a Unified Ecology. Columbia University Press, New York.

Barrett, G.W., J.D. Peles, and S.J. Harper. 1995. Reflections on the use of experimental landscapes in mammalian ecology. In: Landscape Approaches in Mammalian Ecology and Conservation, pp. 157–174. W.Z. Lidicker Jr. (ed.). University Minnesota Press, Minneapolis.

Bissonette, J.A. (ed.). 1997. Wildlife and Landscape Ecology: Effects of Pattern and Scale. Springer-Verlag, New York.

Burke, V.J. 1997. The value of descriptive studies in landscape ecology. US-IALE Newsletter 13(2):6–8.

Christensen, N.L., A.M. Bartuska, J.H. Brown, S. Carpenter, C.D'Antonio, R. Francis, J.F. Franklin, J.A. MacMahon, R.F. Noss, D.J. Parsons, C.H. Peterson, M.G. Turner, and R.G. Woodmansee. 1996. The report of the Ecological Society of America Committee on the scientific basis for ecosystem management. Ecological Applications 6:665–691.

Csuti, B., S. Polasky, P.H. Williams, R.L. Pressey, J.D. Camm, M. Kershaw, A.R. Kiester, B. Downs, R. Hamilton, M. Huso, and K. Sahr. 1997. A comparison of reserve selection algorithms using data on terrestrial vertebrates in Oregon. Biological Conservation 80:83–97.

Forman, R.T.T. 1995. Some general principles of landscape and regional ecology. Landscape Ecology 10:133–142.

Forman, R.T.T., and M. Godron. 1986. Landscape Ecology. John Wiley & Sons, New York.

Franklin, J.F., and R.T.T. Forman. 1987. Creating landscape patterns by forest cutting: ecological consequences and principles. Landscape Ecology 1:5–18.

Golley, F.B. 1996. A state of transition. Landscape Ecology 11:321–323.

Gosz, J.R. 1993. Ecotone hierarchies. Ecological Applications 3:369–376.

Hagen, J.B. 1989. Research perspectives and the anomalous status of modern ecology. Biology and Philosophy 4:433–455.

Haines-Young, R., D.R. Green, and S.H. Cousins (eds.). 1993. Landscape Ecology and Geographic Information Systems. Taylor & Francis, London.

Hargrove, W.W., and J. Pickering. 1992. Pseudoreplication: a *sine qua non* for regional ecology. Landscape Ecology 6:251–258.

Hobbs, R.J. 1994. Landscape ecology and conservation: moving from description to application. Pacific Conservation Biology 1:170–176.

Hobbs, R. 1997. Future landscapes and the future of landscape ecology. Landscape and Urban Planning 37:1–9.

Johnston, C.A. 1990. GIS: More than just a pretty face. Landscape Ecology 4:3–4.

Jones, C.G., and J.H. Lawton (eds.). Linking Species and Ecosystems. Chapman & Hall, London.

Kareiva, P., and U. Wennergren. 1995. Connecting landscape patterns to ecosystem and population processes. Nature 373:299–302.

Kuhn, T.S. 1970. The Structure of Scientific Revolutions, 2nd ed. University Chicago Press, Chicago.

Kuhn, T.S. 1977, Second thoughts on paradigms In: The Structure of Scientific Theories, pp. 459–482. F. Suppe (ed.). University Illinois Press, Urbana.

Lämås, T., and C. Fries. 1995. Emergence of a biodiversity concept in Swedish forest policy. Water, Air and Soil Pollution 82:57–66.

Laudan, L. 1977. Progress and Its Problems. University California Press, Berkeley.

Laurance, W.F., and R.O. Bierregaard Jr. (eds.). Tropical Forest Remnants: Ecology, Management, and Conservation of Fragmented Communities. University Chicago Press, Chicago.

Lidicker, W.Z. Jr. 1995. The landscape concept: something old, something new. In: Landscape Approaches in Mammalian Ecology and Conservation, pp. 3–19. W.Z. Lidicker Jr. (ed.). University Minnesota Press, Minneapolis.

Macfadyen, A. 1975. Some thoughts on the behavior of ecologists. Journal of Ecology 63:379–391.

Margules, C.R., A.O. Nicholls, and R.L. Pressey. 1988. Selecting networks of reserves to maximize biological diversity. Biological Conservation 43:63–76.

McIntosh, R.P. 1987. Pluralism in ecology. Annual Review of Ecology and Systematics 18:321–341.

McIntyre, N.E. 1997. Scale-dependent habitat selection by the darkling beetle Eleodes hispilabris (Coleoptera: Tenebrionidae). American Midland Naturalist 138:230–235.

McKelvey, K., B.R. Noon, and R.H. Lamberson. 1993. Conservation planning for species occupying fragmented landscapes: the case of the northern spotted owl. In: Biotic Interactions and Global Change, pp. 424–452. P.M. Kareiva, J.G. Kingsolver, and R.B. Huey (eds.). Sinauer Associates, Sunderland, MA.
Meentemeyer, V., and E.O. Box. 1987. Scale effects in landscape studies. In: Landscape Heterogeneity and Disturbance, pp. 15–34. M.G. Turner (ed.). Springer-Verlag, New York.
Nadenicek, D.J. 1997. The poetry of landscape ecology: an historical perspective. Landscape and Urban Planning 37:123–127.
Naiman, R.J. 1996. Water, society and landscape ecology. Landscape Ecology 11:193–196.
Nassauer, J.I. (ed.). 1997. Placing Nature: Culture and Landscape Ecology. Island Press, Washington, DC.
Naveh. Z., and A. Lieberman. 1994. Landscape Ecology, 2nd ed. Springer-Verlag, New York.
Nicholls, A.O., and C.R. Margules. 1991. The design of studies to demonstrate the biological importance of corridors. In: Nature Conservation. 2: The Role of Corridors, pp. 49–61. D.A. Saunders and R.J. Hobbs (eds.). Surrey Beatty & Sons, Chipping Norton, NSW, Australia.
O'Neill, R.V., D.L. DeAngelis, J.B. Waide, and T.F.H. Allen. 1986. A Hierarchical Concept of Ecosystems. Princeton University Press, Princeton, NJ.
Peters, R.H. 1991. A Critique for Ecology. Cambridge University Press, Cambridge.
Pickett, S.T.A., and M.L. Cadenasso. 1995. Landscape ecology: spatial heterogeneity in ecological systems. Science 269:331–334.
Pickett, S.T.A., J. Kolasa, and C.G. Jones. 1994. Ecological Understanding. Academic Press, New York.
Pressey, R.L., C.J. Humphries, C.R. Margules, R.I. Vane-Wright, and P.H. Williams. 1993. Beyond opportunism: key principles for systematic reserve selection. Trends in Ecology and Evolution 8:124–128.
Risser, P.G. 1995. The Allerton Park workshop revisited—a commentary. Landscape Ecology 10:129–132.
Risser, P.G., J.R. Karr, and R.T.T. Forman. 1984. Landscape ecology: directions and approaches. Illinois Natural History Survey Special Publication 2:1–18.
Saunders, D.A., G.W. Arnold, A.A. Burbidge, and A.J.M. Hopkins. 1987. Nature Conservation: The Role of Remnants of Native Vegetation. Surrey Beatty & Sons, Chipping Norton, NSW, Australia.
Schreiber, K.-F. 1990. The history of landscape ecology in Europe. In: Changing Landscapes: An Ecological Perspective, pp. 21–33. I.S. Zonneveld and R.T.T. Forman (eds.). Springer-Verlag, New York.
Settele, J., C. Margules, P. Poschlod, and K. Henle (eds.). 1996. Species Survival in Fragmented Landscapes. Kluwer Academic, Dordrecht.
Strong, D.R., D. Simberloff, L.G. Abele, and A.B. Thistle (eds.). 1984. Ecological Communities: Conceptual Issues and the Evidence. Princeton University Press, Princeton, NJ.
Turner, M.G. 1989. Landscape ecology: the effect of pattern on process. Annual Review of Ecology and Systematics 20:171–197.
Turner, M.G. 1998. Landscape ecology: living in a mosaic. In: Ecology, pp. 77–122. S.I. Dodson et al. (eds.). Oxford University Press, New York.

Turner, M.G., R.H. Gardner, and R.V. O'Neill. 1995. Ecological dynamics at broad scales. BioScience Supplement S29–S35.

Ulanowicz, R.E. 1997. Ecology, the Ascendent Perspective. Columbia University Press, New York.

Urban, D.L. 1993. Landscape ecology and ecosystem management. In: Sustainable Ecological Systems: Implementing an Ecological Approach to Land Management, W.W. Covington and L.F. DeBano (eds.). USDA Forest Service Gen Tech Rept. RM-247., Fort Collins, CO.

Wiens, J.A. 1989. Spatial scaling in ecology. Functional Ecology 3:385–397.

Wiens, J.A. 1992. What is landscape ecology, really? Landscape Ecology 7:149–150.

Wiens, J.A. 1995. Landscape mosaics and ecological theory. In: Mosaic Landscapes and Ecological Processes, pp. 1–26. L. Hansson, L. Fahrig, and G. Merriam (eds.). Chapman & Hall, London.

Wiens, J.A. 1997. Metapopulation dynamics and landscape ecology. In: Metapopulation Biology. Ecology, Genetics, and Evolution, pp. 43–62. I.A. Hanski and M.E. Gilpin (eds.). Academic Press, New York.

Wiens, J.A., and B.T. Milne. 1989. Scaling of "landscapes" in landscape ecology, or, landscape ecology from a beetle's perspective. Landscape Ecology 3:87–96.

Wiens, J.A., R.L. Schooley, and R.D. Weeks Jr. 1997. Patchy landscapes and animal movements: do beetles percolate? Oikos 78:257–264.

With, K.A. 1997. Microlandscape studies in landscape ecology: experimental rigor or experimental *rigor mortis*? US-IALE Newsletter 13(1):13–16.

Zonneveld, I.S. 1995. Land Ecology. SPB Academic Publishing, Amsterdam.

Index

A
Abiotic resources, assessing, Open Space Plan example, 190–191
Adaptive Environmental Assessment and Management (AEAM) model, 15
Adaptive management, 15
  in ecosystems, 6, 80
  focus on, as a social option, 8
  integrating with landscape planning, 183–184
  landscaping planning as, 175
Age
  forest age maps, for simulation of harvesting, 122–124
  as a surrogate for merchantability, forest management simulation, 111–112
Agriculture, constituency of, 48
Algyroides mancha (lizard, Iberian Peninsula), 227
Allocation of training activities, Integrated Training Area Management (ITAM) program for, 81–82
Alternatives
  assessment of biological and economic effects of conservation strategies, 162–172
  assessment of future landscapes, 142–148
  in conservation strategy, opportunity cost of, 168–169
  scenarios for developing, 190
  simulating, timber management, 112–114
Americans Outdoors, President's Commission on, 187
Analysis
  of data, 62–63
    biodiversity structure, 130–135
  fractal, 317–329
  of landscape patterns, constrains on, 342–344
  of map patterns, RULE, 293–295
  percolation, 310
  process of, defined, 62
  of scale, tools used in, 211–218
  of self-affine functions, 328–329
  of simulation results, HARVEST, 114–116
  of spatial patterns
    effects of neighborhood rules on, 300–301
    for inferring spatial processes, 80
  statistical treatment of Mandelbrot measures, 322–323
Animal ecology, 226–227
Anisotropy, in spatial patterns, 254
  for plant ecology, 266
Anthracnose, dogwood, 26
Applegate watershed, 28–29
Appraisal, defined, 62
Aquatic boundaries, 28
ARC/INFO, 68
Asymmetry, of road-effect zones, 40–41

Atmosphere, landscape variability effect on, 230–231
Australia, hydrologic changes in, 19
Autocorrelation
relationship between variables, 270–274
spatial
implications of, 253
intensity of, 266–267
Avoidance, to prevent negative ecological impacts, road planning, 43

B
Backcasting, process scenario of, 189
Bark beetle, epidemic of, 26
Barriers, porous, in road planning, 41
Benefits, from forested habitats, 109
Best Linear Unbias Estimator (BLUE), kriging with, 269–270
Binary map, geostatistical properties of, 326–328
Biodiversity
conserving
hierarchical framework for, 127–153
with scarce resources, 154–174
measuring, and level of biological organization, 129
models incorporating, 4–5, 182
Open Space Plan, 191–192, 195, 196
problems of scale in studying, 227–229
suburban examples of, 38
Biogeochemical cycles, 229–230
Biological effects, of alternative conservation strategies, 162–172
Biomass, representation of EDYS, 89
Quadrat Module, 93–94f
Biotic resources, assessing, Open Space Plan example, 191–193
Birds, benefits to, of large patches of natural vegetation, 46
Bison, habitat use assessment, 225
Black grouse, habitat fragmentation effects on, 224–225
Blocking, to define scale, 216
Bonferroni method, for testing significance, 266
Bottlenecks, from interruption of natural patterns by planned roads, 42, 43
Boundaries
disciplinary, 27–28
ecosystem, 28–29
effect on habitat patch description, 295
natural, ecosystem management in terms of, 60
societal, 29–30
Box counting method, fractal analysis, 318–320
Branch and bound methods, to identify an optimal collection of reserve sites, 163
Branching clusters, 307–308
Broad scale, defined, for landscape ecology, 234
BROWNIAN program, for generating self-affine fractals, 315–317
Build-out, in base line development, 189
Butternut canker, 26

C
Calibration, Quadrat Module, EDYS, 92
Canada thistle, invasion of, simulation of controls, 103
Canonical Correspondence Analysis (CCA), partial, 273–274
Capital, cultural, 7
Carrying capacity, earth's, factors defining, 7–8
CART methodology, 130–131
Casco (basic framework), defined, 186
CASE program, for knowledge acquisition management, 61–62
Cattle, habitat use assessment, 225
Cedar Creek experiment, 14
Centers, of biodiversity, identifying, 130–131
Change of Support problem, 258
Chestnut blight, 26
Chorological perspective, in landscape planning, 179

Cladistic analysis, for measuring diversity, 160
ClaraT
  BROWNIAN program, 315–317
  description, 330
  for generating patterns with fractal geometry, 304–332
  iterated function system manipulation in, 312–313
Classification, effects on landscape metrics, 343–344
Classification and regression trees (CART), 352–353
Climate, ecosystem feedback, models for examining, 230–231
CLIPS (C Language Production System), program for organizing information, 62, 68
Clustering
  area weighted average size, RULE, 294
  spatial, for transforming field samples into patches, 257f
Clusters, branching, 307–308
Collaboration, in ecosystem management, 30–31
"Coming Anarchy, The" (Kaplan), 8
Communication, among stakeholders in ecosystem management, 32–33
Communities
  plant, hierarchical scales of processes in, 222
  scenario incorporating consensus in, Open Space Plan, 195
  within society, objectives of, 48–49
  See also Cultural data; Culture, human, and landscape planning
Community-level models
  allocation of training activities, 83–84
  for evaluating the impacts training activities, 85–86
Community Module, EDYS, 93–99
Comparability, as a criterion in policy assessment, 128–129
Compensation, for ecological impacts, road planning, 43
Complementarity, of species distributions, 139–142
Complete spatial randomness (CSR)
  defined, 259
  testing, join-count statistics, 264
Complexity, of environmental systems, 379
  modeling considerations, 65–66
Computer-based decision aids, *See* Decision support systems
Computer-based systems, integrative, 64f
Conceptual unity, and landscape ecology, 373
Conflict, among societal objectives, landscape and regional scales of, 47–49
Connected habitat, defined, 311
Connectivity, and neutral models, 336–342
Consensus building
  for community value expression, Open Space Plan, 193
  spatial concept presentation in, 184
  See also Communities; Community-level models
Conservation biology
  based on taxonomic groups, 157
  constituency of, 48
  goals of, cross-scale conflicts underlying, 232
  hierarchical approach to, 228
  priority of, biodiversity policy, 231–232
Constraints
  on landscape pattern analysis, 342–344
  on policy, fixed budget, 156
Contact rule, for identifying habitat patches, 281
Contaminants, assessing sources of, scale considerations, 230
Correlative studies, 16
Correlograms, spatial, 264, 265–266, 268–269
Corridors, as ecological conduits, 40–41
Cost-effectiveness analysis, 156
Creativity, in spatial concept development, 185

Critical density, percolation map, and infinite cluster appearance, 310–311
Critical resources, community consensus for defining, 193
Cross-boundary issues, for managing forest ecosystems, 24–34
Cross-scale comparison
  in climate analysis, 231
  in habitat analysis, 225
Cultural data
  including in ecology maps, 8, 9
  resource goals and assessments, Open Space Plan, 193–194
Culture, human, and landscape planning, 49, 182
Curdled map, generation with RULE, 286–288
Cycles, biogeochemical, 229–230

D

Data
  analysis and synthesis of, 62–63
  classified by objective, spatial statistics for, 255t
  files produced by RULE, 292–293
  for RULE synthesis, 281
  types and measurements of, spatial statistics for, 256t
  uncertainty in, accounting for, 169–172
Databases, Historical Climate Network, 134
Decision making
  conservation framework for, 155–156
  hierarchical structure of, 128–130
  knowledge available for, 61
    imperfect, 183, 189–190
Decision support systems
  allocation of training activities, U.S. Army, 82–85
  computer based, 58
  ISPBEX-II, for forestry, 66
Defensive planning, strategy of, 188
Delaunay triangulation algorithm, 264
Demographic shifts, effect on public lands, 29
Dendroctonus frontalis (southern pine beetle), 66
Design, constituency of, 48
Design principles
  incorporation of scale, 209–211
  spatially explicit, 19
Detrending a spatial pattern, 270–274, 275
Developmental scenarios, 189–190
Development planning, and political acceptance of open space planning, 193–194
Differentiation diversity, 227–228
Diffusion limited aggregation (DLA), 307–308
Discontinuity, in landscape structure, scale of, 219–220
Disease, forest, in stands after fire-suppression, 26
Dissimilarity, among species, as a measure of diversity, 157
Disturbance, spread of, and landscape heterogeneity, 3
Disturbance regimes, 223–224
  diversity associated with, 228–229
Diversity
  habitat fragmentation effects on, 225
  regional, and environment, 229
  spatial patterns in, 129–130, 227–228
DLA tree program, 308
Dryocopus pileatus (pileated woodpecker), 191
Dynamic models, GIS as a tool in displaying, 214–215
Dynamics
  landscape, 223–224
    modeling of, 355–357
  species-level, 80–81

E

Ecological Applications (journal), special issue on SEMs, 214
Ecological Dynamics System (EDYS), modeling approach, allocation of training activities, 85–101
Ecological economics, 6–8
Ecological Society of America, 33
Economic costs
  of alternative conservation strategies, 162–172

including analysis, in landscape ecology, 9
of preservation, 4
Economics
ecological, 6–8
priority in community planning, 49
Ecosystem, defining, 59–60
Ecosystem management, 5–6
paradigm of, 58
Ecosystem model, systemic planning process based on, 180, 182
Ecosystem scale, limitations of, in landscape planning, 176
Ecosystem services, value of, 161
Ecotone hierarchies, 206
Ecotourism, value of, 161
Edge detection algorithms, for patch delineation, 256–257
Edge effect, correcting for, statistic measures, 262
Education
exercises for teaching landscape ecology, 335–368
instructional software, fractal pattern generation, 304–332
value of small patches of vegetation for, 47
EDYS hierarchical model, 81, 85–101
Efficacy, of computer-based systems, 70
Elicitation of knowledge, defined, 61–62
Elk grazing, simulation of effects of, Community Module, EDYS, 102
Endangered species, effects of efforts expended on, 127
Energy, available, and species richness, 138–139
Enterprise integration, defined, 71
Environment
associations with species richness, 135–139
for ecosystem management, knowledge system, 57–79
incorporating variability in a study area, 258
Environmental management, modeling approach, 81–85
Equal area units, for analysis of biodiversity, 133
ERDAS Toolkit routines, 111
EROS Data Center, USGS, 134
Eutrophication, in suburban water bodies, 44
Everglades, 30
Evolution, of software, 71
Experimentation
controlled, 227
landscape plan implementation as an example of, 182–183
role in landscape ecology, 14
Expert knowledge, for sustainable development, 182
Expert systems, 68–69
Extent
changing, effect on landscape metrics, 343
defined, fractal map, 317
defined, landscape ecology, 234
Extrapolation, among spatial scales, problem of, 211–212

F

Final Environmental Impact Statement (FEIS), Forest Service, simulation based on, 112–113
Fine scale, defined, for landscape ecology, 234
Fire
issue in forest ecosystem health, 24, 27–28
natural cycle of, 25
simulation of effects, Community Module, EDYS, 89, 93, 96–97
First-order statistics, limitations of, 253
Fish and Wildlife Service, U.S., 58
target species selection consultation by, Open Space Plan, 191
Fixed points, in map renormalization, 325–328
Flexibility
in accommodating changing land use, framework model, 186
about sites included in an analysis, 165
Flooding
downstream of suburbs, 44

Flooding (*cont.*)
protection against, with natural vegetation, 45
Flows, directional, planning shapes of objects with consideration of, 38–39
Forecasting, in process scenario development, 189
Forest cover, aggregating a map of, 324
Forest Ecosystem Management Team (FEMAT), 58–59
Forest ecosystems, cross-boundary issues for managing, 24–34
Open Space Plan, 191–192
Forest health, 24
crisis in, 30
monitoring program, 31
Forest management, evaluation and monitoring in, 32
Forestry
constituency of, 48
sustainable development for, government initiatives, 59
Forest Service, 58, 113
ecosystem management policy, 27
fire management policy, 30
history of, 25
Western Forest Health Initiative, 24
FORTRAN 77, 281
FORTRAN 90, 281
code for calculating a fractal, RULE, 292
code for generating a simple random map, 283
RULE source code, 296
Fractal dimension (*D*), 306–307
equation defining, 215
influence of the grid shape on, 293
for scale-invariant measures of structure, 219
Fractal geometry
for analysis of habitat movement, 225–226
ClaraT for generating patterns with, 304–332
RULE algorithm based on, 286–288
as a tool in scaling, 215
Fractals
creation of, 307–317
self-affine, 313–317
Fragmentation
effects on landscape planning, 232
habitat, 224–225
percolation theory, 311
FRAGSTATS, for analyzing HARVEST output, 115–116
Framework
landscape ecology, describing biodiversity in, 228
planning
models of, 178–179, 181f, 183–184
in road and nature conflicts, 42–44
in the Netherlands, 185–186
Functions, of spatial concepts, 184–185
Fusilli puttanesca, distribution of, analysis, 355
Future scenarios, 175–201

G

Game management, constituency of, 48
Gap Analysis Program (USGS), 149, 164
Gap-phase regeneration models, 224
Gap vegetation classes, 136
Gap Vegetation Classification Map, USGS, 135
Geary's c, 265
General Accounting Office (GAO), report of federal initiatives in ecosystem management, 59
Genetic differentiation, 227
Genetic traits, conservation of variety in, 157
Geographic Information System (GIS), 5, 15–16
clumping function of, to generate a forest stand ID, 121
relating biological diversity with habitat degradation, 228
tool for scaling, 214–215
tool in habitat analysis, 225
Geographic location, as a surrogate for nonsampled variables, statistical analysis, 274

Geography, and scaling, 220
Geostatistics, 268–270
  grain problem in, 258
Gliding box sampling
  for determining lacunarity, 294–295
  for producing a Mandelbrot measure, 321
Global change, models of, 230–231
Goals, for conservation at the species level, 156–157
Gradients, environmental, and diversity, 229
Grain size
  changing, effect on landscape metrics, 343
  defined
    for landscape ecology, 234
    fractal map, 317–318
Grasshopper, scale of movement within a habitat, 225–226
Greedy algorithm, 164
Green Heart spatial concept, 185
Greenways, spatial concept of
  North American, 187
  Open Space Plan, 196
Group selection, of stands, HARVEST simulation, 122
Gypsy moth, European, attack on oak, 26

H
Habitat
  analysis of, 224–226
  characteristics of patches of, and diversity, 229
  conservation of, increased land value due to, 169
  fragmentation of
    percolation theory, 311
    from road barriers, 42
HARVEST model, 109–124
Hemlock wooly adelgid, 26
Herbivory, Quadrat Module, EDYS, 89
Heterogeneity
  and flux of organisms, 80
  scale of, and ecosystem-level effects, 3
Hierarchy
  in assessment of agriculture on native biota, 232
  of decision making, 128–130
  in ecology, 129–130
  framework for conserving biodiversity, 127–153, 228
  of landscape structure, 80
    tools for defining scale in, 216
  model design based on, 85–87, 224
    RULE map generation, 286–288
Hierarchy theory
  application to ecological scaling, 206
  and landscape planning, 176, 235
  spatial scale issues in, 217–218, 374
Historical Climate Network database, 134
Holdridge life zone, 5
Holistic approach, in land-use management, 58
Hoosier National Forest (HNF), timber harvest allocation model applied to, 110–124
Horizontal processes, 35–53
Hotspots
  comparison with optimal reserve network, 165, 167f
  versus complementarity, 163
Hubbard Brook watershed, 15
Human context
  human capital and natural capital, 7–8
  intervention as a part of the ecosystems model, 180
  population impacts on ecosystems, and policy considerations, 11–23
Hydrologic structure, integration into a framework model, 186
HyperText Transport Protocol (HTTP), 72

I
Identity, of landscape ecology, 372
Information, analysis and synthesis of, 62–63, *See also* Knowledge
Inheritance relations, in object-oriented programming, 66
Input maps, for RULE, 291

Insect pests
Dendroctonus frontalis (southern pine beetle), 66
grasshopper, 225–226
gypsy moth, 26
hemlock wooly adelgid, 26
infestation in stands after fire-suppression, 26
scale in research on, 227
Integrated pest management (IPM), decision support system for, 66–68
Integrated Training Area Management (ITAM) program, for allocation of training activities, 81–82
Integration
of information, 63
of landscape and regional level analyses, 172
in landscape ecology, 11, 16–18
spatial, framework concept of, 186
of theories in forest health management, 28
Intensity
of harvesting, changes resulting from on simulation, 118–119
spatial, of autocorrelation, 266–267
INTERACTER, for graphic display of maps, 292
Intermediate disturbance hypothesis, regional analogue of, 139
International Biological Program (IBP), 229–230
INTERNET, connecting planners and stakeholders on, 62
Interpolation, with kriging, 268
Interpretation
of information, 63
scale as a tool for, 235–236
Island biogeography theory, 224–225
Isotropy, of spatial pattern, 254
Iterated function systems (IFS), 311–313

J
JAVA, for KSE applications, 72
Jaws model, for maintaining indispensable patterns, 187
Join-count
for data analysis, 257
for qualitative-categorical data, 264

K
KAPPA, program for organizing information, 62
Knowledge
acquisition of, 61–62
for defining an ecosystem problem, 31
and the flow of consumer goods and services, 7–8
imperfect, and planning decisions, 183
for sustainable development planning, 182
Knowledge base, forms of storage, 60
Knowledge engineering, 60–61
Knowledge system, environment for ecosystem management, 57–79
Kriging, spatial model development with, 216, 268

L
Lacunarity index, 219
fractal analysis, 319
RULE output, 281, 294
Lag
defined, 318
renormalization of a binary map described by, 327–328
Lake Tahoe Basin, 28–29
Land and Resource Management Plan Amendment, simulation of plan leading to, 113
Land classification, AVHRR, USGS, 135
Land cover map
for HARVEST simulations, 121
Monroe Country, Pennsylvania, 143–144
quantifying change in, 344–348
scale issues, 220–221
Land Management, Bureau of, 58
Land planning, 231–232
Landsat Thematic Mapper (TM), 366
simulation incorporating data from, 113

Landscape
defined, 176
effects of fire suppression, 25–26
spatially neutral, 308–311
Landscape ecology
defined, 3
spatial statistics in, 274–275
Landscape Ecology (Wiens), 12, 207–208
Landscape metrics
defined, 257
effects of classification on, 343–344
versus spatial statistics, 255–258
Landscape Module of EDYS, 100–101
Landscape planning, defined, 175
Landscape structure, defined, 218–220
Land use
changes in
effect of species habitat, 231–232
effects on biodiversity, 132
classification of, 220–221
decisions about, governmental bodies making, 128–129
Large scale, defined, for landscape ecology, 234
Limiting resources, nitrogen and water as, 89
Linkages
among modules, decision support system, 83–84
Quadrat Module
EDYS, 88f
forming Community Modules, EDYS, 95
of resources, Open Space Plan, 196
Live oaks, California foothills, analysis of distribution, 354–355
Log file
curdled map generated from RULE, 287
fractal map generated from RULE, 288, 289
random map generated from RULE, 283t, 285
Logging
HARVEST model, 109–124
plans for, scale-dependent issues, 232
Longleaf pine ecosystem, 28–29
L statistic, defined, 260

M

Majority rule, for map renormalization, 325
Mammals, small, habitat use assessment, 225
Management Model, allocation of training activities, 83
Mandelbrot measures, 319–323
Mantel's test, partial, 271–273
Map layers, for HARVEST usage, 120–121
Mapping, of natural patterns, in road planning, 42
Maps
random
generation of, 282–289, 338–340
generation of by hand, 298–299
visualization of, 291–292
size of, and reliability of results, 295
Markov model, of succession, 366
Massachusetts, University of, target species selection consultation by Open Space Plan, 191
Maximal coverage problem, 162–168
Metaknowledge, use of, by software systems, 64
Metaphors, spatial concepts as, 184, 187
Metapopulation dynamics, 16
Methodology, of scale definition in landscape ecology, 232–233
Microclimate, effects on, of large patches of natural vegetation, 46
Mineral nutrients, in suburban soil, accumulation of, 44
Mitigation of ecological impacts, road planning, 43
Model-building process, 356–357
Modeling
design for EDYS, 85–87
in landscape ecology, 12, 16
Modifiable Area Unit Problem (MAUP), 258
Monitoring, and evaluation, in forest management, 32

Monroe County, Pennsylvania, assessment of future landscapes, 142–145
Moose, habitat use assessment, 225
Moran's *I*, 264–265
Morphological differences, to measure dissimilarity, 158
Mosaic stability, context for, in a landscape plan, 176
Moving windows technique, for defining scale, 216
MS-DOS operating system, 281
Muddy Creek watershed, assessment of future landscapes, 145–148
Multifractals, 321
Multiple regression, for landscape analysis, 353–355
Multiple variables, scaling using, 212
Multiscale analysis, 227
Multiscale ecological model, for allocation of training activities, U.S. Army installations, 80–108
Mustela vison (mink), 191–193

N

National Environmental Policy Act of 1969, 33, 180
National Forest Plan, simulation using the standards and guidelines of, 110
National Park Service, 58
National Science Foundation, landscape ecology initiative of, 235
National security, and the environment, projection to the future, 8
Native species, preservation in large natural patches of vegetation, 46
Natural boundaries, ecosystem management in terms of, 60
Natural capital, 7
Natural Heritage Programs, 133
Natural processes, horizontal, 36–39
Natural resource management, landscape ecology approach to, 3
Nature Conservancy, The, 133
Nature reserves
  designing, 232, 376
  opportunity cost of, 168–169
Nearest neighbor distance technique, 259
Nearest neighbor rule, 280–281
  for movement by large herbivores, 225
  refined, expected mean nearest distance, 259
Neighborhood rule
  critical percolation density under, 325
  quantifying the effects of, on pattern analysis, 300–301
  specification of, 289–291
Netherlands
  ecological stress in, 177
  Green Heart spatial concept in, 185
  National Ecological Network mapping, 43
  planning framework for addressing road and nature conflicts, 42
Net primary productivity
  estimating, 135
  and scale, 229–230
  and species richness, 138–139
NETSCAPE NAVIGATOR, 72
Net Usable Land Area Process (NULA), 193–194
NETWEAVER, program for organizing information, 62
Neutral simulation models, spatial patterns in, 217, 219
NEXPERT, program for organizing information, 62
Nitrogen cycle, simulation of, Quadrat Module, EDSY, 89
Nitrogen pathways, Quadrat Module, EDYS, 90
Node counting method, for expressing diversity, 157
Nonnative species, effect on forest health, 26
Normalized Difference Vegetation Index (USGS), 135
Nugget effect, 268–269

O

Objectives
  allocation of training activities on Army installations, 81

data classified by, spatial statistics, 255t
maximization of biological diversity conserved, 155–156
Objectives-Oriented Program Planning, for knowledge acquisition, 61–62
Object-oriented programming
defined, 66
design for a knowledge system environment, 65
Object-Oriented Program Planning
object, as defined for, 66
Observation, primary tool in landscape ecology, 12–14
Offensive planning, strategy of, 188
OPENLOOK, 68
Open Space Plan, Orange, Massachusetts, 184, 190–197
Open systems, spatiotemporal scales for, 129
Opportunistic planning, strategy for, 188
Opportunity costs
of alternative conservation strategies, 168–169
of preservation, 4
Optimization
of land use to suit varied communities, 48
in species richness analysis, 131
tools for, 63
Option value
and diversity measurement, 158–159
of genetic prospecting, 161
ORACLE, 68
Orange, Massachusetts, 177
Open Space Plan, 184, 190–197
Order, ecosystem, modes of, 180
Oregon, environmental associations simulation, 135–142
Organic Administration Act of 1897, 25
Organism-based views of the landscape, 349–352
Oscillation of configuration, random map, 327–328
Ownership, of forest land, 29

P

Pacific Northwest Forest Conference, 58
Paradigm shift, in landscape ecology, 371–372
Participatory process, in landscape plan development
framework model, 184
and implementation, 182
Open Space Plan, 193
Patches
identification with a nearest neighbor rule, 280
organizing data into, spatial statistics for, 255–257, 264
size of
defined for landscape ecology, 234
from spatial correlograms, 266
small, of natural vegetation, 47
Patch mosaics, 257
Patch-scale disturbance, 223
Pattern/process paradigm, 60
Patterns/pattern description, 15–16
agents of formation of, 352–355
dynamic relationship with process, in planning, 182
generation by processes, 374
in natural horizontal processes, 36–39
statistical relationships in, 299
surface, sensitivity to the size of sampling units, 258
as a surrogate for process, 130
Percolation
maps, 308–311
renormalization rule, 325
theory, neutral landscape model based on, example, 337–342
Placing Nature: Culture and Landscape Ecology (Nassauer), 375
Planning
attributes of, ecological method, 180–183
consideration of natural horizontal patterns in, 37–38
contribution to, by landscape ecology, 18–20
for ecosystem management, 31–32
strategies for, 175–201
Plan of action, knowledge available for developing, 61

Plant ecology
  quadrat size problem in, 258
  and vegetation analysis, 221–223
Plants
  glucosinolate-producing, example of diversity measures, 159–160
  major associations of, U.S. Army installations, 86t
Point pattern methods, for analysis of discrete data, 257–263
Poisson distribution, point pattern, 259
Policy
  on biodiversity, landscape as the unit of, 231–232
  impact on biodiversity conservation, 155–156
  scientists' participation in making, 6
  spatial consequences to the landscape, scenarios for communicating, 177
  value-based, 128–130
Political hierarchy, interaction with ecological hierarchy, 128–130
Political systems, as variables in landscape ecology, 9
Politics, value of small patches of vegetation in, 47
Population biology, scale dependence in habitat factors affecting, 227
Populations
  abundance of, as a measure of biodiversity, 144
  data, spatial statistics for, 270
  dynamics of, 172
  models of, GIS framework for, 214–215
  viability analyses, 4–5
Power law, coefficient *k* in, as a measure of lacunarity, 319
Preservation, priorities in, techniques for establishing, 4–5
Priorities, setting
  in biodiversity conservation, 154
  for forest management, 31–32
  objective policy analysis for, 128–130
  See also Consensus building; Policy
Private property, concept of, and resource management, 29–30
Probabilities, for multiple habitat types, RULE map generation, 291
Problem solving
  knowledge available for, 61
  landscape level, 29
  See also Decision making; Participatory process, in landscape plan development
Procedural theories of landscape planning, 178
Process, models, linking with spatial databases, 5
Processes
  disturbances driven by, 224
  iterative, in landscape planning, 183
  pattern description as a surrogate for, 130
  patterns produced by, 374
  planning, systemic, 180
  relating to pattern, 16
    in a landscape plan, 175
  spatial, analysis spatial patterns to make inferences about, 80
Process scenario, defined, 189
Product moment correlation coefficient, from the Mantel's statistic, 271
Protective planning, strategy of, 187–188
Pseudorandom number generator, *ran1*, 284
Pseudostationary data, analyzing with geostatistics, 269
Public lands, managing, private land considerations, 6

Q

Quadrat module
  EDYS, 87–92
  size of
    for determining spatial autocorrelation coefficients, 257–258
    effect on quantification of spatial pattern intensity and scale, 263
    and spatial autocorrelation, 267
Qualitative-categorical data, join-count statistics for, 257, 264

Quantitative-numerical data, spatial autocorrelation coefficients for, 264–268
Questions, in landscape ecology, 379–380

R
Randstad (Ring City), 185
Ranges, species
  factors associated with, 225
  and plant community structure, 223
Rarity-based algorithms, 164
Recreation
  constituency of, 48
  scenario incorporating, Open Space Plan, 195
Red-cockaded woodpecker, place in fire management, 28
Reference distribution, for assessing significance of the product moment correlation coefficient, 272
Regression trees, for analyzing spatial structure of biodiversity, 130–131, 135–139
Relevance, of landscape ecology, 11–12, 18–20
Remote sensing
  determining a scale suitable for, 221
  spatial scaling tool, 212–213
Renormalization, spatial, and change of scale, 323–328
RENORM2 program, percolation rule in, 325
Replication, in landscape planning, 182–183
Resampling technique, to define scale, 216
Research, focus of, and scale, 218–236
Reserve network selection, 154, 162–168
  problem of, 169–172
Response variables, biodiversity, 133–134
Rights and responsibilities, in knowledge system relationships, 68
Ripley's *K* plot, 262–263
Ripley's *K* second-order statistic, 259–263
Ripley's K_{12} statistic, 261–262
River Reach File, USGS/EPA, 135
Roadkill, effect on animal populations, 42
Roads, ecology of, 39–44
Root growth, diffusion limited aggregation as a model of, 308
Rothampstead experiment, 14
RULE program, 280–303
  examples, 297–301
  map generator, 284–286

S
SAV, RULE output, area weighted average cluster size, 294
Scale
  changing
    for map generation, 299–300
    to relate process data from diverse observation levels, 5
  components of, fractal analysis, 317–318
  concept of, in landscape ecology, 205–252
  defined, for landscape ecology, 234
  of experiments, in landscape ecology, 14–15, 340–341
  matching process and pattern scales, 233
Scarce resources, for biological diversity conservation, 154–174
Scenario studies
  for explicit projection of policy decisions, 177
  Open Space Plan, 194–197
Science
  of landscape ecology, 373–375
  linkages with the social sciences, 17
  role in landscape ecology, 377–379
Script files, for running RULE, 296–297
Second-order statistics, for distinguishing spatia patterns, 253
Self-affine fractals, generating, 315–317

Self-similarity
of coastlines, 321
fractal patterns, 305–306
and grain size, 319
Semivariance function, spatial, 268
Semivariance renormalization, behavior of, 328f
Sensitivity, of conservation strategies, 170
Shifting-mosaic-steady-state model, 223
Simulation, 217
to assess scale-dependent attributes, 219
of landscapes, 80
GIS framework for, 214–215
of management alternatives, HARVEST model, 109–124
of material and energy fluxes, 229–230
Monte Carlo, for generating a Poisson point pattern process, 260–261
scaling issues, 216–217
Site-scale landscape architecture, 185
Small scale, defined, for landscape ecology, 234
Social forces
and public policy, 8
societal objectives and landscape ecology, 47–49
Socioeconomic context, 17–18
Software, instructional, fractal pattern generation, 304–332
Soil erosion, suburban, 44
Soil formation, catchment-scale processes in, 229
Soil profiles, Quadrat modules, EDSY, 94f
Soil protection, models incorporating, 182
Source-sink relationships, 16
Southern Appalachian ecoregion, 28–29
Southern pine beetle, epidemic of, 26
Spanning cluster, renormalization of, 326
Spatial accounting units, 132–133, 142
Spatial autocorrelation
correcting for, 274
coefficient for quantitative-numerical data, 264–268
Spatial concepts, 175–201
biodiversity, recreation and community values scenarios, Open Space Plan, 196
generic, defining landscape patterns, 186–187
metaphors, 184–187
Spatial distribution, of biodiversity, environmental factors affecting, 130–135
Spatial integration, framework model, 186
Spatially explicit models (SEMs), 172
developing, 213–214
simple simulation models as, 217
Spatially explicit population models (SEPMs), 226–227
Spatial patterns
analysis of
GIS-based, 182–183
to infer spatial processes, 80
landscape ecology as study of reciprocal effects of, 3
in management of training areas, U.S. Army, 85–87
Spatial phase transition, in random maps, 309
Spatial scale, defined, for landscape ecology, 234
Spatial statistics, in landscape ecology, 253–279
Specialization, disadvantages of, in education, 17
Species
abundance of, dependence on quadrat size, 267
movement of, 225
subsets of, and spatial extents for analyzing, 129–130
surrogate for genetic diversity, 133–134
Species richness
dependence on the extent of the study area, 227–228
global scale, conditions associated with, 137–138

joint, locations of, 131
as a measure of species diversity, 156
of rare birds in Tanzania, 228
suburban, 44
maintaining, 46
Spotted owl
effect of old-growth habitat on, 336–342
habitat fragmentation effects on, 224–225
scale effect in research on, 227
Stakeholders, eliciting a information from, 61–62
Standard Normal Deviate (SND), for testing for autocorrelation, 264
Starling, European, scale in research on, 227
State scenario, defined, 189
Stationarity, assumption of, 254
in spatial statistics, 258, 267
Statistical methods, for scale expression, 215–216, 219
Statistics, spatial, in landscape ecology, 253–279
STELLA program, 365
Stochastic models, GIS as a tool in displaying, 214–215
Strategic planning
goals of, 187–188
in harvesting forests, HARVEST software, 118–119
landscape, defined, 183
Study plots, for developing the Community Module, EDSY, 99
Substantive theories of landscape planning, defined, 178
Suburbs
design of, 38–39
patches of nature in, 44–47
Successional dynamics, modeling of, 101, 366
Successional regimes, diversity associated with, 228–229
Support systems, computer-based, 58
Surface pattern methods
field studies using, 263–270
and sampling size, 258
Surface water, horizontal process example, 36–39
Surrogates
age, for merchantability of timber, 111–112
for aquifer protection, 42
geographic location, for nonsampled variables, 274
pattern description, for process, 130
species, for genetic diversity, 133–134
Survival, probability of, and conservation strategy, 157, 169–172
Sustainability
in landscape planning, 175
of plans, and accommodation to natural processes, 38–39
Sustainable development
for forestry, government initiatives, 59
knowledge for, 182
Sweep analyses, of complementarity, 141
Symmetric rule, fixed point under, 325
Synthesis, process of, defined, 62

T

Target species, mapping the range of, 225
Taxonomic groups, habitat fragmentation effects on, 225
Teaching landscape ecology, exercises for, 335–368
Tell City Unit, Hooiser National Forest, 113
Tessellation, hexagonal, framework for a hierarchical spatial structure, 132–133
Timber harvest allocation model (HARVEST), 110–124
Tools, of landscape ecology, 11, 12–16
Topological analysis, in landscape ecological planning, 179
Town planning, constituency of, 48
Toxins, in suburban soil, accumulation of, 44
Training activities, multiscale ecological model for allocation of, U.S. Army installations, 80–108

Transformation
of field samples in homogeneous patches, 257
landscape, relating ecological knowledge and planning, 187
Transition, percolation, 309–310
calculating probabilities, 345–346
Transportation
human processes of, 36–39
as a constituency, 48–49
planning for roads and highways, 41
Triangulation algorithm, Delaunay, 264
Typology, of strategies for landscape planning, 188

U
Uncertainty, planning in the presence of, scenario approach, 189–190
UNIX operating system, 68, 281

V
Values
community, for defining critical resources, 193
ecological
of rural landscapes, 44–45
of small suburban patches of vegetation, 47
policy
hierarchical political levels affecting, 128
relationship with science, 6
Variables, explanatory, accounting for spatial patterns of species diversity, 134–135
Variance stairway, for defining scale, 216
Variogram, for kriging spatial structure, 268
Vegetation
analysis of, and plant ecology, 221–223
interstitial, in suburbs, 44
management prescriptions, Forest Service, 113–114
natural
large suburban patches of, 44–46
small suburban patches of, 47
suitability of current uses for conversion to, 42–43
Vertebrate species, terrestrial, representation, 166f
Vocabulary, for issues of scale, 233–234

W
Wageningen Agricultural University, 177
Water dynamics, bases of, EDSY development, 89
Water quality
models incorporating, 182
protection by natural vegetation, 45
Water resource management, constituency of, 48
Watersheds
Applegate, 28–29
Hubbard Brook, 15
Muddy Creek, assessment of future landscapes, 145–148
Water transport, among quadrats, Community Module, EDSY, 96f
Weight matrix, spatial autocorrelation, 266
Western Forest Health Initiative, 24
triage scheme developed for, 31
White pine blister rust, 26
Window length, fractal analysis, 318
Working method, landscape planning, 180
World Wide Web (WWW), connecting planners and stakeholders on, 62
Worth
of specific species, 161
See also Economic costs; Economics; Values